Bioinformatics

An introduction to programming tools for life scientists

Tore Samuelsson

University of Gothenburg, Sweden

CAMBRIDGE
UNIVERSITY PRESS

CAMBRIDGE UNIVERSITY PRESS
Cambridge, New York, Melbourne, Madrid, Cape Town,
Singapore, São Paulo, Delhi, Mexico City

Cambridge University Press
The Edinburgh Building, Cambridge CB2 8RU, UK

Published in the United States of America by Cambridge University Press, New York

www.cambridge.org
Information on this title: www.cambridge.org/9781107008564

First published 2012

Printed in the United Kingdom at the University Press, Cambridge

A catalogue record for this publication is available from the British Library

Library of Congress Cataloguing in Publication data
Samuelsson, Tore, 1951–
Genomics and bioinformatics : an introduction to programming tools for life scientists / Tore
Samuelsson.
pages   cm
Includes bibliographical references and index.
ISBN 978-1-107-00856-4
1. Genomics – Data processing.   2. Bioinformatics.   I. Title.
QH447.S26   2012
572.8′6 – dc23   2012006477

ISBN 978-1-107-00856-4 Hardback
ISBN 978-1-107-40124-2 Paperback

Additional resources for this publication at www.cambridge.org/9781107008564

Genomics and Bioinformatics

With the arrival of genomics and genome sequencing projects, biology has been transformed into an incredibly data-rich science. The vast amount of information generated has made computational analysis critical and has increased demand for skilled bioinformaticians.

Designed for biologists without previous programming experience, this textbook provides a hands-on introduction to Unix, Perl and other tools used in sequence bioinformatics. Relevant biological topics are used throughout the book and are combined with practical bioinformatics examples, leading students through the process from biological problem to computational solution. All of the Perl scripts, sequence and database files used in the book are available for download at the accompanying website, allowing the reader to easily follow each example themselves. Programming examples are kept at an introductory level, avoiding complex mathematics that students often find daunting. The book demonstrates that even simple programs can provide powerful solutions to many complex bioinformatics problems.

Tore Samuelsson is a Professor in Biochemistry and Bioinformatics at the Institute of Biomedicine, University of Gothenburg, Sweden. He has been active in bioinformatics research for more than 15 years and has over 10 years' experience of teaching in the field, including the development of web resources for molecular biology and bioinformatics education.

Genomics and

CONTENTS

PREFACE

We currently see a vast amount of information being generated as a result of experimental work in biomedicine. Particularly impressive is the development in DNA sequencing. As a result, we are now facing a new era of genomics where a lot of different species, as well as many different human individuals, are being analysed. There are many important biological questions being addressed in such genome-sequencing projects, including questions of medical relevance. A critical technical part of all these projects is computational analysis. With the large amount of sequence information generated, computational analysis is often a bottleneck in the pipeline of a genomics project. Therefore, there is great demand for individuals with the appropriate computational competence. Ideally, such individuals should not only be proficient in the relevant mathematical and computer scientific tools, but should also be able to fully understand the different biological problems that are posed. This book was partly motivated by the urgent need for bioinformatics competence due to recent developments in genomics.

A student or scientist may enter into bioinformatics from different disciplines. This book is written mainly for the biologist that wants to be introduced to computational and programming tools. There are certainly books out there already for that type of audience. However, I was attracted by the idea of assembling a book that would cover a large number of relevant biological topics and, at the same time, illustrate how these topics may be studied using relatively simple programming tools. Therefore, an important principle of the book is that it will attempt to convince the reader that relatively simple programming is sufficient for many bioinformatics tasks and that you need not be a programming expert to be effective. Another important principle of the book is that I wanted the bioinformatics examples to be very practical and explicit. Thus, the reader should be able to follow all the details in a procedure all the way from a biological problem to the results obtained through a technical approach. As one demonstration of this principle, all files and scripts mentioned in this book are available for download at www.cambridge.org/samuelsson. This means the reader is able to try it all out on his/her own computer. I also wanted this book to illustrate the interdisciplinary nature of bioinformatics. Therefore, I have chosen to include a substantial amount of biological motivation as well as programming technology. As a result, the book has a number of rather sudden transitions from descriptions of biological topics to very technical computing matters.

This book is intended as a guide to Perl and Unix-based computing tools for students with a background in molecular biology, biochemistry, cell biology or genomics who have no previous background in this type of computing. In addition, PhD students and scientists at all levels in these fields who want to be introduced to such programming tools should hopefully benefit from the book. The computational parts of the text should be easy to understand for a student lacking a background in computer science; the programming examples presented are at a fairly basic level. The book is complemented by exercises for further study, with mixed levels of difficulty. For the benefit of the student without a mathematical background, the book is by and large non-mathematical, avoiding topics such as probability theory. In summary, the reader of this book should be any student or scientist with some insight into biology, but who also wants to learn about bioinformatics at a more technical level. I also think of the book as being of potential interest to the student or scientist with a background in computer science or programming, but who seeks biological motivation and wants to know more about biological problems that are typically addressed in genomics and bioinformatics.

I present a number of biological and medical topics related to DNA and protein sequences and show how they may be exploited using bioinformatics tools. The book inspires from a biological point of view by selecting relevant and interesting examples; some of the examples will be understood also by non-biologists. Many of the biology topics presented in the book are related to human genomics or human disease, emphasizing the importance of bioinformatics in human medicine. The examples chosen are mainly in the field of sequence bioinformatics. This is a classic area of bioinformatics that has been described previously in many textbooks, but is enjoying renewed interest following current developments in genomics. 'Personal genomics', as touched upon in the final chapters, will be an important area in biomedicine and clinical medicine.

The material in this book is divided into a number of major biological or bioinformatical topics; gene technology, human disease, evolution, gene function, information resources, gene identification methods and personal genomes. Within each of these topics there are different examples of problems that require bioinformatics tools.

The material is organized from the perspective of sequence bioinformatics. First, simple sequence operations such as translation and pattern matching are presented in Chapters 1–5. Chapter 6 deals with RNA secondary structure, and there is a discussion of pairwise alignments and sequence similarity searches in Chapters 7 and 11. Multiple alignments and molecular phylogeny are covered in Chapters 8–10. Different methods of functional assignment are discussed in Chapters 11–13, while molecular sequence databases are discussed in Chapter 14. Finally, gene prediction methods are covered in Chapters 15–17.

From a computational point of view the book focuses on the Unix operating system and the Perl programming language as these are the predominant bases of computational tools in the area of bioinformatics. The Perl content is also organized in a specific fashion, with new concepts introduced in each chapter. For this reason, it is a good idea to read the chapters of the book in the order they are presented. Should a reader contemplate studying the chapters in a different order, Appendix III, providing a short reference guide to Perl, might be helpful. The Perl examples are at a fairly simple level throughout the book, although the Perl code tends to get somewhat more complex towards the end. As mentioned above, a major principle in the design of the book is to convince the reader that relatively simple programming is sufficient to handle many common biological problems. It should also be pointed out that this book is not a complete Unix or Perl reference. In addition, there are more advanced areas of Perl programming that are not covered, such as references and object-oriented programming. A reader seeking information on such topics should consult additional books, such as those listed in Appendix III.

ACKNOWLEDGEMENTS

Nick Lane says in his book *Power, Sex, Suicide: Mitochondria and the Meaning of Life* that 'writing a book sometimes feels like a lonely journey into the infinite, but that is not for lack of support...'. In the same vein there are a number of people that I am indebted to in the context of my own journey into the infinite. They are listed below in a (partly) random order.

A number of people provided help on specific chapters. For the sections on NCBI Entrez and BLAST, I received information and comments from Peter Cooper, Dennis Benson and Eric Sayers, all at the NCBI. Marie–Claude Blatter of the Swiss–Prot communication team provided information about Swiss–Prot. Sean Eddy, at HHMI Janelia Farm Research Campus, provided helpful information regarding HMMER and Infernal. Gunnar Hansson, University of Gothenburg, with whom I collaborated on mucin bioinformatics, had helpful comments on the chapter dealing with these proteins. I'm grateful to Magnus Alm Rosenblad of the Department of Cell and Molecular Biology, University of Gothenburg, for discussions about chloroplast RNAs and many other topics that unfortunately would not fit into this book. Stefan Washietl allowed me to use his code generating double-stranded DNA shown in Appendix III. I'm grateful to Russell Doolittle for feedback on the chapter about blood clotting and to Joe Felsenstein for information about Dnapars. For the story on the HIV criminal case, my sources of information included an article by Pam Lambert in *People* (http://www.people.com) and one article by Stephen G. Michaud at truTV.com (http://trutv.com). I'm also indebted to a large number of anonymous Wikipedia authors.

For the chapter on thylacine, I had much help from Robert Paddle of the Australian Catholic University. In addition, his book, *The Last Tasmanian Tiger*, was a great source of information. Caroline Freeman of the University of Tasmania provided comments on the thylacine chapter, and also supplied a copy of the Burrell photograph. Thanks also to Ellen Alers at the Smithsonian Institution Archives, Washington, for sending me the photograph of the Washington thylacines. Jacqui Ward of the Tasmanian Museum and Art Gallery provided the photograph of two thylacines in Beaumaris zoo. The image of the Darwin termite in the chapter about termites was obtained courtesy of Katja Schultz of the Tree of Life Project (http://tolweb.org) and Smithsonian Institution, National Museum of Natural History.

For the chapters on personal genomes I received help from a number of people. I'm grateful to Adam Siepel and Melissa Jane Hubisz, Department of Biological

Statistics and Computational Biology at Cornell University, for sharing their SNP data from a number of human individuals. I gratefully acknowledge help regarding the Bushmen data from Stephan Schuster and Webb Miller at the Center for Comparative Genomics and Bioinformatics, Penn State University. They also supplied information on the thylacine story. In addition, Stephan Schuster generously provided photographs of the Bushmen individuals. With regard to the chapter on the family quartet, I received comments from Jared Roach, Gustavo Glusman and David Galas at the Institute of Systems Biology, Seattle. In particular, I'm grateful to Gustavo Glusman, who produced simulated data for chromosome 4 and provided a lot of helpful information concerning his processing of genotype data.

A number of people at the University of Gothenburg and at Chalmers University of Technology read my draft manuscript and had very useful comments: Marina Axelson-Fisk, Per Elias, Graham Kemp and Ka-Wei Tang. In particular, I'm also grateful to Marcela Dávila López for her detailed comments and ideas for improvement. Katrina Halliday, Hans Zauner, Lynette Talbot, Jonathan Ratcliffe and other staff at Cambridge University Press were very positive and helpful during the compilation of this book. I also thank Gary Smith for careful copy-editing. In addition, I'm very grateful to the Hasselblad Foundation for awarding me a stipend to spend two months in Grez-sur-Loing in France to work on the book. Special thanks to Birgitta Bergenholtz at the Foundation and Bernadette Plissart at Hotel Chevillon in Grez. I take this opportunity to apologize to the students of a bioinformatics course that I notoriously neglected while I was in France. I also sincerely thank my three 'A's, Annika, Anders and Anna for their contributions, including a set of Lego pieces, but most of all I acknowledge their great support and patience during the time I put together this book.

Finally, as an important source of inspiration I would like to mention my mentor and former supervisor Ulf Lagerkvist, who tragically passed away in 2010. He was an inspiration to all his students, not only because of his scientific achievements and attitude towards science, but also because he authored a number of highly readable books in the areas of life sciences and scientific history.

DESIGN AND CONVENTIONS OF THIS BOOK

This book is designed such that it covers a number of biological topics, one in each chapter. The topics are arranged in the following major categories:

- introduction to genetic information (one chapter)
- gene technology (three chapters)
- human disease (three chapters)
- evolution (three chapters)
- gene function (three chapters)
- information resources (one chapter)
- gene identification methods (three chapters)
- personal genomes (three chapters).

In each of the chapters one or more specific problems are addressed in a bioinformatics section where Perl, Unix or other bioinformatics software are used. The Unix or Perl topics that are novel to the chapter are listed in a box at the beginning of the bioinformatics section. In the bioinformatics section of each chapter the following conventions are used. Some text is presented in a coloured `fixed-width font`. These are (1) Unix command lines; (2) Perl code; and (3) names of files, programs or Unix utilities. Whenever something is to be typed at the Unix command line, this is indicated with a % symbol, to represent the Unix command line prompt. Thus, a reader trying these commands at his/her computer should *not* type the % symbol. An example would be:

```
% uname
```

which means that by typing 'uname' the program `uname` (a Unix utility to print system information) will be executed.

Complete Perl scripts are present within specifically highlighted boxed areas. The Perl scripts are, to some extent, explained and commented within the actual scripts, but mainly in the text preceding or following the script. All files that are used in the examples of this book, including all Perl scripts, are available for download at the supplementary website (www.cambridge.org/samuelsson).

For more background and practical information on Unix and Perl, the reader is referred to Appendices I and III. Appendices I and III also contain suggestions for further reading. A selection of bioinformatics software that is used in a Unix environment is presented in Appendix II. The web resources provided with this book have more information, such as solutions to the Perl exercises of the book, Python examples and a listing of bioinformatics resources.

Some of the figures in this book were created with R, a free software environment for statistical computing and graphics (http://www.r-project.org). In such cases, the R scripts are available for downloading from the web resource for this book. The scripts are not explained in any detail, but a short R reference is provided in Appendix IV.

Throughout this book it is assumed that the reader has access to a computer running a Unix operating system and that Perl is installed on this system. For more general background and technical information about Unix and Perl, see Appendices I and III.

This book assumes the reader has a basic knowledge of molecular biology, biochemistry or cell biology. In case the reader needs more background information in these areas, the following are all examples of excellent textbooks:

Alberts, B. (2008). *Molecular Biology of the Cell: Reference Edition*. New York, Garland Science.
Barton, N. H., D. E. G. Briggs, J. A. Eisen, D. B. Goldstein and N. H. Patel (2007) *Evolution*. Cold Spring Harbor, NY, Cold Spring Harbor Laboratory Press.
Berg, J. M., J. L. Tymoczko and L. Stryer (2010). *Biochemistry*. New York, W. H. Freeman.
Lodish, H. F. (2008). *Molecular Cell Biology*. New York, W. H. Freeman.

Introduction
Working with the molecules of life in the computer

[Jim Kent] embarked on a four-week programming marathon, icing his wrists at night to prevent them from seizing up as he churned out computer code by day. His deadline was June 26, when the completion of the rough draft was to be announced.

(*DNA: The Secret of Life*, describing Jim Kent's efforts in the human genome sequencing project in 2000; Watson and Berry, 2003)

I t was a somewhat historic event when President Bill Clinton announced, on 26 June 2000, the completion of the first survey of the entire human genome. We were able for the first time to read all three billion letters of the human genetic make-up. This information was the ground-breaking result of the Human Genome Project. The success of this project relied on advanced technology, such as a number of experimental molecular biology methods. However, it also required a significant contribution from more theoretical disciplines such as computer science. Thus, in the final phase of the project, numerous pieces of information like those in a giant jigsaw puzzle needed to be appropriately combined. This step was critically dependent on programming efforts. Adding further tension to the programming exercises was the fact that a private company, Celera, was competing with the academic Human Genome Project. This competition was sometimes referred to as 'the Genome War' (Shreeve, 2004). While computationally talented people like Jim Kent 'churned out computer code', other gifted bioinformaticians, such as Gene Myers at Celera, worked on related jigsaw-puzzle problems. Ideally, scientists should not war against each other; however, there was an important conclusion from these projects in which important genetic information was generated: *computing is an essential part of biological research*.

This book will present a number of examples from the world of biology and biomedicine, where programming can significantly help us out. This first chapter will discuss very basic biology concepts and will introduce fundamentals of genetic information at the molecular level. We will then go on to examine some simple programming code to represent and process that genetic information. The code will not be the advanced work of the programmers mentioned above, but it is a start.

Life on Earth and evolution

What are the characteristics of life on Earth? We observe a wealth of living organisms in a large number of ecological niches; on land and in fresh- and seawater; in the soil and deep beneath the Earth's surface. There is a variety of organism types, all the way from small viruses and bacteria to plants and animals.[1] As we compare many of the life forms, they seem very different. For example, an oak tree, a cow and a slime mould do not look that similar to the human eye. But in a sense these differences are somewhat superficial. In fact, a closer look at organisms at a microscopic or molecular level reveals that they are strikingly similar. For instance, all living forms (with the exception of viruses) are built from cells. All cells have certain characteristics in common. For instance, each cell is surrounded by a membrane. Passage of molecules across the membrane is rigorously controlled, and as a result the environment within a cell is very different from that of the outside world. All kinds of cells have a general biochemical machinery in common. Thus, the basic metabolism of living cells is very similar in all kinds of organisms. As one example, a large majority of cells have the capacity to extract energy by oxidation of carbohydrates. Yet another principle shared by all organisms is a basic system for storage and processing of genetic information. That information is stored in the form of a DNA molecule, and the genetic message is a long sequence of chemical units referred to as *bases* in DNA. Furthermore, there is a universal system for production of RNA and protein molecules as specified by this genetic information, as will be explained in some detail later in this chapter.

Why are species similar at the molecular level? The simple answer is one major principle of life: species are related by *evolution*. Evolution is a major key to the understanding of life. Or, as expressed by Theodosius Dobzhansky in his famous quotation: 'Nothing in biology makes sense except in the light of evolution', where he argues against creationism and intelligent design (Dobzhansky, 1964).

What are the basic concepts of evolution? One key element is that all species on our planet are related and have a common ancestor. Our planet is now about 4500 million years old; the first living forms appeared when it was about

---

[1] To be more strict about this, there are three major kingdoms of life: eubacteria, archaea and eukaryotes. The eubacteria and archea are single-cell organisms, whereas eukaryotes typically are multicellular organisms (e.g. plants and animals). Eukaryotes have a more organized intracellular structure, with a membrane-surrounded nucleus and organelles such as mitochondria. In addition to the organisms of the three kingdoms of life, there are viruses. These are not free-living organisms as they are dependent on their host organism for propagation.

1000 million years old. It is believed that life came about as a result of favourable conditions, such as the presence of water and other critical compounds, together with a suitable temperature range. Early life forms were very primitive compared to many of the living organisms that we observe today. Eventually, further evolution gave rise to diversification and new, more 'complex' species. The total number of species on our planet today is not known exactly, but exceeds two million. Although a systematic inventory of species was started by Carl von Linneaus and others in the eighteenth century, new species are still being identified today.

For evolution to proceed there must be a certain degree of change in the genetic message as it is being transmitted between generations. Such changes are referred to as *mutations*. Another key element to understanding evolution is the process of *selection*, or 'survival of the fittest', as was one early wording. This selection determines what mutations will stay in a population. To put it simply, the evolution of species is determined by factors such as environmental conditions and by competition between different individuals and species. For instance, bacteria that are able to adapt rapidly to a change in their environment may be more likely to survive compared to bacteria that do not have this property. As an example from more recent evolution, humans with enhanced brain functions presumably had a selective advantage compared to primates that did not.

What is the scientific evidence of evolution? What is, for instance, the evidence that species such as man developed from more 'primitive' species? We will not go into detail about this here, but Charles Darwin accumulated an overwhelming amount of data in support of his theory of evolution. Today, we also have detailed information about the genetic information of many species and are able to monitor changes during the course of evolution in remarkable detail. With this recent information there is even more overwhelming support of Darwinian theory.

There is no doubt that the concept of evolution is fundamental in genomics and bioinformatics, and many of the issues brought up in this book are directly or indirectly related to this concept. For instance, consider a common situation where biologists have identified a specific sequence of bases in a DNA molecule present in humans and they want to attribute this sequence to a biological function. This sequence, or a similar one, might be present in another species – let's say mice. This is because sequences of bases, just as species, are related by evolution. We may also find that the function of the sequence in mice has been elucidated. In such a case we may infer a function of the human sequence, based on what we have collected by studying mice. We will see a few cases of this sort later in the book (Chapters 11–13). These are examples of situations where we are making use of the concept of evolution without necessarily being interested in evolution as such. Then again, we may actually be interested in evolutionary events; topics like these are dealt with in Chapters 8–10.

In summary, evolution is fundamental in bioinformatics.[2] If we are to understand more about the actual mechanisms of evolution, we need to know more about how mutations occur and what they mean in terms of biological information. For this, in turn, we need to know more about the molecular genetic machinery.

The machinery of genetic information: more about DNA

We have already seen that the DNA (deoxyribonucleic acid) molecule carries genetic information, i.e. the information that will be transmitted from one generation to the next. DNA is copied in a process called *replication*. The information in DNA ultimately guides the production of different proteins. First in this process, an RNA (ribonucleic acid) copy is made in a process known as *transcription*. Non-informational portions of the primary product of transcription (introns) are removed in a process called *splicing*. The RNA then directs the *synthesis of protein* (a process also called *translation*). See Fig. 1.1 for an overview of transcription, splicing and translation.

Each protein has a specific function in the cell. In a human cell there are in the order of 21 000 different proteins. These come in a variety of functional classes. One important category is *enzymes*. These are biocatalysts responsible for catalysing thousands of chemical reactions within the cell, reactions that would not be possible without enzymes. Other important functions of proteins are to act in the transport of molecules and to act as a 'skeleton', maintaining a specific cell architecture. Proteins are the molecules that directly determine all properties of a living cell. DNA does not do anything like that; it simply carries the genetic information that specifies which proteins are to be made.

We will now examine how the flow of information occurs from the DNA to the protein in more detail. First, we have to understand the basic chemistry of the DNA molecule. It is a very long polymer with repeating sugar and phosphate units. Attached to the sugars are bases. The bases are the variable units in DNA, and it is the sequence of bases that constitutes the actual genetic information.

A DNA molecule has a distinct chemical polarity. One end of the DNA is referred to as the 5′ end, because the hydroxyl group in that end is attached to

---

[2] If, for some reason, you prefer not to believe in this basic Darwinian concept (and there are in fact people that tend to support other ideas about the development of species) genomics and bioinformatics will not make sense to you at all and there is no point in reading further.

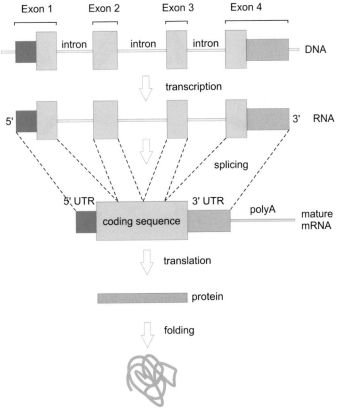

Fig. 1.1 *Flow of genetic information in a eukaryotic cell.* A primary RNA transcript is first produced in the process of transcription. This transcript is subject to splicing, where the exons are joined. The mature mRNA contains the exons as well as a polyA tail at its 3′ end. This RNA has a 5′ untranslated region (5′ UTR, green), a coding sequence (yellow) and a 3′ UTR (blue). The coding sequence determines the sequence of amino acids that are incorporated into the protein during translation. Finally, the amino acid sequence of the protein determines the folding of the protein into a three-dimensional structure. This structure has specific biological properties. Cells from all kingdoms of life are characterized by this flow of genetic information, although bacterial genes do not have introns and are therefore not subject to splicing.

a carbon of the sugar and that carbon has the number five, according to the numbering scheme for the sugar structure. Similarly, the other end of the DNA molecule is referred to as 3′. The polarity of DNA means that the sequence of bases read from one end of the molecule is not equivalent to the sequence of bases read from the other end.

The bases are referred to as adenine, guanine, cytosine and thymine, abbreviated as A, G, C and T (Fig. 1.2). The A and G bases have a double ring structure and are referred to as *purines*, whereas C and T have a simple ring structure

Fig. 1.2 *Three-dimensional structure of the DNA double helix.* The sugar–phosphate backbones of the two strands are shown as red and blue ribbons. Base-pair interactions between guanine and cytosine – with three hydrogen bonds – and between adenine and thymine – with two hydrogen bonds – are shown in the lower panel. Structures are from the Protein Data Bank entry 7BNA. The sequence of both DNA strands is CGCGAATTCGCG. The figure was produced with the UCSF Chimera software (Pettersen *et al.*, 2004).

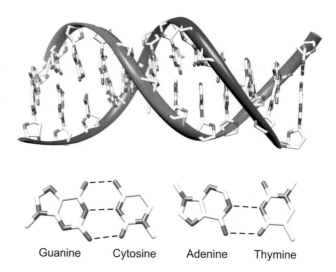

Guanine   Cytosine   Adenine   Thymine

and are called *pyrimidines*. The sugar–phosphate–base unit in DNA is referred to as a *nucleotide*. From an information perspective a nucleotide is equivalent to a base because the base is the only variable unit in DNA.

We now arrive at one basic concept of sequence bioinformatics. The genetic message in DNA can be represented as a simple string of the letters A, G, C and T. For example:

```
5' - AGGACACGACGACTATTGG - 3'
```

Normally you see the sequence written in the direction shown here, i.e. with the 5' end to the left and the 3' end to the right.

DNA may sometimes occur as a *single-stranded* molecule (as in certain viruses), but normally we find DNA as a *double-stranded* unit. It has a double helical structure with two antiparallel strands. In this case the word 'antiparallel' means that the two strands run in 'parallel', but have opposite polarity.

The structure of double-stranded DNA was elucidated by Francis Crick and James Watson and was presented in a famous paper in *Nature* in 1953, an *annus mirabilis* for science[3] (Watson and Crick, 1953) (Fig. 1.2). The paper was a landmark in molecular biology, and in 1962 the two authors received the Nobel Prize in Physiology or Medicine for their discovery. The double-stranded DNA is held together by pairing between bases. Chemical bonds known as hydrogen bonds are formed between bases and the pairing is always such that adenine pairs with thymine and cytosine pairs with guanine. This distinct pairing is referred to as *complementarity*, and one strand of DNA is said to be the complement of the other. The double-stranded DNA may thus be represented as two strings, like this:

[3] http://www.nature.com/nature/dna50/archive.html.

```
5′ – AGGACACGACGACTATTGG – 3′
3′ – TCCTGTGCTGCTGATAACC – 5′
```

Note that the two strands have opposite polarity. As realized early by Watson and Crick, the base complementarity is the basis for replication of DNA. In this process one strand of DNA serves as the template for the synthesis of the other.

When sequences of DNA are deposited in databases, there is no need to put both strands there, as one of them may easily be inferred from the other. Similarly, only one strand is provided as input to many computer programs for sequence analysis, although the other strand is implicit. There is a somewhat confusing terminology as to the two strands of DNA (the reference strand and its complement), where you can encounter expressions such as plus/minus, forward/reverse, top/bottom, sense/antisense or even Watson/Crick.

The fact that the information in DNA may be represented as a long string of characters is of fundamental importance in sequence bioinformatics. It is actually quite surprising how useful the string representation is, as the DNA molecule has a specific three-dimensional structure, and as such is far from one-dimensional.

Genes and genomes

The example above shows a relatively short piece of DNA sequence. In reality, DNA molecules are very long. For instance, the genetic material of humans is distributed among *chromosomes*. There are 22 human chromosomes in addition to the X and Y chromosomes. Each chromosome is, in essence, a very long DNA molecule. The typical length of a human chromosome is about 100 million base pairs. We refer to the base or nucleotide sequence of all chromosomes of an organism as the *genome* of that species. Thus, when we say 'the human genome' we should think of the genetic information in *all* the human chromosomes. Genomes and the biological signals they contain are studied in the science referred to as *genomics*.

More than ten years ago the complete base sequence of all human chromosomes, more than three billion bases, was determined. This was indeed a significant advance in biological research (Lander *et al.*, 2001; Venter *et al.*, 2001). Among many different important medical applications, it helps us to better understand all diseases with a genetic background. Knowing the complete sequence of the human genome, we come to the obvious question: what exactly is the information carried within the genome? What biological signals are found as we examine a human chromosome from one end to the other? To begin with, one important category of elements are *genes*. A gene is a portion of the DNA molecule that contains all the information for production of a protein. The length of human genes is highly variable; they can be as small as a few thousand and as large as one million bases. Some genes are specified by one of the strands of DNA, and others are specified by the other (complementary) strand.

Genes at work: transcription and translation

What matters to the physiology of a cell are the processes in which DNA serves as a template for production of RNA and protein products. The first step in the flow of genetic information is transcription, in which the information in DNA is copied into an RNA molecule. RNA is a nucleic acid, in the same way as DNA. However, RNA has a sugar unit that is ribose, instead of deoxyribose. In addition, RNA contains the base uracil (abbreviated as U) instead of thymine (T). Like DNA, RNA has 5′ to 3′ polarity.

One of the strands in DNA serves as the template for the synthesis of RNA. The RNA produced is, unlike DNA, single-stranded. Note that the sequence of the RNA produced is complementary to one of the DNA strands and it is equivalent in sequence (although U replaces T) to the other DNA strand (Fig. 1.3).

Transcription takes place with the help of the enzyme *RNA polymerase*. The biochemical machinery of transcription is to some extent similar to DNA replication, because in both of these processes a new strand of nucleic acid is formed by copying information from a template strand of DNA.

The different RNAs produced by transcription have different functions in the cell. However, a very important functional class of RNAs is *messenger RNA (mRNA)*. These RNAs contain information for the production of proteins. Proteins are large polymers like DNA and RNA, but the building blocks are amino acids rather than nucleotides. There are 20 different amino acids. These are presented along with some of their properties in Table 1.1.

During protein synthesis the sequence of bases in RNA guides the incorporation of amino acids into proteins. Protein synthesis is also named *translation* because a 'translation' occurs from the language of nucleic acids (DNA/RNA) to the language of proteins. Information contained in a protein may be represented as a string of letters, the same as it can be for RNA and DNA. As there are 20 different amino acids, the protein alphabet contains 20 characters (Table 1.1), compared to four in the nucleic acid alphabet (A, T, C and G in the case of DNA).

A set of well-defined rules determine the relationship between the RNA sequence and the sequence of amino acids in proteins. A sequence of three bases, known as a *codon*, specifies a distinct amino acid. The *genetic code* is a table showing

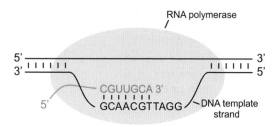

Fig. 1.3 *Transcription.* Double-helical DNA is unwound, allowing RNA polymerase to synthesize RNA in 5′ to 3′ direction (in orange) using one of the strands of DNA as a template. The RNA polymerase moves from left to right.

Table 1.1 *The amino acids*

Amino acid	Physicochemical property	Three-letter abbreviation	One-letter abbreviation
Alanine	Non-polar	Ala	A
Glycine		Gly	G
Cysteine	Non-polar, sulfhydryl group	Cys	C
Isoleucine	Non-polar, aliphatic	Ile	I
Leucine		Leu	L
Methionine		Met	M
Valine		Val	V
Proline	Aliphatic, side chain joined to both α carbon and amino group	Pro	P
Serine	Aliphatic – hydroxyl group	Ser	S
Threonine		Thr	T
Phenylalanine	Aromatic	Phe	F
Tryptophan		Trp	W
Tyrosine		Tyr	Y
Asparagine	Polar	Asn	N
Glutamine		Gln	Q
Aspartic acid	Acidic	Asp	D
Glutamic acid		Glu	E
Arginine	Basic	Arg	R
Histidine		His	H
Lysine		Lys	K

what codons correspond to what amino acids (Fig. 1.4). The genetic code was elucidated in the 1960s. Later, we will see how this code may be represented in a Perl program.

There are 64 codons in the genetic code and 61 of these specify an amino acid. The other three, UAA, UAG and UGA, are termination signals during protein synthesis. Because there are 61 codons that specify an amino acid and there are only 20 amino acids, for most amino acids there is more than one codon; because of this, the genetic code is said to be *degenerate*. The codon AUG, coding for methionine, is also used as a start codon, i.e. the start site of translation will (almost) always be the codon AUG.[4] Fig. 1.5 shows a region of bacterial mRNA

---

[4] The codon AUG is also used to encode internal methionines in a protein. Therefore, there are additional criteria to distinguish a start AUG from an internal AUG. For instance, in bacteria there is a specific sequence upstream of the start codon that helps in initiating protein synthesis.

UUU Phe F	UCU Ser S	UAU Tyr Y	UGU Cys C
UUC Phe F	UCC Ser S	UAC Tyr Y	UGU Cys C
UUA Leu L	UCA Ser S	UAA Stop	UGA Stop
UUG Leu L	UCG Ser S	UAG Stop	UGG Trp W
CUU Leu L	CCU Pro P	CAU His H	CGU Arg R
CUC Leu L	CCC Pro P	CAC His H	CGC Arg R
CUA Leu L	CCA Pro P	CAA Gln Q	CGA Arg R
CUG Leu L	CCG Pro P	CAG Gln Q	CGG Arg R
AUU Ile I	ACU Thr T	AAU Asn N	AGU Ser S
AUC Ile I	ACC Thr T	AAC Asn N	AGC Ser S
AUA Ile I	ACA Thr T	AAA Lys K	AGC Arg R
AUG Met M	ACG Thr T	AAG Lys K	AGG Arg R
GUU Val V	GCU Ala A	GAU Asp D	GGU Gly G
GUC Val V	GCC Ala A	GAC Asp D	GGC Gly G
GUA Val V	GCA Ala A	GAA Glu E	GGA Gly G
GUG Val V	GCG Ala A	GAG Glu E	GGG Gly G

Fig. 1.4 *The universal genetic code.* Every three-base word (codon) corresponds to an amino acid (in three- and one-letter abbreviations), except UAA, UAG and UGA, which are termination signals during protein synthesis.

encoding a protein; a region starting with an AUG codon and ending with a stop codon.

What is the machinery responsible for carrying out the instructions as presented in the genetic code? Critical molecules are *transfer RNAs* (tRNAs), which act as adaptors between the languages of nucleic acids and proteins. This is possible because every tRNA carries a sequence, the *anticodon*, which is complementary to the codon on the mRNA, and to the amino acid corresponding to the codon.

The protein formed in the process of translation is a linear sequence of amino acids. That sequence will govern folding of the protein molecule into a distinct three-dimensional shape. Even more importantly, that shape is associated with one or more specific biological functions.

In summary, all cells are characterized by a flow of genetic information where DNA is first copied to RNA, which in turn is used to guide the production

```
AAAAUUCCUAUGAAGGUGAAUUUAGAGUGGAUAAUUAAACAGUUACAAAUGAUAGUUAAA
         M  K  V  N  L  E  W  I  I  K  Q  L  Q  M  I  V  K
AGAGCAUAUACUCCCUUUUCUAACUUUAAAGUUGCAUGUAUGAUUAUUGCUAACAACCAA
 R  A  Y  T  P  F  S  N  F  K  V  A  C  M  I  I  A  N  N  Q
ACUUUUUUUGGAGUUAACAUUGAAAAUUCUUCCUUUCCAGUAACUUUGUGUGCUGAAAGA
 T  F  F  G  V  N  I  E  N  S  S  F  P  V  T  L  C  A  E  R
AGCGCCAUUGCUAGCAUGGUUACAAGUGGUCAUAGGAAAAUUGAUUAUGUUUUUGUUUAC
 S  A  I  A  S  M  V  T  S  G  H  R  K  I  D  Y  V  F  V  Y
UUCAAUACUAAAAAUAAGAGUAACUCACCCUGUGGAAUGUGCAGACAAAACUUACUGGAA
 F  N  T  K  N  K  S  N  S  P  C  G  M  C  R  Q  N  L  L  E
UUUUCCCAUCAAAAAACAAAGCUUUUUUGUAUUGAUAAUGAUAGUAGUUAUAAACAAUUU
 F  S  H  Q  K  T  K  L  F  C  I  D  N  D  S  S  Y  K  Q  F
UCCAUUGAUGAAUUAUUAAUGAAUGGUUUUAAAAAGAGCUAAUGGAUAAACUUAGAUUA
 S  I  D  E  L  L  M  N  G  F  K  K  S  STOP
```

Fig. 1.5 *Open reading frame (ORF).* A region of the genome of the bacterium *Mycoplasma genitalium* that encodes the protein cytidine deaminase is depicted. The start site of translation, the codon AUG, is indicated as well as the stop codon UAA. The sequence in orange is the amino acid sequence of the deaminase protein, with one-letter symbols for the amino acids as listed in the genetic code in Fig. 1.4 and Table 1.1.

of a protein. Finally, the protein folds up into a functional unit based on its sequence of amino acids (Fig. 1.1). The processes of gene transcription and translation are collectively referred to as *gene expression*. An important aspect of gene expression is that not all genes are expressed at the same level; rather, transcription and translation are regulated processes.

Organization of the human genome

Adding to the complexity of eukaryotic genes is the fact that most of them are mosaics of *exons* and *introns* (Fig. 1.1). This genomic structure of the gene is maintained in the primary RNA transcript, but in the process of *splicing*, all exons will be joined, while the introns are removed as a kind of waste material (for more about splicing, see Chapter 16). The splicing process thus generates the mRNA sequence as it is presented to the protein-synthesizing machinery. In a typical human gene, most introns are very much longer than exons; in addition, many intergenic regions are very long. Taken together, this has the effect that only a minor portion of the human genome is composed of coding sequences. The human genome is also characterized by a large number of repetitive sequences, typically placed in introns and intergenic regions, which make up nearly half of the human genome. The human genome is distributed between 22 different linear chromosomes in addition to the X and Y chromosomes. About 21 000 different protein-coding genes have been identified in these chromosomes. They all have names – for example, ABL1, BRCA1 or TAS2R38 – following a nomenclature as decided by the HUGO Gene Nomenclature Committee (www.genenames.org).

BIOINFORMATICS

We turn now to the basic flow of genetic information from DNA to RNA and from RNA to proteins, and to some computational aspects of this information flow. How can we computationally represent the different elements of genetic information storage and expression – DNA replication, transcription and translation? We will turn to some Perl code for this.

For the following sections of this chapter, as well as throughout the book, it is assumed that the reader has access to a

Tools introduced in this chapter		
Perl	basic syntax	
	defining variables	
	scalars and hashes	
	the functions `print`, `reverse`, `substr` and `length`	
	the transliteration `tr///` operator	
	`for` loops	

computer running a Unix operating system and that Perl is installed on this system. For more general background and technical information about Unix and Perl, see Appendices I and III.

Inferring products of DNA replication

We will first consider some Perl code related to the process of DNA replication. More precisely, we will address the problem of producing the reverse complement of a DNA sequence. If we have the sequence GCAATGG as one of the two strands in a length of double-stranded DNA, we want to infer the sequence of the complementary antiparallel strand. For this problem, we create a file with the following lines:

```
$dna = 'GCAATGG';
$rev = reverse($dna);
$rev =~ tr/ATCG/TAGC/;
print "$rev\n";
```

In the first line we define a sequence of bases and we store it in the variable referred to as $dna. All variables of this type in Perl have the dollar ($) character in front of them.[5] In this case, the variable $dna is specified as containing a string. A string is typically enclosed by single (') or double (") quotes.

We now want to convert the sequence in $dna to its reverse complement. In Perl we could do that in different ways, but in this simple example we do the conversion in two steps. First, we reverse the order of bases in the DNA, and then we replace each base with its complement.

The second line of the code does the reversing of characters using the function reverse. A function in Perl is some operation carried out on one or more variables, shown within parentheses following the name of the function. You can define functions yourself, but in this case the reverse function is built into the Perl language. This function will reverse the order of characters in a string – in this case, the bases of DNA. This is to say, reversing the string GCAATGG will result in GGTAACG. It is convenient to have the result of the operation stored as yet another variable. In this case the new string will be stored in $rev.

The third line performs an operation to replace bases on the basis of complementarity:

```
$rev =~ tr/ATCG/TAGC/;
```

This statement is using something called the *transliteration* operator. (This operator is distantly related to the world of *pattern-matching* operations to be

---

[5] In this respect, Perl is different from most other programming languages.

discussed in the next chapter.) The transliteration or `tr///` operator has the form `tr/SEARCHLIST/REPLACEMENTLIST/`, meaning that the characters found in `searchlist` are to be replaced with the characters in `replacementlist`. The `=~` symbol is the *binding operator* and the two / symbols are delimiters. The way to interpret the statement `$rev =~ tr/ATCG/TAGC/;` is 'the variable `$rev` is subject to the `tr` operation, i.e. the content of `$rev` will change as a result of this operation'. Replacements are done such that the first character in the search list is replaced with the first character in the replacement list, while the second character in the search list is replaced with the second character in the replacement list, and so on. In this case, all As will be replaced by Ts, all Ts by As, etc. The whole statement therefore means, in plain language, 'take the sequence stored in `$rev`, replace characters according to the base-pair rules and store this new sequence in `$rev`'.

Finally, with the fourth line we print the contents of `$rev` to the screen. The `"\n"` expression refers to a new line character, i.e. we want to have a new line following the contents of the `$rev` variable.

In our statement above that reads `$dna = 'GCAATGG'`, the string is enclosed within single quotes, whereas in the line reading `print "$rev\n"` the `$rev` variable is within double quotes. What is the difference between single and double quotes? The double quotes are special because they indicate that we are to do *variable interpolation*, meaning that `"$rev"` refers to the *contents* of that variable. If we had used `'$rev'` instead, exactly `$rev` would have been printed, i.e. the *name of the variable and not its contents*.

All of the lines above are *statements*, and as such they need to end with a semicolon (;). The line breaks make the code more readable. However, they are not strictly required and we could also have put all of the code in one line or with one line break, like this:

```
$dna = 'GCAATGG'; $rev = reverse($dna); $rev =~ tr/ATCG/TAGC/;
print "$rev\n";
```

We now store the lines of code in a file named `complement.pl`. Perl programs often have the `.pl` extension, but this is not necessary; you may call your Perl code file almost anything. This Perl program can now be executed by typing the following at the Unix command line:

```
% perl complement.pl
```

You should now see the following output from this program:

```
CCATTGC
```

And that was our first, very simple, Perl program!

Inferring RNA products of transcription

Next, let us consider transcription. We can modify the code above to produce the RNA product from transcription:

```
$dna = 'GCAATGG';
print "The DNA sequence is $dna\n";
$rna = $dna;
$rna =~ tr/T/U/;
print "and the RNA sequence is $rna\n";
```

This is very similar to the code in the reverse complement exercise, but we are now assuming the original DNA sequence is the same as the RNA sequence, except that we replace T (thymine) with U (uracil). Again, we are using the transliteration operator tr/// to do the replacement.

Inferring protein products of translation

For the final part of the genetic information flow we want to have a piece of code to translate an RNA sequence into a sequence of amino acids (Code 1.1). The problem of translation is somewhat more complex than the previous examples, because we need to read a sequence of codons from the DNA (or RNA) and we need to handle the genetic code, i.e. the conversion from codons to amino acids.

First, we think about how to represent the genetic code. There are different kinds of variables in Perl, including *scalars*, *arrays* and *hashes* (see Appendix III). Scalars are the fundamental data type, typically strings or numbers. The genetic code is most conveniently represented as a hash. You can think of a hash in this case as a two-column table in which each row contains two connected elements. For this example all rows contain a codon, as well as the amino acid represented by it. In a hash table we refer to *keys* and *values* (in this example the codons and amino acids, respectively). Names of hashes are always preceded by '%'. We may call the hash table in this case %code, and we can write it as:

```
%code = ('UUU' => 'Phe',
         'UUC' => 'Phe',
         ... );
```

You can think of the symbol '=>' as meaning 'corresponds to' or 'is linked to'. The hash may also be written as below, but typing it in this way, it is perhaps less clear what values belong to what keys.

```
%code = ('UUU', 'Phe',
         'UUC', 'Phe',
         ... );
```

Once the hash table is created, we can use it to find out which amino acid corresponds to a specific codon. For example, for UUU we use the expression:

```
$code{'UUU'}
```

This expression evaluates to 'Phe'. Please note that while the whole hash table is referred to as %code, a value of the hash table is referred to using $code and the key enclosed within curly brackets. This means that we could also have defined the whole hash using this form:

```
$code {'UUU'} = 'Phe';
$code {'UUC'} = 'Phe';
# etc ...
```

We now turn to the problem of reading the codons from the initial sequence. To do this we make use of a *loop statement*, as well as the function substr to extract substrings.

A loop statement is typically used whenever some operation is to be repeated a number of times. A simple example is the following loop, in which a number is increased:

```
for ($i = 0; $i < 5; $i++) {
    print "$i ";
}
```

In this code $i = 0 describes the starting condition and $i < 5 tells the loop to continue as long as $i is less than 5. $i++ is shorthand for $i = $i + 1, meaning we increase $i by 1 with each round of the loop. The code above will print:

```
0 1 2 3 4
```

The substr function is used to extract portions from a string variable. It may have the form substr (variable, offset, length). The first argument is the string variable name, the second is the position of the first character to be extracted, and the third argument is the number of characters to be extracted from the position specified by the second argument. For example:

```
print substr ('AGCTT', 3, 2);
```

This will print 'TT'. Please note that the positions in the string are numbered from 0. Hence, in the string 'AGCTT', 'A' has position 0, 'G' position 1, etc.

Consider these principles as applied in Code 1.1. This code will extract codons from a sequence and identify the corresponding amino acids.

Code 1.1 translation.pl

```perl
# The genetic code is represented as a hash. Each codon (hash key)
# is associated with an amino acid (hash value).
# A stop codon is shown as '*'

%code = (
    'UUU' => 'F', 'UUC' => 'F', 'UUA' => 'L', 'UUG' => 'L',
    'CUU' => 'L', 'CUC' => 'L', 'CUA' => 'L', 'CUG' => 'L',
    'AUU' => 'I', 'AUC' => 'I', 'AUA' => 'I', 'AUG' => 'M',
    'GUU' => 'V', 'GUC' => 'V', 'GUA' => 'V', 'GUG' => 'V',
    'UCU' => 'S', 'UCC' => 'S', 'UCA' => 'S', 'UCG' => 'S',
    'CCU' => 'P', 'CCC' => 'P', 'CCA' => 'P', 'CCG' => 'P',
    'ACU' => 'T', 'ACC' => 'T', 'ACA' => 'T', 'ACG' => 'T',
    'GCU' => 'A', 'GCC' => 'A', 'GCA' => 'A', 'GCG' => 'A',
    'UAU' => 'Y', 'UAC' => 'Y', 'UAA' => '*', 'UAG' => '*',
    'CAU' => 'H', 'CAC' => 'H', 'CAA' => 'Q', 'CAG' => 'Q',
    'AAU' => 'N', 'AAC' => 'N', 'AAA' => 'K', 'AAG' => 'K',
    'GAU' => 'D', 'GAC' => 'D', 'GAA' => 'E', 'GAG' => 'E',
    'UGU' => 'C', 'UGC' => 'C', 'UGA' => '*', 'UGG' => 'W',
    'CGU' => 'R', 'CGC' => 'R', 'CGA' => 'R', 'CGG' => 'R',
    'AGU' => 'S', 'AGC' => 'S', 'AGA' => 'R', 'AGG' => 'R',
    'GGU' => 'G', 'GGC' => 'G', 'GGA' => 'G', 'GGG' => 'G'
);
$dnaseq = 'GAACTGGGT';      # the DNA sequence

print "$dnaseq\n";

$rnaseq = $dnaseq;
$rnaseq =~ tr/T/U/;         # the RNA sequence

print "$rnaseq\n";

for ( $i = 0 ; $i < length($rnaseq) - 2 ; $i = $i + 3 ) {
    $codon = substr( $rnaseq, $i, 3 );
    $amino_acid = $code{$codon};
    print " $amino_acid ";
}
print "\n";
```

In this case our starting point is a DNA sequence of nine bases. We convert the sequence from DNA to RNA, and the `for` loop towards the end of the Perl code has the procedure for the actual translation. One codon at a time is extracted with the `substr` function. We start out with `$i = 0` and increase `$i` by three in each step. Because we move in steps of three bases, the values of `$i` will be 0, 3,

6, etc. How do we know when to end the codon extraction and translation? We make use of another function, `length`, which returns the length of a string. In this case the length of the `$rna` string is nine. As the last codon will be extracted using `$i = 6`, there is no point in going on with larger values of `$i`. Therefore, `$i < length($rna) - 2` means that `$i` is less than the (length of `$rna`) minus 2 (9 − 2 = 7). The loop is continued until this condition is no longer met, i.e. when `$i` is equal to or greater than 7.

Running the script in Code 1.1 will result in the following being printed to the screen:

```
GAACTGGGT
GAACUGGGU
 E  L  G
```

This is, of course, a ridiculously short piece of DNA sequence to be translated, and this problem could more easily have been handled manually. However, if we were to do this with a DNA sequence containing thousands of bases, manual translation would be daunting.

This first chapter presented a method for manipulating and translating nucleotide sequences. If you did not find these Perl examples too discouraging, remember that Perl is not necessarily more advanced than this, and you can do wonders in bioinformatics by using code at this level of difficulty!

EXERCISES

1.1 Write a piece of Perl code that will print a specific DNA sequence as well as its complementary strand. Thus, assuming we start out with a DNA strand with the sequence GAACTGGGT, the code should elucidate the complementary strand and print like this:

```
5′ GAACTGGGT 3′
3′ CTTGACCCA 5′
```

1.2 Modify Code 1.1 so the complementary strand of the original strand sequence GAACTGGGT is translated. As with the original strand, we only consider the first reading frame of the nucleotide sequence, i.e. the first codon of the complementary strand is ACC.

1.3 Modify Code 1.1 so that not only the first reading frame, but also the other two are presented. Thus, the modified code should present the following type of output:

```
GAACTGGGT
 E  L  G
  N  W
   T  G
```

REFERENCES

Dobzhansky, T. (1964). Biology, molecular and organismic. *Am Zool* **4**, 443–452.

International Human Genome Sequencing Consortium (2004). Finishing the euchromatic sequence of the human genome. *Nature* **431**(7011), 931–945.

Lander, E. S., L. M. Linton, B. Birren, *et al.* (2001). Initial sequencing and analysis of the human genome. *Nature* **409**(6822), 860–921.

Pettersen, E. F., T. D. Goddard, C. C. Huang, *et al.* (2004). UCSF Chimera: a visualization system for exploratory research and analysis. *J Comput Chem* **25**(13), 1605–1612.

Shreeve, J. (2004). *The Genome War: How Craig Venter Tried to Capture the Code of Life and Save the World*. New York, Alfred A. Knopf.

Venter, J. C., M. D. Adams, E. W. Myers, *et al.* (2001). The sequence of the human genome. *Science* **291**(5507), 1304–1351.

Watson, J. D. and A. Berry (2003). *DNA: The Secret of Life*. New York, Alfred A. Knopf.

Watson, J. D. and F. H. Crick (1953). Molecular structure of nucleic acids: a structure for deoxyribose nucleic acid. *Nature* **171**(4356), 737–738.

Gene technology
Cutting DNA

2

When I come to the laboratory of my father, I usually see some plates lying on the tables. These plates contain colonies of bacteria. These colonies remind me of a city with many inhabitants. In each bacterium there is a king. He is very long, but skinny. The king has many servants. These are thick and short, almost like balls. My father calls the king DNA, and the servants enzymes . . . My father has discovered a servant who serves as a pair of scissors. If a foreign king invades a bacterium, this servant can cut him in small fragments, but he does not do any harm to his own king . . .

> (Silvia, daughter of Werner Arber and ten years old at the time of the quote. From *The Tale of the King and his Servants*; Lindsten and Nobelstiftelsen, 1992)

In the last 30 years we have seen a dramatic development in molecular biology research. Genetic information has been mapped in great detail for many different living organisms. We are able to examine gene expression, biochemical reactions and molecular interactions within the cell in a manner that was quite impossible 50 years ago. This basic research has had a great impact on many areas, including medicine and biotechnology. For instance, molecular details of many diseases such as cancer have been worked out, making new methods of diagnosis and therapy possible. In addition, pharmaceutically important proteins such as insulin may be produced in high yield. In the world of plants, crops have been genetically modified to achieve increased crop yields and resistance to insects, or to make them produce specific substances in large quantities.

This revolution in biology would not have been possible without *gene technology*. Critical methods were invented in the 1970s and 1980s. In *cloning*, a specific piece of DNA is isolated genetically so it may be reproduced in large amounts to allow further study. The DNA is typically propagated in a convenient host for such work, such as the bacterium *Escherichia coli*.

To allow cloning we need methods to construct new DNA molecules in the test tube. One important category of tools is *restriction enzymes*. These are used to cut DNA in a sequence-specific manner and will be discussed in this chapter. To construct new DNA molecules we also need to join fragments that were produced with restriction enzymes; this is carried out with an enzyme named *DNA ligase*. In the following chapters we will deal with other important tools: *RNA interference*, used to examine the function of genes; and the *polymerase chain reaction* (PCR), used to amplify specific regions of DNA.

Early days of restriction enzymes

The history of restriction enzymes began long before the days of cloning, back in the early 1950s. The first clues were obtained during experiments with *phages*, i.e. viruses that specifically target bacteria. It was observed by Salvador Luria and Guiseppe Bertani that a phage that grew well in one bacterial strain often grew poorly in a second strain (Luria and Human, 1952; Bertani and Weigle, 1953). However, when a small number of phages that survived this second strain were used in a new infection experiment they grew well in the same bacterial strain, but at the same time they had lost their ability to infect the original strain. These unexpected findings indicated that the phage had acquired a property related to the strain where it had last been replicated. The phenomenon was later explained with a model of 'restriction-modification'. According to this model, which was later proved to be correct, certain bacterial strains contain an enzyme (a restriction enzyme) that is able to cleave DNA ('restrict' infection or growth of the phage). At the same time, there is a modification system that protects the bacterial host DNA from being degraded by its own restriction enzyme. In essence, the restriction/modification system is there to protect bacteria from foreign DNA.

The first restriction enzymes were isolated and characterized in the late 1960s. These were later classified as 'type I', meaning that they cleave DNA at random positions far away from the actual recognition site. For this reason, they are not that useful in gene technology. However, in 1970 Hamilton Smith and colleagues described the first 'type II' restriction enzyme, an enzyme isolated from the bacterium *Haemophilus influenzae* (HindII) (Smith and Wilcox, 1970). Each type II enzyme cut DNA in a sequence-specific manner and at a precise location in the DNA. Werner Arber (Linn and Arber, 1968) and Daniel Nathans (Danna and Nathans, 1971) also made important contributions in the discovery of restriction enzymes and, together with Hamilton Smith, they shared the 1978 Nobel Prize for Physiology or Medicine. The quote of Silvia Arber's, which opened this chapter, was used by her to explain why her father received the Nobel Prize.

Soon after the isolation and characterization of the first restriction enzyme, the potential of this type of enzyme in gene technology was realized. Researchers have systematically hunted for restriction enzymes in a large number of bacterial

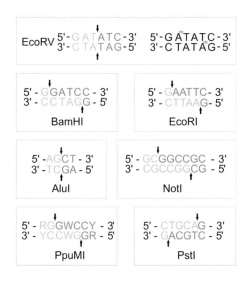

Fig. 2.1 *Restriction enzymes*. Seven examples of restriction enzymes are shown together with their recognition sequences and sites of cleavage. The sequence cleaved by the enzyme EcoRV is shown on top. It cuts as indicated to generate 'blunt ends'. When adenines are methylated (lower-case 'm' in orange) cleavage by EcoRV is prevented. EcoRV and AluI both generate blunt ends. BamHI, EcoRI, NotI and PpuMI cleave to generate a 5′ single-stranded end and PstI cleaves to generate a 3′ single-stranded end. Most of the enzymes identify a sequence six base pairs in length, except for AluI, PpuMI and NotI, which recognize a sequence of four, seven and eight base pairs, respectively. The recognition sequence of PpuMI is less specific than for the other enzymes of this figure (R = A or G; Y = C or T; W = A or T).

species. Today, approximately 3000 different restriction enzymes have been identified and more than 600 of these are commercially available (Roberts *et al.*, 2010).

Properties of restriction enzymes

The molecular action of restriction enzymes has, in some cases, been worked out in great detail. We also know the three-dimensional structure of different restriction enzymes. One example is EcoRV, isolated from *Escherichia coli* (Kostrewa and Winkler, 1995). The EcoRV enzyme recognizes a portion of double-stranded DNA with the sequence GATATC (Fig. 2.1).

Note that the recognition sequence of EcoRV is an *inverted* repeat or a *palindrome*, which means the sequence read from 5′ to 3′ is the same in both of the DNA strands. This is in fact a property of most sequences recognized by restriction enzymes. There is also a corresponding symmetry in the restriction enzymes as such. Thus, the EcoRV enzyme has two subunits related by a two-fold rotational symmetry (Fig. 2.2).

The *E. coli* bacterium needs to be protected against the activity of its own enzyme, EcoRV. Otherwise, the host DNA would be cut at every position where the sequence GATATC occurs. The protection is achieved through a DNA modification. Thus, when one of the As is methylated, it is no longer cleaved by EcoRV (Fig. 2.1). This modification is carried out by a specific enzyme. In fact, for each restriction enzyme there is a corresponding methylating enzyme.

The cleavage at the site GATATC results in 'blunt' ends. However, many restriction enzymes cleave asymmetrically in the manner shown in Fig. 2.1. This means that 'sticky ends' will be produced. Such sticky ends may easily

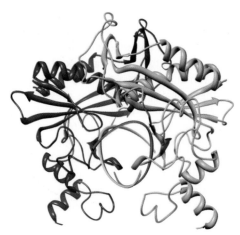

Fig. 2.2 *Three-dimensional structure of the restriction enzyme EcoRV bound to DNA*. The backbone structures of the two protein subunits are shown in blue and green. The phosphate–sugar backbone of double-helical DNA is shown, with the two strands in red and yellow. This figure was produced with the UCSF Chimera software (Pettersen *et al.*, 2004) using the entry 1RVB in the database Protein Data Bank (PDB) (Kostrewa and Winkler, 1995).

base-pair to other DNA molecules cleaved with the same enzyme. After pairing, the DNA ends may be covalently joined by the enzyme DNA ligase. This is of practical importance in gene technology whenever new DNA molecules are constructed in the test tube[1] (Fig. 2.3).

The names of restriction enzymes are composed of letters and roman numerals. The first three letters refer to the bacterial species which is the source of the enzyme. For instance, for the enzyme named EcoRI, the first three letters refer to *Escherichia coli*. Restriction enzymes differ in the nature of substrate specificity. First of all, these enzymes differ with respect to the length of the recognition sequence. Most commonly used restriction enzymes recognize a sequence either four or six bases in length. For instance, the enzyme AluI recognizes the sequence AGCT, and EcoRI recognizes GAATTC. 'Four-cutters' cleave more often than 'six-cutters' in a DNA molecule. In a 'random' sequence of DNA you would expect a specific sequence of four bases to occur with a probability of $(1/4)^4 = 1/256$. The corresponding probability for a 'six-cutter' is $(1/4)^6 = 1/4096$. These numbers tell us about the approximate size of fragments that are the result of cleavage with four- and six-cutters.

Enzymes also differ in their degree of specificity. For instance, whereas BamHI recognizes, in a highly specific manner, the sequence GGATCC (Fig. 2.1), an enzyme such as BstX2I cleaves in the same manner but is less stringent as it requires a sequence 'RGATCY', where R represents either A or G (= puRine), and Y is either C or T (= pYrimidine).

BIOINFORMATICS

Tools introduced in this chapter		
Perl	pattern matching and the `m//` and substitution `s///` operators	
	conditional statements	
	`foreach` loops	

For cloning and gene technology experiments, molecular biologists want to identify restriction enzyme cleavage sites in DNA molecules. Computer programs were developed for this type of analysis early in the development of restriction enzyme technology. Here we will see how Perl may be used

---

[1] Also, blunt ends may be joined, but less effectively and not with the specificity associated with sticky ends.

to develop a simple tool to identify restriction enzyme cleavage sites. Perl is a good choice in this case, as it is very powerful when it comes to pattern matching, as will be discussed below.

Pattern matching

Patterns and *regular expressions* are very useful in Perl. Consider a simple example. We want to identify the recognition sequence for the enzyme EcoRI, 'GAATTC', in a nucleotide sequence. We have stored the sequence to be analysed in a variable:

```
$dna = 'AACGGAATTCCCTCTC';
```

Remember from the first chapter that we used the `substr` function to extract words from a DNA sequence. In this case we could use that function to extract every possible six-letter word in the sequence stored in `$dna` and for each of the words we could test if it exactly matches 'GAATTC'. In this case we find that there is one such match in this sequence. However, another way of finding the site 'GAATTC' is to make use of a regular expression. Thus, to find out whether there is a match, we could use the following:

```
if ($dna =~ /GAATTC/) {print "match";}
```

In this case, the regular expression is the string `GAATTC` and the expression /GAATTC/ tells us that we are dealing with the *pattern-matching* operator.[2] In the expression above, the symbols `=~` represent the *binding operator*, meaning that the variable preceding that operator (`$dna`) is to be tested with respect to the pattern. Thus, the way to interpret the expression `$dna =~ /GAATTC/` is: 'Search for a match of the regular expression `GAATTC` in the sequence stored in `$dna`. If there is a match, return "true", otherwise return "false".' The testing mechanism above involves an *if conditional statement*; if there is a match to the regular expression, then print 'match'.

But regular expressions are more versatile than this. We may allow more ambiguity in the pattern. For instance, if we want to find out whether a certain string matches all sequences recognized by the enzyme Bfm1, which recognizes CTRYAG, we can use the expression

```
/CT[AG][CT]AG/
```

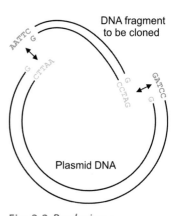

Fig. 2.3 *Producing a recombinant DNA molecule.* A circular DNA plasmid vector molecule has been cut with the enzymes EcoRI and BamHI to generate two different 'sticky ends'. A DNA fragment to be introduced into the plasmid vector has been cut with the same enzymes. A sticky end of EcoRI or BamHI in the fragment will base-pair with the end of the plasmid cut with the same enzyme. Once the DNA molecules have been annealed, they can be covalently joined with the aid of the enzyme DNA ligase.

DNA fragment to be cloned

Plasmid DNA

---

[2] Incidentally, the expression /GAATTC/ could also have been written m/GAATTC/, but the m (for 'match') is commonly left out. However, you need the m// syntax whenever you want to use other symbols that use slashes as delimiters for the pattern. See Chapter 14 for an example.

Here, the characters [and] are not used in the actual matching. Rather, they have a special meaning. (There are many other symbols that have a special meaning in regular expressions, see Appendix III.) In the pattern above, the expression [AG] means that *either* A or G will match the character following CT. Therefore, the whole pattern above will, in words, specify 'CT, followed by either A or G, followed by either C or T, followed by AG'.

In this chapter we encounter a Perl script showing another example of a pattern-matching operator, the *substitution* operator, s///. It is different from the matching // operator above (as well as the transliteration operator encountered in the previous chapter). It has the form s/PATTERN/REPLACEMENT/. This is to say that the pattern specified within the first two slashes will be replaced by the string specified between the second and third slashes. Consider this code:

```
$seq = 'GCAATAT'
$seq =~ s/T/U/;
```

It has the effect that $seq ends up as 'GCAAUAT'. We note that the first T from the left was replaced by a U. But if we want all the Ts to be replaced by Us? In such a case we use the *global modifier* to the s/// operator:

```
$seq =~ s/T/U/g;
```

The matching operation is now changed so that matching and replacement is carried out in all possible instances within the string.

In summary, the important difference between the match and substitution operators is that the match operator is used to find out if there is a match to a regular expression, whereas the substitution operator is used whenever we want to do replacements in a string based on regular expressions. We will return to other examples where regular expressions are used later in this book, as they are a very useful part of Perl and are used for many operations with biological sequences. The reader is also referred to Appendix III for a summary of the most common syntax.

Identifying restriction enzyme sites with Perl

The script used in this chapter is shown in Code 2.1. We will go through the details of this code. First, we need a list of restriction enzymes and their corresponding recognition/cleavage sites. These may be downloaded from databases like REBASE, which store this type of information (Roberts *et al.*, 2010). For the following example we use a very small collection of enzymes that was extracted from a much more comprehensive list. We have also simplified things by not considering the actual cleavage sites, only the recognition sequences. We store the restriction enzyme data as a hash table, just as we did with the genetic code in the previous chapter. In this hash, the enzyme names are the keys, and the recognition sequences are the values:

```
%enzymes = (
      'BclI'    => 'TGATCA',
      'BfmI'    => 'CTRYAG',
      'Cac8I'   => 'GCNNGC',
      'EcoRI'   => 'GAATTC',
      'HindIII' => 'AAGCTT'
);
```

We also want to modify the list in %enzymes. For this reason we make a copy of it and name it %enzymes_mod:

```
%enzymes_mod = %enzymes;
```

In the definition of %enzymes we see that the enzyme BfmI recognizes 'CTRYAG'. As explained above, R is either A or G, and Y is either C or T. The symbols R and Y are examples of *ambiguity codes* for nucleotides; there are more of them, as shown in Code 2.1. There is a simple way to represent the ambiguities in Perl, because we can make use of regular expressions. We will use this idea for all the ambiguity codes in our script. For instance, 'R' has to be assigned to [AG]. We do this mapping of ambiguity to regular expressions with yet another hash table:

```
%amb = (
      'R' =>  '[AG]',
      'Y' =>  '[CT]',
      'N' =>  '[AGCT]',
      'W' =>  '[AT]',
      etc . . .
      };
```

Using this table we can substitute all restriction enzyme sites as listed in the hash %enzymes_mod (which is the copy we made from %enzymes). We do this with the following lines:

```
foreach $val (values %enzymes_mod) {
      foreach $key (keys %amb) {
            $val =~ s/$key/$amb{$key}/g;
      }
}
```

This may look a little complex, but let's examine the different pieces of this code. First, the line

```
$val =~ s/$key/$amb{$key}/g;
```

is an example of pattern matching with the substitution operator, s///, as explained above.

We want to have the replacement done for all keys of the %amb hash. Thus, if we have the string 'CTRYAG', which is one of the values in the %enzymes_mod hash, we want it to be transformed to 'CT[AG][CT]AG'. We do this with a foreach loop. A foreach loop is used when you want to have the same code executed for a range of scalars. (There are other loop constructs in addition to foreach. We encountered for in the previous chapter. Regarding loop constructs, including while and until, see Appendix III.) We use the following code to go through all keys of the %amb hash and do substitutions of a value ($val) of the %enzymes_mod hash:

```
foreach $key (keys %amb) {
        $val =~ s/$key/$amb{$key}/g;
}
```

We want to do this for all values of the %enzymes_mod hash, so we need yet another foreach loop, foreach $val (values of %enzymes_mod) {}. The whole operation has the effect that we have modified the hash %enzymes_mod so that it now has the following key–value pairs:

```
BclI     TGATCA
BfmI     CT[AG][CT]AG
Cac8I    GC[AGCT][AGCT]GC
EcoRI    GAATTC
HindIII  AAGCTT
```

Having constructed this hash, we finally use a loop construction to go through all possible six-letter words to test if they match the different restriction enzyme recognition sequences:

```
for ( $i = 0 ; $i < length($seq) - 5 ; $i++ ) {
        $testseq = substr( $seq, $i, 6 );
        foreach $key ( keys %enzymes_mod ) {
                if ( $testseq =~ /$enzymes_mod{$key}/ ) {
                        $pos = $i + 1;
                        print
                "$key\t$pos\t$testseq\t$enzymes{$key}\n";
                }
        }
}
```

The first two lines of this code construct words starting from the 5′ end of the sequence. For instance, if the sequence to be analysed is 'GATCTGACTAGC-GAGC...', then the three first words tested will be 'GATCTG', 'ATCTGA' and 'TCTGAC'. Each of these words (stored in $testseq) are tested against all the values of the `%enzymes_mod` hash table. We make use of the match operator mentioned above:

```
$testseq =~ /$enzymes_mod{$key}/
```

The variable we test for matching is `$testseq`. The slash symbols `//` delimit the pattern (regular expression). The expression as a whole is to be interpreted as: 'Is there a match in the sequence `$testseq` to the pattern as defined by `$enzymes_mod{$key}`?' The whole expression within parentheses returns true if there is a match. If true is returned, then we print (1) the name of the enzyme, (2) the position of the enzyme site, (3) the sequence recognized and (4) the sequence/pattern associated with the enzyme:

```
if ( $testseq =~ /$enzymes_mod{$key}/ ) {
    $pos = $i + 1;
    print "$key\t$pos\t$testseq\t$enzymes{$key}\n";
}
```

The position of the first base in the DNA sequence has index 0 when we are using the `substr` function. But we want to think of this as position number 1 in the nucleotide sequence, and this is why we use a new variable `$pos`, which is `$i + 1`. The `\t` in the print command is a symbol to represent tabs, so the output will be tab-delimited columns.

Code 2.1 cut.pl

```
# very simple cutter

%enzymes = (
    'BclI'    => 'TGATCA',
    'BfmI'    => 'CTRYAG',
    'Cac8I'   => 'GCNNGC',
    'EcoRI'   => 'GAATTC',
    'HindIII' => 'AAGCTT'
);

%enzymes_mod = %enzymes;

# modify the ambiguity letters
%amb = (
    'R' => '[AG]',
    'Y' => '[CT]',
    'N' => '[AGCT]',
    'W' => '[AT]',
    'M' => '[AC]',
    'S' => '[CG]',
    'K' => '[TG]',
```

```
    'V' => '[ACG]',
    'H' => '[ACT]',
    'D' => '[AGT]',
    'B' => '[CGT]'
);

foreach $val ( values %enzymes_mod ) {
    foreach $key ( keys %amb ) {
        $val =~ s/$key/$amb{$key }/g;
    }
}
$seq = 'GATCTGACTAGCGAGCGTGATCAAGCTTGTGTAGGAATTCCTTGATGCTGTAGCGCGAGC';

for ( $i = 0 ; $i < length($seq) - 5 ; $i++ ) {
    $testseq = substr( $seq, $i, 6 );
    foreach $key ( keys %enzymes_mod ) {
        if ( $testseq =~ /$enzymes_mod{$key}/ ) {
            $pos = $i + 1;
            print "$key\t$pos\t$testseq\t$enzymes{$key}\n";
        }
    }
}
```

The output from Code 2.1 shows that a number of restriction enzyme sites are identified:

```
Cac8I    11    GCGAGC GCNNGC
BclI     18    TGATCA TGATCA
HindIII  23    AAGCTT AAGCTT
EcoRI    35    GAATTC GAATTC
BfmI     48    CTGTAG CTRYAG
Cac8I    55    GCGAGC GCNNGC
```

The script therefore illustrates that restriction enzyme cleavage sites may be effectively identified using Perl and regular expressions. You will also notice, however, that there are a few limitations to our script if we were to use this as a more serious application to identify cleavage sites. Some examples are presented below, but it is relatively easy to improve on the script in these respects:

(1) We have analysed only a few enzymes but want to use a more comprehensive list; for such a list of restriction enzymes, see the web resources for this book.

(2) We have analysed a short DNA sequence, specified within the script. In a real situation, we would probably want to analyse a much longer sequence

of DNA, and we would typically want to read that from a file. We will see in the following chapter how to perform that sort of operation.

(3) Some restriction enzymes are not symmetrical, i.e. they do not recognize a fully palindromic sequence. One example shown above is Cac8I, which has the recognition sequence GCGAGC. For this reason we need to analyse also the complementary strand of the DNA. However, we saw in the previous chapter how to produce the complementary strand of a DNA sequence.

(4) We have considered only 'six-cutters', but need to consider a range of different categories of enzymes, including 'four-cutters'.

(5) We may want to have more sophisticated output; for example, we may wish to print the exact point of cleavage, or to print the original DNA sequence along with the recognition sites.

EXERCISES

2.1 Assume you want to analyse the same sequence as in Code 2.1, but instead you are interested in identifying recognition sites of the two enzymes AluI and DpnI. These enzymes recognize sequences AGCT and GATC, respectively. Modify Code 2.1 to achieve this analysis.

2.2 In Code 2.1 we have used a `for` loop to go through every possible word six letters in length. We do it like this because we want to find out about the positions of the different enzyme recognition sequences. Show how the code may be simplified if we do not care about the positions and we only want to know whether a certain enzyme cleaves or not in the sequence under investigation (stored in $seq).

2.3 Write some Perl code to generate the complement of a DNA sequence that contains ambiguity symbols. For instance, the symbol R is either A or G. The complementary bases in that case are T and C, and these may be represented by Y. Therefore, the 'complement' of R is Y. As another example, the symbol B is either C, G or T. The complementary bases are G, C and A, which are represented by V. Thus, the complement of B is V. Write a piece of Perl code that prints the complement of the following restriction enzyme recognition sequences: GTMKAC, GDGCHC and ACNNNNGTAYC.

REFERENCES

Bertani, G. and J. J. Weigle (1953). Host controlled variation in bacterial viruses. *J Bacteriol* **65**(2), 113–121.

Danna, K. and D. Nathans (1971). Specific cleavage of simian virus 40 DNA by restriction endonuclease of *Hemophilus influenzae*. *Proc Natl Acad Sci U S A* **68**(12), 2913–2917.

Kostrewa, D. and F. K. Winkler (1995). Mg2+ binding to the active site of EcoRV endonuclease: a crystallographic study of complexes with substrate and product DNA at 2 A resolution. *Biochemistry* **34**(2), 683–696.

Lindsten, J. E. and Nobelstiftelsen (1992). *Nobel Lectures, Including Presentation Speeches and Laureates' Biographies: 1971/1980*. Amsterdam, Elsevier.

Linn, S. and W. Arber (1968). Host specificity of DNA produced by *Escherichia coli*, X: in vitro restriction of phage fd replicative form. *Proc Natl Acad Sci U S A* **59**(4), 1300–1306.

Luria, S. E. and M. L. Human (1952). A nonhereditary, host-induced variation of bacterial viruses. *J Bacteriol* **64**(4), 557–569.

Pettersen, E. F., T. D. Goddard, C. C. Huang, *et al*. (2004). UCSF Chimera: a visualization system for exploratory research and analysis. *J Comput Chem* **25**(13), 1605–1612.

Roberts, R. J., T. Vincze, J. Posfai and D. Macelis (2010). REBASE: a database for DNA restriction and modification – enzymes, genes and genomes. *Nucleic Acids Res* **38** (database issue), D234–D236.

Smith, H. O. and K. W. Wilcox (1970). A restriction enzyme from *Hemophilus influenzae*: I – purification and general properties. *J Mol Biol* **51**(2), 379–391.

Gene technology
Knocking genes down

3

Restriction enzymes, one of many tools of the molecular biology toolbox, were introduced in the previous chapter. The present chapter is devoted to another important method designed to examine the function of specific genes.

Interfering with gene expression

In order to study the biological function of a gene, the molecular biologist needs methods that allow alteration of that gene – for example, an alteration that reduces the production of the protein specified by the gene. The behaviour of a normal cell may then be compared to a cell where the gene of interest has been manipulated, thereby allowing a conclusion regarding the function of the gene of interest. In the yeast *Saccharomyces cerevisiae*, for instance, there are methods that allow the complete removal of genes (such a removal is often referred to as a *gene knock-out)*. If, for instance, the yeast gene named PSY3 is removed or inactivated, it gives rise to an increased level of mutations. This observation suggests to us that this particular gene is related to the repair of DNA damage.

For studies of human genes, you may want to examine a species more closely related to man than is yeast. You can delete or inactivate genes in mammals such as mice, but this is more technically involved than in yeast. On the other hand, there are other methods that make possible a reduction in gene expression in animals. One such important method involves *RNA interference* (*RNAi*) (also known as *RNA silencing*). In one common type of experiment, an mRNA is inactivated by the introduction of a small synthetic RNA which is complementary to the mRNA. In the bioinformatics example below we will see how we can design a small RNA for such an experiment. RNAi is an example of a *gene knock-down* method, as the expression of the target gene is reduced (compared to the gene knock-out, where a specific gene is completely removed or inactivated[1]).

There are significant biotechnology applications to RNAi, but it is important to point out that such RNAs also occur in nature and are part of complex pathways to regulate gene expression. The phenomenon of RNA interference was first

---

[1] A third category of experiments is a *gene knock-in*, meaning that a new gene is introduced in the species under study.

discovered in the 1990s. Critical experiments were carried out by Andrew Fire and Craig Mello. They injected double-stranded RNA (dsRNA) into cells of the worm *Caenorhabditis elegans* and found that if the dsRNA contained a sequence matching a specific gene, the expression of that particular gene was dramatically reduced (Fire *et al.*, 1998). In 2006 Fire and Mello received the Nobel Prize in Physiology or Medicine for their pioneering work on RNA silencing.

Small silencing RNAs

RNA silencing occurred early during eukaryotic evolution and is therefore found in a significant majority of eukaryotic species.[2] Different categories of small RNAs are involved. The first two categories to be described were *microRNA* (miRNA) and *siRNA* (small interfering RNA) (Siomi and Siomi, 2009). Both of these RNAs are 21–25 nucleotides long and are part of mechanisms that act to reduce the expression of genes by at least three different mechanisms: (1) cleavage of mRNA; (2) repression of translation; and (3) promotion of mRNA degradation (Fig. 3.1).

One role of miRNA is to control the expression of endogenous genes. In contrast, siRNA downregulates exogenous sequences, like those resulting from viral infection. The synthesis of miRNA and siRNA occur by different pathways. All miRNA is encoded in the genome, and as a first step precursor miRNA is produced by transcription. Such a precursor is processed by the enzyme Dicer to produce a small duplex RNA. The miRNA is one of the strands of this duplex and it eventually forms a complex with a protein named Argonaute.[3] The miRNA–protein complex (also called RISC, for 'RNA-induced silencing complex') will in turn target an mRNA which has a sequence complementary to the miRNA. The siRNAs, on the other hand, originate from double-stranded RNAs (dsRNAs). These dsRNAs could be the result of a viral infection. As with miRNA, the siRNAs are generated by processing with Dicer.

The miRNAs and siRNAs will both target mRNAs with a sequence complementary to that of the small RNA. In the case of siRNA the complementarity will be perfect, whereas for miRNA there are mismatches in the hybrid formed. An example of non-identical matching between an miRNA and an mRNA in the worm *C. elegans* is shown in Fig. 3.2. Because of the complementarity between an miRNA and its mRNA target, a number of computational methods have been developed that are designed to predict mRNA targets based on the sequence of the miRNA (Maziere and Enright, 2007).

---

[2] Strangely, it has been lost in the widely studied yeast *S. cerevisiae*, but is present in closely related yeasts.

[3] Argonaute proteins are named after an *Arabidopsis thaliana* mutant that resembles argonauts.

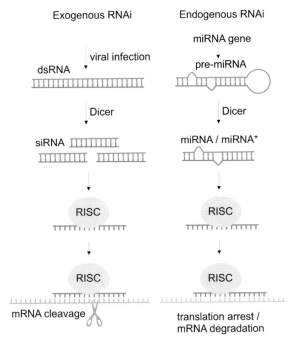

Exogenous RNAi   Endogenous RNAi

Fig. 3.1 *Two pathways of RNA interference*. The synthesis of miRNA and siRNA occur by different pathways. For miRNA, a precursor referred to as pri-miRNA (not shown) is first formed by transcription. This RNA is processed to give pre-miRNA. This RNA is in turn transported to the cytosol and processed by the enzyme Dicer to produce a small duplex RNA. The miRNA is one of the strands of this duplex (and the other is referred to as miRNA*). An Argonaute protein and other proteins associate with the duplex and the miRNA* is degraded. The miRNA–protein complex (also called RISC, for 'RNA-induced silencing complex') will finally target an mRNA which has a sequence complementary to the miRNA. The siRNAs originate from double-stranded RNAs (dsRNAs). These dsRNAs could be the result of a viral infection. As with miRNAs, the siRNAs are generated by processing with Dicer. They are also part of a RISC complex, but with Argonaute proteins that are specific to siRNA. The miRNAs and siRNAs will both target mRNAs based on sequence complementarity. In the case of siRNA, the complementarity is perfect, whereas for miRNA there are mismatches in the hybrid formed. The siRNA-mediated mechanism effectively cuts mRNA, whereas the miRNA pathway gives rise to translation arrest or enhanced mRNA degradation.

More recently, additional small RNAs have been discovered that are relatives of miRNA and siRNA (Siomi and Siomi, 2009). One example is the family of *piRNAs* with a role in germ cell development (Lau *et al.*, 2006). They are named piRNAs because they are bound to an Argonaute protein with the name PIWI. The piRNAs, 24–31 nucleotides long, are produced by a mechanism independent of Dicer, as they are produced from single-stranded precursors. Yet another class of small RNAs, endogenously produced siRNAs (endo siRNAs, eiRNAs)

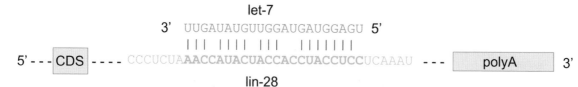

Fig. 3.2 *Targeting of mRNA by RNAi in* C. elegans. The base-pairing interaction between the RNAi let-7 and the mRNA lin-28 is shown. Note mismatches in the hybrid formed. The lin-28 gene product is important during development in *C. elegans*. The RNA let-7 regulates the expression of lin-28 as well as other genes, such as lin-41 (Kaufman and Miska, 2010).

have been identified, suggesting that the borders between the different classes of small RNAs are not so distinct and in the future we may see other means of classifying silencing RNAs.

RNAi: functions and applications

Silencing RNAs have a number of important functions. The miRNAs down- or upregulate a number of genes; this type of regulation is particularly important in development. There are also examples where miRNAs are linked to human tumour formation (He *et al.*, 2005; Lu *et al.*, 2005). Some miRNAs may act as oncogenes and others as tumour suppressors. The siRNAs have important functions in immunity as they protect during viral infection, while piRNAs protect cells from harmful expression of *transposon* DNA, a kind of mobile DNA element (Senti and Brennecke, 2010).

In addition to these significant biological effects of small RNAs, there are important applications in biotechnology. Below, we will deal with the details of designing RNA sequences for experiments of RNAi-based gene knock-downs. Such experiments are widely used in molecular biology to examine the function of genes.

RNAi is also being exploited in therapeutics. One of the first to be used in clinical trials was an siRNA named Bevasiranib[4] for treatment of a macular degeneration disease (a visual impairment). This RNA is targeted against the mRNA of the vascular endothelial growth factor, a protein responsible for stimulation of blood vessel growth. RNAi is additionally being tested as a therapeutic agent in a variety of diseases such as Parkinson's disease, HIV infection, rheumatoid arthritis and cancers.

Plant biotechnology is another interesting area where RNA silencing is used. One example is concerned with cotton, a well-known source of fibre. Not so well known is that cottonseeds are very rich in wholesome protein. For every

---

[4] Bevasiranib was originally developed by OPKO Health, Inc.

kilogram of fibre, there is nearly 0.4 kg of protein produced in the cottonseed. In fact, there is a yearly production of cottonseed protein amounting to nine million metric tons! This would be a useful resource as food protein, but there is a problem in that cottonseed also produces the compound *gossypol*, which is toxic to animals, including humans. In order to make cottonseed available for consumption, scientists have used RNAi to decrease the production of gossypol (Sunilkumar *et al.*, 2006). This was possible by designing a small RNA directed against the enzyme δ-cadinene synthase, which is responsible for the first step in a pathway leading from (+)-δ-cadinene to gossypol. This RNAi technology in plants also has the advantage that the silencing is stable and heritable, as is the case for a number of different species, including *C. elegans*.

For any experiment where we want to introduce a silencing RNA, we first have to think about how to design the silencing RNA. This will be described in more detail below. Once we know the appropriate sequence(s), these are used to produce corresponding dsRNAs. The strands of the dsRNA are produced chemically. To obtain the dsRNA a molecular biologist will typically turn to a company specialized in synthesizing short oligonucleotide sequences. The dsRNA will then be introduced into the cell under study and it will then activate the RNAi pathway. Sometimes the RNA is not directly supplied to cells, but delivered as a plasmid DNA encoding the desired RNA. In *C. elegans*, RNA interference experiments are particularly simple. Bacteria such as *E. coli* that carry the desired dsRNA may be fed to the worms, and as a result the RNA is transferred to the worm.

BIOINFORMATICS

Silencing RNAs and design principles

We now turn to the design of the silencing RNA. In order to reduce the expression of a gene, we need to supply a short dsRNA molecule to the cell under study. Can we select just any short sequence matching the mRNA of that gene? Well, not really

Tools introduced in this chapter		
Perl	reading from a file	
	more on regular expressions	
	concatenating strings	
	logical and comparison operators	
	function chomp	

(Birmingham *et al.*, 2007). First of all, it is common practice to make use of the protein-coding sequence of the mRNA. This is because coding sequences are better conserved than untranslated regions and they are less likely to contain polymorphisms (variations that are observed between individuals in a population). Splice variation is also a matter to consider. If you want to knock-down a specific gene you may want to effectively knock-down all transcripts of that gene. In such a case you need to consider mRNA sequences that are common to all transcript variants.

There are also issues of specificity that are important. You want the short RNA to be specific to the mRNA under consideration, but by chance a short RNA sequence may match an unrelated mRNA. Therefore, design of siRNA includes database searches to exclude those RNA sequences that non-specifically match other mRNAs. It is common practice to use the program BLAST for this task. This step of siRNA design will not be dealt with here, but BLAST is discussed in Chapters 7 and 11.

There are a number of additional criteria for an siRNA to be effective. In many instances siRNA duplexes are designed with a protruding UU at the 3′ end as duplexes of this nature were early found to be more effective. Thus, an example of a duplex being used in an RNAi experiment is the following:

```
5′      CCAGGGAUGCCAAAUCAGUUU    3′  (sense)
        |||||||||||||||||||||
3′   UUGGUCCCUACGGUUUAGUCA       5′  (antisense)
```

Note the protruding 3′ ends with the UU sequence. The lower strand will be targeted to mRNA. The overhang on the sense strand is not critical, but for the coming example we will make use of that feature in the design of the siRNA. Thus, we are making use of the sense sequence of the mRNA, which in this case is:

```
5′ AACCAGGGAUGCCAAAUCAGUUU 3′
```

We now can define our first rule in a strategy to identify suitable siRNAs:

(1) Identify sequences in the coding sequence of the mRNA, 23 nucleotides in length where the first two positions are AA and the last are UU (or TT if we think about the corresponding DNA sequence).

The siRNA should also have a suitable content of GC base pairs. Such base pairs increase the stability of double-stranded DNA. Unwinding of the siRNA duplex is an important part of the siRNA mechanism and this unwinding will be more difficult the more stable the duplex. On the other hand, a too low GC content and low stability of the duplex is not effective, perhaps because of too low stability in the mRNA–siRNA hybrid. We can therefore invent a second rule:

(2) GC content of the siRNA sequence should be 30–50%.

Sequences with any base repeated four or more times are avoided because such sequences tend to reduce selectivity. Therefore, we arrive at our next rule:

(3) Avoid sequences where either A, U, C or G is repeated four times.

Furthermore, stretches with either G or C in every position have been shown to negatively influence siRNA activity. They tend to inhibit duplex dissociation, just like any siRNA sequence with a very high GC content. Our fourth and last rule is therefore:

(4) Avoid sequences with six consecutive positions that are either C or G.

We will now design a Perl script to identify the siRNA sequences that meet all four of these criteria. A few additional criteria are also helpful but will not be dealt with here (Birmingham *et al.*, 2007). As mentioned above, a BLAST search should be carried out to exclude candidates that match sequences of undesired mRNAs. Furthermore, sequences that may form internal base-pairing should be avoided as they disturb the formation of the RNA duplex. An example is:

```
5' GCUACAGGUUUCCUGUAGCG 3'
```

This RNA is able to form the structure

```
5'      GCUACAGGU
        ||||||||  U
3'      GCGAUGUCCU
```

Identifying siRNA candidates

We will now examine some Perl code that may be used to implement the four rules as above (Code 3.1). For this example we will be using an mRNA sequence derived from the human BRCA1 gene, which is an important genetic factor in breast cancer.

In the previous chapter we defined a nucleotide sequence explicitly within the script. This is fine, but not convenient when it comes to longer sequences. Furthermore, a sequence we want to make use of is often stored in a file and we want to be able to read the sequence from that file. So how do we read from a file in Perl? To begin with, here is a procedure to print all lines in a file, i.e. the output of the program will simply be the contents of the file:

```perl
open(IN, 'mrna.fa');
while (<IN>) {
    print;
}
close IN;
```

The IN used here is referred to as a *filehandle*. You can regard it as a kind of nickname for the file. We need not call it IN, we could have named it BRCAFILE, OBAMA or just about anything, but you typically see a filehandle with upper-case letters so it may be more easily spotted in a script.

As expected, the `open` and `close` statements refer to the opening and closing of the file. The statement `open(IN, 'mrna.fa')` opens the file named `mrna.fa` for *reading* (for *writing* to a file, see Appendix III). In this example the `mrna.fa` file contains an mRNA sequence.

There is a danger in the statement `open(IN, 'mrna.fa');`, because if for some reason this file is not available in the current directory, or not available for the `open` operation, there will be no particular error message from Perl. In this case nothing will be printed and you do not know whether the file was missing or whether it was empty. A better programming practice is therefore to add a construct to give an error message whenever the file could not be opened. We can use something like this:

```
open(IN, 'mrna.fa') or die "Could not open file mrna.fa\n";
```

Here, `or` is the logical OR operator (you may also use `||`; for more on logical operators, see Appendix III). The statement has the effect that either the `open` command is successful *or* the `die` function will come into play. The `die` function is an instruction to tell Perl to exit after having reported the error message provided as the argument to `die`.

Let us return to the code above reading the file. What does the condition '`while (<IN>)`' mean? In Perl it is assumed that we read a file one line at a time. The way to interpret the condition is therefore: 'as long as we read lines from the file'. As long as we do that, we carry out things within the `while` loop, which in this case is the simple instruction: `print;`. What does `print` mean in this case? This is actually an example of Perl shorthand. In principle the `print` function has an argument, so if we were to say '`print ('OK, I see')`' then the expression within parentheses will be printed to the screen. However, when there is *no* argument to the `print` function, Perl assumes that the argument is a special variable with the somewhat strange name `$_`. In this case `$_` is the default input, which is the line read from the file. Therefore, for every loop of the `while` condition, `print` will use the line currently read as its argument.[5]

The file `mrna.fa` not only contains a sequence. It is in the *FASTA* format,[6] which means the first line is a 'definition' line:

```
>gi|237757283:233-5824 Homo sapiens breast cancer 1,
early onset (BRCA1), transcript variant 1, mRNA
```

In general, the definition line has a '>' character, followed by an identifier (in this case `gi|237757283:233-5824`) and an optional description of the sequence. The lines that follow after the definition line specify the actual nucleotide sequence.

---

[5] To be more precise, the condition (`<IN>`) assumes `$_`, as it is equivalent to ($_ = `<IN>`).

[6] FASTA format is named after its use in the FASTA software developed by William Pearson.

For our script we want to analyse the mRNA sequence. To do this we first have to store the sequence in a variable. We could therefore read all lines, excluding the definition line. However, before we merge all lines to form the complete sequence, we have to be aware that all lines end with an end-of-line character. Before merging the lines, therefore, we need to remove those end-of-line characters with the `chomp` function. This is a function that removes an end-of-line character in a string, if present.

Each line of sequence could be added to any previous sequence with:

```
$seq = $seq . $_;
```

Here, the dot '.' is the operator that is used in Perl for concatenating two strings. A shorthand for the previous statement is:

```
$seq .= $_;
```

After reading all lines in the file, we have stored the complete sequence in the variable $seq. We now analyse it with respect to possible RNAi sequences. As a suitable length for RNAi is 23 nucleotides (nt), we will break up the sequence into pieces of that size. We will start by taking the first 23 nt from the 5′ end and then move one position to the right until we come to the 3′ end. We do this with a `for` loop like this:

```
for (  $i = 0 ; $i < length($seq) - 22 ; $i++ ) {
    $testseq = substr( $seq, $i, 23 );
    # some processing of $testseq . . .
}
```

This construct is very similar to what we used in the previous chapters to extract substrings from a sequence.

How do we then analyse each 23 nt piece for the four properties as we have defined them above? For rule 1 we use:

```
if ( $testseq =~ /^AA.*TT$/ )
```

This is an example of a pattern match where we are using the matching operator as in the previous chapter. Here, enclosed within the delimiters is `^AA.*TT$`, a regular expression where the `^AA` to the left means that AA should match at the *beginning* of the sequence in $testseq. The '.' (dot) refers to any symbol and the '*' (star) following that is a special *quantifier* meaning that there should be zero or more of the preceding item. Therefore, `.*` means 'zero of more of any characters'.[7] The $ at the end of the regular expression indicates that the

---

[7] There are other types of quantifier. The most common is an expression of the form with numbers within curly brackets. An example is {3,10}, which means at least three but at most ten. For more on this topic, see below in this chapter, Chapter 5 and Appendix III.

preceding item (TT) should be at the very *end* of the string $testseq. In summary, we are looking for strings that begin with AA and end with TT.

For rule 2 we use:

```
$gc_content = ( $testseq =~ tr/GC// ) / 21;
if ( ( $gc_content >= 0.3 ) && ( $gc_content <= 0.5 ) )
```

First, we define a variable $gc_content to store the sum of G and C over the total length of the sequence. Therefore, we first need to count the number of Gs and Cs. Counting characters may be done in different ways in Perl. Here we use the tr/// operator that we used in Chapter 1 to replace characters in a string. As a simple example, consider this code

```
$dna = 'GCAT';
$count = ($dna =~ tr/GC//);
```

The operation $dna =~ tr/GC// has the effect that nothing will be substituted in $dna because there are no characters between the two right-most slashes. So what is the point in doing such an operation? Well, the expression ($dna =~ tr/GC//) will evaluate to the number of attempted replacements, which is two in this case. In essence, the number of Gs and Cs will be counted. We could also have used the tr expression tr/GC/GC/, but many Perl programmers are lazy typists and prefer a shorter expression.

In a condition like $gc_content >= 0.3 we are testing whether $gc_content is greater than or equal to 0.3 (the >= is an example of a *comparison operator*; for more on these, see Appendix III).

The symbol && is the logical AND operator. The condition (($gc_content >= 0.3) && ($gc_content <= 0.5)) therefore means that the variable $gc_content should be in the range 0.3–0.5.

For rule 3 we use:

```
unless ( ( $testseq =~ /A{4}/ )
      || ( $testseq =~ /T{4}/ )
      || ( $testseq =~ /G{4}/ )
      || ( $testseq =~ /C{4}/ ))
```

A regular expression such as A{4} means A repeated four times. In plain language the whole statement is interpreted as 'unless ($testseq contains AAAA or TTTT or GGGG or CCCC)'. An unless statement is related to if statements such that any unless statement can be rephrased as an if statement. For instance, if you say

```
unless ($gc_content >= 0.3)
```

that is equivalent to

```
if (!($gc_content >= 0.3))
```

where the exclamation point indicates a negation (actually yet another logical operator, NOT). But you could in this case also say the following, which is more straightforward and easy to understand:

```
if ($gc_content < 0.3)
```

Finally, there is rule 4:

```
unless ( $testseq =~ /[GC]{6}/ )
```

The number within curly brackets is a quantifier, specifying in this case that we are matching a series of six characters that are either G or C.

Code 3.1 sirna.pl

```perl
# First read the sequence from a file named 'mrna.fa'
$seq = '';
open(IN, 'mrna.fa') or die "Could not open file mrna.fa\n";
while (<IN>) {
    unless (/>/) {
        chomp;
        $seq .= $_;
    }
}
close IN;

# Now analyse the sequence read from file

# Step through each position of the sequence

for ( $i = 0 ; $i < length($seq) - 22 ; $i++ ) {

    $testseq = substr( $seq, $i, 23 );

    # check if first two positions are AA and
    # last are TT
    if ( $testseq =~ /^AA.*TT$/ ) {

        # test GC content

        # count the number of G's and C's
        $gc_content = ( $testseq =~ tr/GC// ) / 23;

        # is the GC content within the range 30-50?
        if ( ( $gc_content >= 0.3 ) && ( $gc_content <= 0.5 ) ) {
```

```
            # does the sequence contain stretches of As, Ts, Cs or Gs?

        unless ( ( $testseq =~ /A{4}/ )
             || ( $testseq =~ /T{4}/ )
             || ( $testseq =~ /G{4}/ )
             || ( $testseq =~ /C{4}/ )

             # avoid also regions of six positions with G or C
             || ( $testseq =~ /[GC]{6}/ ))
        {
             print "pos $i $testseq\n";
        }
      }
    }
}
```

When the script in Code 3.1 is executed the first lines of the output will be:

```
pos 165 AAAGGGCCTTCACAGTGTCCTTT
pos 223 AAAGTACGAGATTTAGTCAACTT
pos 522 AAGACGTCTGTCTACATTGAATT
pos 776 AAAGTATCAGGGTAGTTCTGTTT
```

Hence, we have obtained a number of candidate siRNA sequences that could be used in an RNAi experiment. As indicated above, we also should make sure by a separate investigation that a sequence by chance does not match an mRNA other than the one of interest. For the RNAi experiment it will also be useful to have access to some random sequence for the purposes of a negative control. Such a sequence could be constructed by taking the original sequence and shuffling it, i.e. take all the characters of that string and put them in a random order (see the web resource for this book for an example script).

In summary, we have obtained a list of suitable RNAi sequences. As the next step we may request these RNAs from a company producing oligonucleotides. As a rule, companies also offer a service to computationally design suitable RNAi sequences, using, for instance, an mRNA as a starting point. Now, however, you know how to do it yourself! This could be useful if you wanted to introduce some rule of your own in the RNAi design scheme.

EXERCISES

3.1 In Code 3.1 we used `open(IN, 'mrna.fa')`; to read from the file `mrna.fa`. If you want to *write* to a file you may use `open(OUT, '>newfile')`; where `newfile` is the name of the file you want to write to. Try this method to print the output of Code 3.1 to a file instead of to the screen. (For more on filehandles, see also Appendix III.)

3.2 In Code 3.1 the DNA sequence of the siRNA candidate is printed. Modify the code such that the translation product is also printed below the DNA sequence. The translation product should be from the relevant reading frame, i.e. it should be a portion of the BRCA1 protein.

3.3 Modify Code 3.1 so it also checks whether the first codon in the input sequence is AUG and the last codon is any of the three stop codons.

REFERENCES

Birmingham, A., E. Anderson, K. Sullivan, *et al.* (2007). A protocol for designing siRNAs with high functionality and specificity. *Nat Protoc* **2**(9), 2068–2078.

Fire, A., S. Xu, M. K. Montgomery, *et al.* (1998). Potent and specific genetic interference by double-stranded RNA in *Caenorhabditis elegans*. *Nature* **391**(6669), 806–811.

He, L., J. M. Thomson, M. T. Hemann, *et al.* (2005). A microRNA polycistron as a potential human oncogene. *Nature* **435**(7043), 828–833.

Kaufman, E. J. and E. A. Miska (2010). The microRNAs of *Caenorhabditis elegans*. *Semin Cell Dev Biol* **21**(7), 728–737.

Lau, N. C., A. G. Seto, J. Kim, *et al.* (2006). Characterization of the piRNA complex from rat testes. *Science* **313**(5785), 363–367.

Lu, J., G. Getz, E. A. Miska, *et al.* (2005). MicroRNA expression profiles classify human cancers. *Nature* **435**(7043), 834–838.

Maziere, P. and A. J. Enright (2007). Prediction of microRNA targets. *Drug Discov Today* **12**(11–12), 452–458.

Senti, K. A. and J. Brennecke (2010). The piRNA pathway: a fly's perspective on the guardian of the genome. *Trends Genet* **26**(12), 499–509.

Siomi, H. and M. C. Siomi (2009). On the road to reading the RNA-interference code. *Nature* **457**(7228), 396–404.

Sunilkumar, G., L. M. Campbell, L. Puckhaber, R. D. Stipanovic and K. S. Rathore (2006). Engineering cottonseed for use in human nutrition by tissue-specific reduction of toxic gossypol. *Proc Natl Acad Sci U S A* **103**(48), 18054–18059.

4

Gene technology
Amplifying DNA

My little silver Honda's front tires pulled us through the mountains. My hands felt the road and the turns. My mind drifted back into the lab. DNA chains coiled and floated. Lurid blue and pink images of electric molecules injected themselves somewhere between the mountain road and my eyes.

(Kary Mullis about his invention of PCR in *Dancing Naked in the Mind Field*; Mullis, 1998)

In order to work with DNA in the laboratory we often need to produce that DNA in a larger quantity. In addition, a longer DNA molecule like a whole human chromosome is difficult to work with, and we rather want to focus on a smaller region of DNA. In a classic cloning experiment a smaller DNA fragment is introduced into a circular plasmid DNA, which is allowed to replicate within bacterial cells. Once the bacterial cells have grown to a certain density, these cells may be isolated by centrifugation and the plasmid DNA purified from the bacterial cells. This is a somewhat time-consuming procedure, but there is an alternative method in gene technology that is more convenient. Thus, our third example of gene technology is the *polymerase chain reaction* (PCR), invented by Kary Mullis in 1984 (Mullis *et al.*, 1986).

What is PCR?

PCR is one of the most widely used techniques in DNA technology. As in traditional cloning it is a means to amplify a shorter region of DNA, i.e. to produce a distinct region of DNA in a large quantity. However, PCR is carried out in a simple test tube reaction where DNA is amplified with the help of the enzyme DNA polymerase. Another technical advantage of PCR compared to conventional cloning in some bacterial host is that the DNA product is relatively pure.

How is a region of interest amplified using PCR? The method assumes you know a piece of sequence towards the ends of the region you want to amplify.

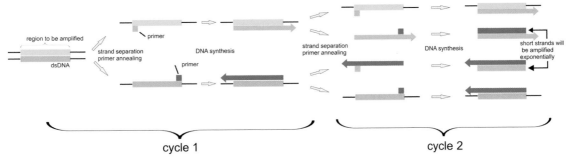

Fig. 4.1 *The polymerase chain reaction (PCR).* This reaction is carried out in a test tube. The following critical ingredients are required: a double-stranded DNA with a region to be amplified; short oligonucleotide sequences known as primers; a DNA polymerase to produce new DNA strands; and deoxyribonucleotide triphosphates as substrates for this enzyme. The DNA is first heated to about 95 °C to allow the strands to separate. The temperature is then lowered to 50–65 °C to allow annealing (by base-pairing) of two different primers to the DNA strands. One primer pairs with the 3′ end of the forward strand and the other with the 3′ end of the reverse strand. Next is the elongation step, where DNA polymerase produces new DNA by extending the primers in 5′ to 3′ direction. For this step the temperature is raised to a level suitable for the DNA polymerase being used. A thermostable enzyme like Taq polymerase is typically used such that it will survive the high temperatures of the strand separation step. The strand separation, primer annealing and chain elongation steps form one cycle of PCR. This cycle may now be repeated any number of times by simply adjusting the temperature appropriately. The first two cycles are shown. After many more cycles the short fragments indicated in the figure have by far outnumbered all other DNA molecules in the reaction mixture.

Primers for DNA replication are constructed on the basis of these sequences such that the primers match sequences that flank the region you want be amplify (Fig. 4.1). The first step in PCR is that the strands in the DNA to be amplified are separated by heating. The next step is an annealing step where the temperature is lowered, and as a result, primers associate with DNA strands on the basis of sequence complementarity. Two primers are used and they associate with the target DNA as shown in Fig. 4.1. The third step is one in which DNA synthesis occurs, extending from the primers. The three steps – (1) strand separation; (2) annealing of primer to strands; and (3) elongation – form one cycle in PCR. Multiple cycles may now be performed in the same test tube by appropriately adjusting the temperature. Because a high temperature is used in the strand separation step, it is necessary to use a thermostable DNA polymerase. Such enzymes may be isolated from bacteria thriving at high temperatures, like those isolated from hot springs. In each PCR cycle the amount of DNA will ideally be doubled. For example, after 30 cycles the amount of DNA has in theory increased by a factor of 2^{30} (approximately 10^9).

The quite dramatic amplification achieved with PCR has the effect that it is extremely sensitive. In principle, a single molecule of DNA may be amplified and analysed.[1] This sensitivity is useful whenever the amount of DNA in a sample is restricted. This may be the case at a crime scene, or in the analysis of very old biological samples where the DNA has been severely degraded.

PCR applications

Not only is PCR used in all kinds of biomedical research, but there are a very large number of practical applications, such as important clinical uses. It is used in diagnosis of viral diseases such as HIV, and in diagnosis of bacterial infections like tuberculosis. The responsible agent in tuberculosis is *Mycobacterium tuberculosis*. This bacterium is difficult to identify without molecular techniques such as PCR because it grows slowly in the laboratory. Furthermore, PCR is used to monitor cancer therapy as well as to detect cancer-specific chromosomal aberrations of a type we will discuss in Chapter 7. It is used in paternity testing and in many forensic applications. A single hair root or a faint blood stain at a crime scene is enough to obtain an informative PCR product. Finally, PCR has been important when analysing ancient DNA that might be tens of thousands of years old. The results allow conclusions as to the nature and evolution of species that are now extinct. Examples include the woolly mammoth, which went extinct about 4000 years ago. Thus, a Siberian mammoth specimen about 20 000 years old was recently subject to sequence analysis, allowing reconstruction of a significant portion of its genome (Miller *et al.*, 2008). Another example is analyses of Neanderthal humans that went extinct around 30 000 years ago (Green *et al.*, 2008, 2010). Yet another example, the extinct Tasmanian tiger (Miller *et al.*, 2009) will be discussed in more detail in Chapter 10. In Chapter 12 we will encounter another success story of PCR, the identification of the remains of the Romanovs, the Russian tsar family killed in 1918 by Bolsheviks.

Primer design

From this brief account of rather striking applications of PCR we will leap back to the more technical problem of appropriately designing PCR primers. There are many different methods that have been designed in order to computationally identify suitable primers for PCR. They all assume that you know the sequence of the region you want to amplify, or at least the regions where the primers are. Examples of criteria used in the design of PCR primers are:[2]

---

[1] In terms of mass this is very little; the weight of a single human chromosome is in the order of picograms.

[2] Based on information at http://premierbiosoft.com/tech_notes/PCR_Primer_Design.html and http://biochem218.stanford.edu/Projects%202001/Burpo.pdf.

(1) primer length should be in the range 18–22 nucleotides;
(2) melting temperature (the temperature at which half of the DNA duplex will dissociate to single-stranded entities) should be 52–58 °C;
(3) GC content should be 40–60%;
(4) sequences with stable secondary structure should be avoided;
(5) repeats should be avoided;
(6) long runs of single bases should be avoided;
(7) primer should be specific to target DNA (check with BLAST).

You may note that some of these criteria resemble those we considered for design of RNAi in the previous chapter. There are many programs developed for PCR primer design, and we could also think of implementing the design ideas in the form of a Perl script. However, for our programming example we will instead consider a somewhat less common application of PCR. This application addresses the problem of wanting to amplify a DNA region but in a case where we do not have access to any nucleotide sequence at all for primer construction. For example, we may be facing an organism where we yet do not have access to any nucleotide sequences.[3] How could we design PCR primers in this case? One solution is that we make use of a gene which is strongly conserved in sequence, like a ribosomal RNA gene. In this case we could predict conserved sequences assuming that the species of interest has close relatives where the sequence is known. But what if we wanted to amplify a protein-coding gene? We consider first a simple example to better understand this problem.

Assume we want to use PCR to amplify a specific protein-coding region from some organism where there is no sequence information, but we do have access to an amino acid sequence as shown below, derived from one or more closely related organisms.

```
Gly-Val-Thr-Lys-Trp-Lys-Met
```

We assume that this sequence is also found in our species of interest. For PCR we then need to derive a nucleotide sequence from the amino acid sequence. We can do this computationally, but there is of course the problem that the genetic code is degenerate, i.e. many amino acids have more than one codon. For instance, the amino acid serine is encoded by six different codons: UCU, UCC, UCA, UCG, AGU and AGC. This means that for any amino acid sequence we start out with there will typically be a large number of possible nucleotide sequences. For the experimental PCR method we therefore need to make use of

---

[3] And there are still quite a few organisms that have not been subject to any sequence analysis, some two million or so.

a mixture of primers where all possible sequences are represented. We can infer all possible nucleotide sequences with the help of the genetic code:

```
Protein: Gly-Val-Thr-Lys-Trp-Lys-Met
DNA:     GGN-GTN-ACN-AAR-TGG-AAR-TAG
```

where N represents any nucleotide, and R is either A or G. For the PCR reaction we need to construct all possible oligonucleotides based on this sequence. The total number of sequences will be $4 \times 4 \times 4 \times 2 \times 2 = 256$, because we have three 'N's and two 'R's.

For the PCR reaction we want to avoid too large a number of oligonucleotides because the risk of mispriming (priming in the wrong site in DNA) increases with the number of sequences. Therefore, we want to make use of regions in the protein sequence that are rich in amino acids that are encoded by relatively few codons. Amino acids like Trp and Met, which both have only one codon, seem ideal.

BIOINFORMATICS

Tools introduced in this chapter	
Perl	shebang
	warning messages with `perl -w`
	`strict` module
	declaring variables with `my`
Unix	`sort`
	pipes

We now arrive at the programming part of this discussion. We want to construct a piece of code that allows a systematic investigation of the amino acid sequence of interest. Thus, we want to infer all possible nucleotide sequences based on the amino acid sequence. We also want to know about the degeneracy of the nucleotide sequence, i.e. the total number of oligonucleotide sequences that need to be produced to cover all possible alternatives.

Reverse translation

The code to identify nucleotide sequences that are of interest for a PCR experiment as described above is shown in Code 4.1. You will notice how in this script, when variables are introduced, they are preceded by the word `my`. There are a few other changes that are new to this chapter. We will return to a discussion of these towards the end of this chapter. Otherwise, the Perl script uses many of the techniques from the scripts in the previous chapters, as described in the following.

We first define a hash %reverse_code, which is the 'reverse' genetic code, i.e. for each amino acid it is specified by the different codons that encode that amino acid. But the different codons are represented as a single string, using ambiguity codes as in the previous chapter examples with restriction enzymes. For example, the value for the key 'L' (for the amino acid leucine) will be YUN because the leucine codons are CUU, CUC, CUA, CUG, UUA and UUG. We are reading the sequence from a file, as in the previous chapter and we store the protein sequence in the variable $pep. Then we analyse the protein sequence using a for loop when we extract a sequence of seven amino acids (corresponding to 21 codons at the level of DNA sequence). Every sequence of seven amino acids is then 'reverse translated' into a DNA sequence and we calculate the extent of degeneracy.

We use a measure of degeneracy which is the number of sequences that need to be synthesized in order to cover all possible DNA sequences. For instance, the amino acid lysine ('K') has two different codons, AAA and AAG. The degeneracy associated with lysine is therefore two. We multiply the individual numbers to obtain the degeneracy associated with the DNA sequence of seven codons. In the script in Code 4.1 an expression like $degen *= 2; is shorthand for $degen = $degen * 2;.

As in the previous chapter, we are making use of pattern matching. For instance, in the line

```
if ($base =~ /[RYWMSK]/) {$degen *= 2;}
```

we are testing whether the nucleotide symbol stored in the variable $base matches the pattern /[RYWMSK]/, i.e. we are testing whether that symbol matches any of the nucleotide ambiguity symbols R, Y, W, M, S and K. Each of these symbols denotes two different bases. For instance, W is either A or T. Therefore, all these ambiguity symbols are associated with a degeneracy of two.

Code 4.1 pcr.pl

```perl
#!/usr/bin/perl -w

use strict;

# the 'reverse' genetic code
# This hash has amino acid symbols as keys,
# and the degenerate codons as values
my %reverse_code = (
    'L', 'YUN', 'F', 'UUY', 'I', 'AUH', 'M', 'AUG', 'V', 'GUN',
    'S', 'WSN', 'P', 'CCN', 'T', 'ACN', 'A', 'GCN', 'G', 'GGN',
    'Y', 'UAY', 'H', 'CAY', 'Q', 'CAR', 'N', 'AAY', 'K', 'AAR',
    'D', 'GAY', 'E', 'GAR', 'C', 'UGY', 'W', 'UGG', 'R', 'MGN'
);
```

```perl
# reading the protein sequence from a file
my $pep = '';
open(IN, 'brca1.pep') or die "Could not open file\n";
while (<IN>) {
    unless (/>/) { chomp; $pep .= $_; }
}
close IN;

for ( my $i = 0 ; $i < length($pep) - 6 ; $i++ ) {
    print "pos $i ";
    my $test = substr( $pep, $i, 7 );
    my $degen = 1;

    for ( my $j = 0 ; $j < 7 ; $j++ ) {
        my $aa = substr( $test, $j, 1 );
        my $codon = $reverse_code{$aa};
        print "$codon";

        # calculate degeneracy
        for ( my $k = 0 ; $k < 3 ; $k++ ) {
            my $base = substr( $codon, $k, 1 );
            if ( $base =~ /[RYWMSK]/ ) { $degen *= 2; }
            if ( $base =~ /[VHDB]/ ) { $degen *= 3; }
            if ( $base =~ /[N]/ ) { $degen *= 4; }
        }
    }
    print "\t$degen\n";
}
```

The output from Code 4.1 will be a list of amino acid positions along with the degeneracy that was calculated for the reverse-translated DNA sequence derived from that position:

```
1 AUGGAYYUNWSNGCNYUNMGN 65536
2 GAYYUNWSNGCNYUNMGNGUN 262144
3 YUNWSNGCNYUNMGNGUNGAR 262144
4 WSNGCNYUNMGNGUNGARGAR 65536
5 GCNYUNMGNGUNGARGARGUN 16384
6 YUNMGNGUNGARGARGUNCAR 8192
7 MGNGUNGARGARGUNCARAAY 2048
8 GUNGARGARGUNCARAAYGUN 1024
9 GARGARGUNCARAAYGUNAUH 768
```

For the PCR experiments we do not want to consider a sequence like that in position 2, with a degeneracy of 262 144, but rather one like that in position

8, with degeneracy 768. In the output from the Perl program we may want to identify the positions associated with the smallest degeneracy. We could do a sorting of the positions within the Perl script (see Appendix III). We could also direct the output from Code 4.1 to the Unix program sort in this manner:

```
% perl pcr.pl | sort -k3 -n -r
```

The pipe (|) symbol is used to indicate that the output coming from the left of the pipe is directed to the program specified to the right of the pipe. In this case the Perl program output is used as input to the Unix sort command. We do a sorting based on column 3, as specified by -k3. The -n parameter specifies a *numerical* sort; the -r parameter indicates that we reverse the order of the sorted output, so the listing will be from higher to lower numbers. For more on the Unix sort command, see Appendix I. From our sorting procedure above we identify two pairs with a degeneracy of only 64:

```
pos 652 AARAARAARUAYAAYCARAUG    64
pos 351 AARGARUGGAAYAARCARAAR    64
```

Good manners during Perl programming

Now some words about executing a Perl script and also some important notes about good common practice when it comes to writing Perl code. There are elements of such practice in Code 4.1. First, the top line is:

```
#!/usr/bin/perl -w
```

The first characters, #!, are known as a *shebang*.[4] When a file in a Unix operating system contains a shebang the operating system will understand that whatever follows this line is a script. The script is interpreted by the program that is specified immediately after the shebang. In this particular case the program is the Perl executable, located in the directory /usr/bin/. One consequence of the whole shebang line is that in order to have the script executed you do not need to type

```
% perl pcr.pl
```

Instead, you can type

```
% ./pcr.pl
```

---

[4] Also known as hashbang, hashpling, pound bang or crunchbang! If you want to know about the etymology, 'shebang' comes from the words 'sharp' describing the exclamation point, and the word 'hash' describing the '#' character.

where the `./` means you are referring to `pcr.pl` in the current directory. This assumes the file named `pcr.pl` has an executable flag. If it does not you have to make it executable by using the Unix `chmod` command (see also Appendix I):

```
% chmod +x pcr.pl
```

Finally, what is meant by the `-w` flag to `/usr/bin/perl`? This is one of many possible parameters to Perl and it means that warning messages will be issued. This is useful, for example, in a situation where you have made a typo. Consider this code stored in the file `test.pl`:

```
$dna = 'GCCAT';
print $dan; # typo
```

If you try this code out:

```
% perl test.pl
```

nothing will be printed, because `$dan` does not have any contents. However, if you use

```
% perl -w test.pl
```

warning messages will result:

```
Name "main::dan" used only once: possible typo at strict_w.pl line 3.
Name "main::dna" used only once: possible typo at strict_w.pl line 2.
Use of uninitialized value $dan in print at strict_w.pl line 3.
```

In this case Perl warns that both `$dna` and `$dan` are used only once, and that the contents of `$dan` have not been defined. These kind of messages are very useful for spotting typing errors.

Now to other characteristics of the script of this chapter; `use strict` and `my`. These concepts will not be fully explained here, but practical consequences of them are outlined. The line `use strict;` specifies that the Perl *module* named `strict` is to be used in the script. We have not previously encountered modules, but they are frequently used in Perl. You may regard them as pieces of code that are added to what is already within the core of the Perl programming language (see also Appendix III). The application of `use strict` has the consequence that all variables *must be declared.* The declaration may be done with `my`. If you use

```
use strict;
$codon = 'GCA';
print "$codon\n";
```

the code will not be executed, but there will be an error message saying 'Global symbol "$codon" requires explicit package name.' In other words, the strict module will spot that you did not declare $codon with my. But all will be fine if you write:

```
use strict;
my $codon = 'GCA';
print "$codon\n";
```

The use of use strict and my will also help to spot typos:

```
use strict;
my $dna = 'GCCAT';
print $dan;
```

This code will give rise to the error message:

```
Global symbol "$dan" requires explicit package name at strict_w.pl line 3
```

The practice of using use strict and my also helps to keep track of the different variables in your script. An example of where things may go wrong is the following, where the same variable name has been used in two different places in the same script. First without my:

```
for ( $i = 0 ; $i < 2 ; $i++ ) {
    print "outer loop $i ";
    # some code here ....
    for ( $i = 0 ; $i < 2 ; $i++ ) {
        print "inner loop $i ";
    }
    print "\n";
}
print "\n";
```

This will print

```
outer loop 0 inner loop 0 inner loop 1
```

because when you enter the second (inner) for loop, $i will reach the value 1 and back in the first (outer) loop it will reach the value 2; as a result this loop will be exited. But with my, the $i in the second loop will be local to that loop and not confused with the $i of the first loop. The output will instead be:

```
outer loop 0 inner loop 0 inner loop 1
outer loop 1 inner loop 0 inner loop 1
```

Declaring a variable with `my` has the effect that the variable is local to the enclosing block, in this case the inner `for` loop. We want to behave well and from now on in this book we will therefore make use of the `#!/usr/bin/perl -w` line, as well as `use strict` and `my` declarations.

EXERCISES

4.1 When designing primers for PCR you may want to avoid *dinucleotide repeats*, such as ATATATAT. Write a Perl script to analyse the sequence `mrna.fa` in the previous chapter. Identify all regions in the mRNA with at least four tandemly repeated dinucleotides.

4.2 See Appendix II for information on the Unix `head`, `tail` and `cut` commands. What is the `head` command to print the first two lines of the file `brca1.pep`? The output of Code 4.1 may be piped to the `head`, `tail` and `cut` commands. What is the command to print the first ten lines of the output from the code? What is the command to print only the reverse-translated sequences that are in the second column of the output?

4.3 An important issue in the design of PCR primers is the temperature of the primer/template annealing. A simple estimate of the melting temperature is to assign 2 °C to each A–T pair and 4 °C to each G–C pair. Design Perl code that will analyse the nucleotide sequence in `mrna.fa`, and that will list the annealing temperature for each possible sequence of 21 nucleotides.

REFERENCES

Green, R. E., J. Krause, A. W. Briggs, *et al.* (2010). A draft sequence of the Neandertal genome. *Science* **328**(5979), 710–722.

Green, R. E., A. S. Malaspinas, J. Krause, *et al.* (2008). A complete Neandertal mitochondrial genome sequence determined by high-throughput sequencing. *Cell* **134**(3), 416–426.

Miller, W., D. I. Drautz, J. E. Janecka, *et al.* (2009). The mitochondrial genome sequence of the Tasmanian tiger (*Thylacinus cynocephalus*). *Genome Res* **19**(2), 213–220.

Miller, W., D. I. Drautz, A. Ratan, *et al.* (2008). Sequencing the nuclear genome of the extinct woolly mammoth. *Nature* **456**(7220), 387–390.

Mullis, K. (1998). *Dancing Naked in the Mind Field*. New York, Pantheon Books.

Mullis, K., F. Faloona, S. Scharf, *et al.* (1986). Specific enzymatic amplification of DNA in vitro: the polymerase chain reaction. *Cold Spring Harb Symp Quant Biol* **51**(1), 263–273.

Human disease
When DNA sequences are toxic

## 5

It consists essentially in a spasmodic action of all the voluntary muscles of the system, of involuntary and more or less irregular motions of the extremities, face and trunk. . . . The first indications of its appearance are spasmodic twitching of the extremities, generally of the fingers which gradually extend and involve all the involuntary muscles. This derangement of muscular action is by no means uniform; in some cases it exists to a greater, in others to a lesser, extent, but in all cases gradually induces a state of more or less perfect dementia.

(Charles Oscar Waters, 1841, describing what is now known as Huntington's disease)

A fter a discussion of gene technology methods in the preceding chapters, we now turn to a few topics of more immediate medical interest. We will be dealing with different kinds of human diseases that have a genetic component and see how some aspects of these diseases may be examined using bioinformatics tools such as Perl.

Inherited disease and changes in DNA

In Chapter 1 we saw how genetic information flows from DNA to proteins. The sequence of bases in DNA determines the sequence of amino acids in protein and that sequence in turn determines the biological function of the protein. The relationship between genetic information in the form of DNA sequences and biological function in proteins has been demonstrated by numerous experiments carried out in molecular biology laboratories. At the same time it is intriguing to note that this relationship is also elegantly demonstrated in nature. Thus, we know of numerous examples of naturally occurring changes in DNA that have marked effects on the function of the corresponding protein. Such changes often give rise to disease. In discussing this further we need to clarify what types of alterations are observed in DNA. We may distinguish two major categories.

Glu

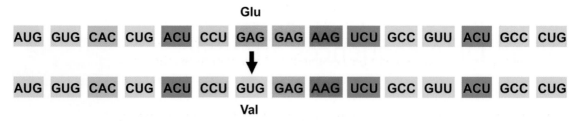

Fig. 5.1 *Sickle-cell mutation.* N-terminal part of β-globin chain is shown. In sickle-cell anaemia a mutation A to T causes a change from a glutamic acid codon (GAG, above) to a valine codon (GUG, below). This amino acid is much more hydrophobic than glutamic acid and causes aggregation of haemoglobin molecules.

First, there are highly local changes, such as point mutations (single nucleotide changes) and addition or deletion of a smaller number of nucleotides. Second, there are large-scale *rearrangements* of DNA sequences that result from *DNA recombination* events.

DNA recombination is a process in which one portion of DNA moves to another DNA molecule. This happens, for example, during crossing-over (which is covered in more detail in Chapter 20). During a chromosomal *translocation*, parts of chromosomes are moved between non-homologous chromosomes. Yet another result of DNA recombination is *gene amplification*, in which one region of a chromosome, like a single gene, is produced in many copies.

We will cover the category of rearrangement in the form of chromosomal translocation in Chapter 7. Here we will focus on the category of local changes. Numerous diseases with a genetic background are the result of such changes. A classic example of point mutation is *sickle-cell anaemia*, affecting one of the polypeptide chains in haemoglobin, referred to as the β-subunit. The disease got its name from the abnormal shape of red blood cells observed in people affected by this disease. Haemoglobin molecules in people with sickle-cell anaemia tend to aggregate and form fibres that extend throughout blood cells. The altered shape of the blood cells prevents blood flow, typically resulting in pain in different parts of the body and an increased risk of strokes.

The molecular background of sickle-cell anaemia was discovered in 1956. Vernon Ingram demonstrated that the amino acid glutamic acid, normally present in position 7 of the β-globin chain, had been changed to valine (Ingram, 1956, 1957) (Fig. 5.1). This residue is located on the surface of the globin molecule and the mutation has the effect of forming a hydrophobic patch. This patch causes globin chains to aggregate.

Humans have two copies of each chromosome, and thus two copies of each gene (for more on this subject, see Chapters 18–20). In cases of sickle-cell anaemia, both copies of the β-globin gene are mutated, whereas individuals that have one normal gene and one copy of the mutated gene are more or less healthy. In certain parts of the world the fraction of the population that carries the sickle-cell gene is fairly large. In Western Africa the gene is present in about 1% of the population. The reason it is so common in this part of the world is that the sickle-cell gene protects against malaria because the responsible parasite, *Plasmodium falciparum*, spends part of its life cycle in red blood cells.

Another disease with a genetic background is Lesch–Nyhan syndrome (LNS), a disease with neurological symptoms in which affected individuals are often characterized by bizarre behaviour such as self-mutilation (Sculley *et al.*, 1992; Nyhan, 1997). This disease is the result of mutations in the gene encoding hypoxanthine-guanine phosphoribosyltransferase (HGPRT). This enzyme is part of a pathway in which purines are converted to the corresponding nucleotides. In the absence of the enzyme, as in LNS, an excess of purines give rise to an overproduction of uric acid. However, it is still not clear why the absence of the enzyme gives rise to the strange neurological symptoms. Compared to sickle-cell anaemia, LNS is a relatively rare disease, with approximately 1 in 380 000 live births affected. Sickle-cell anaemia is the result of a very specific point mutation; LNS may result from a variety of changes, including deletions, insertions and point mutations in coding regions, as well as mutations that affect splicing.

Huntington's disease and CAG repeat expansion

We now turn to another genetic disease, *Huntington's disease*, which we are going to cover in more detail. The mechanism giving rise to this disease is different from those of sickle-cell anaemia and LNS as the fatal mutations in this case involve an *expansion of short repeats*. Huntington's disease (Lawrence, 2009) was first described in 1841 by Charles Oscar Waters, who had recently graduated from the Jefferson Medical College in Philadelphia.[1] Waters noted not only the symptoms, but also the hereditary nature of this disorder. Another American physician, George Huntington, later examined the disorder more carefully; he was only 22 years old when he described his first results in a paper published in 1872.

The disease was described in the nineteenth century, but it was much later that we came to a better understanding of the disease at the molecular level. A major breakthrough occurred in 1993 when the responsible gene was identified. This was possible by careful analysis of families affected by the disease. It was found that the gene encodes a protein 3144 amino acids in length, referred to as *huntingtin*. Unfortunately, we still do not fully understand the function of the

---

[1] The opening quotation to this chapter is from a letter he wrote to his former professor.

```
HD_HUMAN   MATLEKLMKAFESLKSFQQQQQQQQQQQQQQQQQQQQQQQQ-----------------------------------------PPPPPPPPPPPQLPQP
var1       MATLEKLMKAFESLKSFQQQQQQQQQQQQQQQQQQQQQQQQQQQQQQQQQQQQ----------------------------PPPPPPPPPPPQLPQP
var2       MATLEKLMKAFESLKSFQQQQQQQQQQQQQQQQQQQQQQQQQQQQQQQQQQQQQQQQQQQQQ------------------PPPPPPPPPPPQLPQP
var3       MATLEKLMKAFESLKSFQQQQQQQQQQQQQQQQQQQQQQQQQQQQQQQQQQQQQQQQQQQQQQQQQQQQQQQQQQQPPPPPPPPPPPQLPQP
```

Fig. 5.2 *Repeat region in huntingtin.* N-terminal portion of the huntingtin protein with poly-glutamine (polyQ) region shown. The two sequences on top are normal variants, whereas the longer polyQ inserts in the bottom sequences are more likely to give rise to Huntington's disease.

Fig. 5.3 *Slippage during DNA replication.* Replication slippage occurs at repetitive sequences, such as a series of CAG triplets. In backward slippage, as shown here, the newly synthesized strand loops out, and as a result the new DNA strand contains more CAG triplets than the template strand. In forward slippage a region of the template strand loops out and this gives rise to a smaller number of CAG triplets.

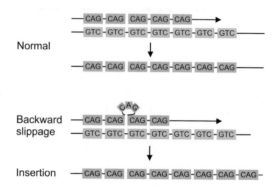

protein. Important for the disease, however, is a region where the amino acid glutamine is repeated a large number of times (Fig. 5.2).

The length of the poly-glutamine (polyQ) region is subject to variation among human individuals. Normally there are less than 28 glutamine residues, but in rare cases when the number increases to more than 40 it gives rise to Huntington's disease. The exact mechanism of pathology is not known, but the version of the protein in which the polyQ region has been expanded is known to be toxic to cells. The polyQ region is thought to stimulate protein aggregation, a process that might be involved in the progression of disease.

At the level of mRNA, the polyQ region is a result of the codon CAG being repeated many times. Such a repeat region in a gene is rather prone to mutational events that change the number of repeats. It is believed that triplet expansion occurs by *slippage* during DNA replication. In such a case the newly synthesized strand will loop out during DNA replication while base-pairing is maintained (Fig. 5.3).

Huntington's disease is not the only disease caused by a CAG triplet repeat expansion; a number of other hereditary neurodegenerative diseases involve such an expansion.[2] Furthermore, a number of diseases have been described that involve repeats of other triplets. The first triplet expansion disease to be identified was *fragile X syndrome*. Its symptoms are severe mental retardation combined with certain physical characteristics such as large ears and a long, narrow face. The affected gene is FMN1 ('fragile X mental retardation 1'), located

---

[2] Examples are: spinal bulbar muscular atrophy (SBMA); dentatorubropallidoluysian atrophy (DRPLA); and six different types of spinocerebellar ataxia (SCA1, 2, 3, 6, 7 and 17).

on the X chromosome. Its mRNA contains a region in the 5′ UTR with a CGG triplet repeat. The triplet is normally found in less than 55 copies, but patients with the fragile X syndrome may have up to 230 repeats. With this large number of repeats the transcription of this gene is turned off.

BIOINFORMATICS

All genes encoding mRNAs with triplet repeats are 'dangerous' in the sense that they may be expanded as a result of DNA replication errors, and this expansion may give rise to disease. Furthermore, a signifi-cant amount of genetic variation between human individuals may be attributed to repeat-length polymorphism. We may take a bioinformatics approach to examine these issues (for one example, see Molla *et al.*, 2009). For instance, we will consider the

Tools introduced in this chapter		
Perl	regular expressions and quantifiers	
	capturing in patterns	
	defining functions	
	array @_	
Unix	redirecting output to a file	
	wc command	

question: how common are mRNAs with triplet repeats in the human genome? We return to the wonderful world of programming tools and design a Perl script to find the answer to that question (Code 5.1).

Identifying mRNAs with CAG repeats

In Perl we can conveniently identify repeated elements in a nucleotide sequence. As in the previous chapters, regular expressions play an important role in the code. The repeated element in this case is CAG. This sequence is obviously included in the pattern, but we will have to specify how many times this sequence is to occur. We therefore need to make use of a regular expres-sion with quantifiers. Examples of possible regular expressions with CAG using quantifiers are:

```
/(CAG){6}/     CAG must match exactly six times
/(CAG){6,9}/   CAG must occur at least six times but at most nine times
/(CAG){6,}/    CAG must occur at least six times
/(CAG)*/       CAG occurs zero or more times
/(CAG)+/       CAG occurs one or more times
```

(For more on quantifiers, see Appendix III.) We need to group the characters CAG by using parentheses, because if we do not, i.e. the expression used is:

```
/CAG{6}/
```

then the quantifier will refer only to the preceding character, G, and this is not what we want.

For the task of finding CAG repeats when we require at least six consecutive CAGs, either of the expressions /(CAG){6}/ or /(CAG){6,}/ will work. For instance, if we assume that our sequence to be tested is in the variable $seq, then

```
$seq =~ /(CAG){6}/;
```

will return true whenever we have CAG repeated six times.

But we also want to identify the number of repeats. Therefore, we make use of yet another useful Perl function related to regular expressions. Perl is able to store in memory exactly what has been matched in a regular expression. All we need to do is to put the relevant expression(s) within parentheses. The contents may then be recalled using the variables $1, $2, $3, etc. First, a simple example:

```
$seq = 'GAACT';
$seq =~ /([AG]{3})CT/;
print $1;
```

The match within parentheses will be captured as $1, which is 'GAA' because the regular expression matches three consecutive characters that are either A or G. We may use any number of parentheses. Consider this example:

```
$seq = 'GAACT';
$seq =~ /([AG]{3})(.*)(.)$/;
print "$1 $2 $3";
```

Here $1 will contain GAA and $3 will contain T because it is the character at the end of the string (as represented by the $ symbol) and .* can match any string of length zero or more. In this case it will capture the 'C' in 'GAACT'.

We now exploit this memory function for our example with CAG repeats. Assume we want to report regions with at least six CAG triplets. We also want to match as many consecutive CAGs as possible. Therefore, we prefer /(CAG){6,}/ because a Perl regular expression will by default match in a *greedy* manner, i.e. it will try to match as much as possible. This type of behaviour is seen in this example:

```
$seq = 'GCCCCATG';
$seq =~ /(C{1})/;      # match one C
print "$1\n";          # will print C
$seq =~ /(C{1,})/;     # match one or more Cs
print "$1\n";          # will print CCCC
```

In Code 5.1 we have defined a *function* named find_cag_repeat that will find CAG repeats in a sequence. What is a function? Functions do things to variables that are used as input; functions typically produce an output. We already mentioned functions that are built into the Perl language, like `reverse` and `substr`. But we can also construct our own functions. We typically do it whenever we suspect that the same procedure or code is to be used repeatedly in a script. We may also want to have functions to better organize our script and to make it more readable.

As an example of a function defined by the user, consider a function to produce the reverse complement of a DNA sequence. A function is defined in a section listed as `sub` (for subroutine), followed by the name of the function and some code within curly brackets. The code below is an example; it will achieve the same thing as a piece of code in Chapter 1, except that it uses a function:

```perl
use strict;
my $dna = 'GCAATGG';
my $rev = revcomp($dna);
print "$rev\n";
sub revcomp {
    my $str = $_[0];
    $str = reverse($str);
    $str =~ tr/ATCG/TAGC/;
    return $str;
}
```

Here we defined and used a function called `revcomp`. In the definition of that function, the second line looks a bit odd. What is `$_[0]`? It is the first element of an array named `@_`. This array is a list of the arguments passed on to the function. The individual elements of this array are referred to by `$_[0]`, `$_[1]`, etc. In this case there is only one element in that array, the `$dna` variable. Therefore, the second line means that the variable `$str` receives its contents from the first element in the `@_` array, which is `$_[0]`. The last line of the `revcomp` function 'return $str;' means that the contents of the variable `$str` will be the output of the function.

We now examine the code for identifying CAG repeats (Code 5.1).

Code 5.1 find_cag_short.pl

```perl
#!/usr/bin/perl -w

use strict;

my $myid = 'short test sequence';
my $myseq = 'CGGATACTGGGGACTAAGCAGCAGCAGCAGCAGCAGCAGTTT';

find_cag_repeat( $myid, $myseq );

sub find_cag_repeat {
    my ( $id, $seq ) = @_;
    if ( $seq =~ /((CAG){6,})/ ) {
```

```
        my $len = length($1);      # the string matched within
                                   # the outer parentheses is stored
                                   # in memory and recalled
                                   # with $1
     $id = substr( $id, 0, 20 );
     print "Repeat with length $len found in $id\n";
   }
}
```

The function `find_cag_repeat` has two arguments, `$myid` and `$myseq`. Remember, the arguments passed to a function are stored in the array `@_`. In the function (subroutine) code these values are in turn passed to `$id` and `$seq` with the statement `my ($id, $seq) = @_;`.

In this case, our expression for testing the match to CAG repeats is:

`$seq =~ /((CAG){6,})/`

Note that there are two nested parentheses. The match corresponding to the outer parentheses, which is the match we are interested in, is recalled with `$1`. The inner parentheses (CAG) is recalled with `$2`. The output from the script will be:

```
Repeat with length 21 found in short test sequence
```

In Code 5.1 we analysed a single and very short sequence that was invented for this purpose only. The result is therefore fairly uninteresting. But why not scale up considerably and modify the code to read a whole database of sequences, like all human mRNAs? This is done in the script shown in Code 5.2. The function `find_cag_repeat` is exactly the same; the only difference is the input sequence. In this case we read the sequence from a file named `refseq_human`, which contains 46 044 mRNA sequences.[3]

To read this file we use a routine similar to that described in Chapter 3. A difference is that here we have a file with *multiple* concatenated sequences in a FASTA format, and we have to treat these sequences separately. We do it in such a way that whenever we encounter a line starting with a '>' symbol, i.e. the FASTA sequence definition line, it is read as a signal that the last line of the previous sequence has been read. Thus, when we have encountered the '>' symbol we carry out the `find_cag_repeat` operation on the sequence most recently read from the file. Having done this, we again set the `$seq` variable to

---

[3] These sequences are available from the web supplement to this book, but were originally downloaded using the NCBI Entrez tool, which is discussed in Chapter 14. Alternatively, they may be downloaded from ftp://ftp.ncbi.nih.gov/refseq/H_sapiens/mRNA_Prot/human.rna.fna.gz.

the null string (`' '`) and we set the variable $id as the current line. All lines that do not have a '>' character contain a sequence, and these are concatenated with any previous sequence in the variable $seq.

We also have to take into account that the very first time we encounter the '>' character we do not have a sequence to process, hence the condition `if ($id ne ' ')`. In addition, when we have reached the end of the file, we are left with one sequence to process. Therefore, we need to call the function `find_cag_repeat` one last time after having read the whole file.

Code 5.2 find_cag.pl

```perl
#!/usr/bin/perl -w

use strict;

my $infile = 'refseq_human';    # all human refseq
                                # sequences Dec 2010.

open(IN, $infile) or die "Oops, could not open $infile\n";
my $id = '';
my $seq = '';
while (<IN>) {
    chomp;
    if (/^>/) {
        if ( $id ne '' ) {
            find_cag_repeat( $id, $seq );
        }
        $id = $_;
        $seq = '';
    }
    else {
        $seq .= $_;
    }
}
close IN;

find_cag_repeat( $id, $seq );

sub find_cag_repeat {
    my ( $id, $seq ) = @_;
    if ( $seq =~ /((CAG){6,})/ ) {
        my $len = length($1);
        $id = substr( $id, 0, 20 );
        print "$id\trepeat length $len\n";
    }
}
```

The first lines of output from Code 5.2 will be:

```
>gi|157151758|ref|NM    repeat length 36
>gi|116875847|ref|NM    repeat length 18
>gi|223646108|ref|NM    repeat length 39
>gi|114431247|ref|NM    repeat length 33
>gi|125346191|ref|NM    repeat length 27
>gi|154350223|ref|NM    repeat length 21
```

We may execute Code 5.2 and have the results printed to the screen. However, we may also want to save the results in a file. We can redirect the output to a file:

```
% perl find_cag.pl > find_cag.out
```

The > symbol is used to indicate that the output from the program or command to the left of the symbol is directed to a file (see also Appendix I). As an alternative to this operation, we could have written code in the Perl script to specify that the output should go to a specific file. However, the operation shown here at the level of the command line is often more convenient and requires less code.

So, how many mRNAs were found with CAG repeats? In case you want to know how many lines there are in the output file, you may use the Unix command `wc -l`:[4]

```
% wc -l find_cag.out
```

It turns out that more than 200 mRNAs in the human genome have repeats of this kind!

Although we may be very happy about this Perl code, let us think for a moment about possible problems. One important thing to note is that even though there is more than one CAG repeat region, only one of these will be reported. Consider this simple example with two separate CAG repeats:

```
$dna = 'CAGCAGAAACAGCAGCAG';
if ($dna =~ /(CAG{2,})/) {print $1;}
```

In this case only the first two CAGs of the $dna string will be captured. Therefore, the script needs to be modified in case we want all CAG repeat regions in a sequence to be reported. We could, for example, make use of this construct with the global modifier to the match operator:

```
while ($dna =~ /(CAG){2,}/g) {print " $1";}
```

---

[4] `wc` is short for 'word count' and, combined with the parameter `-l`, it produces the number of lines.

When there is no longer a match to a CAG repeat region the `while` loop will be terminated.

An alternative method is based on the substitution operator:

```
while ($dna =~ s/((CAG){2,})// ) {print " $1";}
```

Here, we are replacing all regions of the sequence where CAG is repeated twice or more.

EXERCISES

5.1 Using Code 5.2 as a starting point, implement a method to analyse human mRNAs so that not only one region of CAG repeats is reported, but all such regions. You may use constructs based on the matching or substitution operators:

```
while ($dna =~ /(CAG){6,}/g)    {print " $1";}
while ($dna =~ s/((CAG){6,})//) {print " $1";}
```

How many mRNAs are you able to identify that have more than one CAG repeat region?

5.2 In the collection of RefSeq sequences used in Code 5.2 (`refseq_human`) many of the mRNAs have a description of function. Modify the code so that this information, instead of the identifier, is reported in the output. Are there proteins of the output that are described as 'huntingtin'?

5.3 Modify Code 5.2 so that the *nucleotide positions* of the different CAG repeats are also reported.

REFERENCES

Ingram, V. M. (1956). A specific chemical difference between the globins of normal human and sickle-cell anaemia haemoglobin. *Nature* **178**(4537), 792–794.

Ingram, V. M. (1957). Gene mutations in human haemoglobin: the chemical difference between normal and sickle cell haemoglobin. *Nature* **180**(4581), 326–328.

Lawrence, D. M. (2009). *Huntington's Disease*. New York, Chelsea House.

Molla, M., A. Delcher, S. Sunyaev, C. Cantor and S. Kasif (2009). Triplet repeat length bias and variation in the human transcriptome. *Proc Natl Acad Sci U S A* **106**(40), 17095–17100.

Nyhan, W. L. (1997). The recognition of Lesch–Nyhan syndrome as an inborn error of purine metabolism. *J Inherit Metab Dis* **20**(2), 171–178.

Sculley, D. G., P. A. Dawson, B. T. Emmerson and R. B. Gordon (1992). A review of the molecular basis of hypoxanthine-guanine phosphoribosyltransferase (HPRT) deficiency. *Hum Genet* **90**(3), 195–207.

6

Human disease
Iron imbalance and the iron responsive element

An inherited disease affecting the iron-binding protein ferritin

Some genetic disorders are comparatively common, such as sickle-cell anaemia, discussed in the previous chapter. However, there are also disorders that are extremely rare. Here we will deal with one such example, *hyperferritinaemia cataract syndrome* (Beaumont *et al.*, 1995; Girelli *et al.*, 1995, 1996, 1997; Kato *et al.*, 2001). It results in high levels of the protein *ferritin* in the blood. Another symptom is cataracts, i.e. clouding of the lens of the eye. The first cases of this disorder were described in 1995 (Beaumont *et al.*, 1995; Girelli *et al.*, 1995). One of the reports was concerned with two families that had lived in northern Italy for many generations (Girelli *et al.*, 1995). The affected family members already had symptoms of glare and impaired visual acuity in childhood. They also had high serum levels of ferritin. It was soon found that the molecular basis of the disease is a mutation in the ferritin mRNA. In the previous chapter we encountered inherited diseases that are the result of a mutation in the coding sequence of the gene, as exemplified by sickle-cell anaemia, Lesch–Nyhan syndrome and Huntington's disease. However, in the case of hyperferritinaemia cataract syndrome the mutation is not in the coding sequence. Instead, it affects a regulatory region in the 5′ untranslated region (UTR) of the ferritin mRNA (for a description of UTR, see Fig. 1.1).

Many proteins of iron metabolism are regulated at the level of translation

What is the function of ferritin? Free iron ions are toxic to cells and an important protective function of ferritin is to bind these ions. The production of the ferritin protein is carefully regulated in order to maintain a suitable level of free ions and to optimize the interaction with other proteins involved in iron metabolism. (For more on the intricate and complex pathways of iron metabolism, see Outten and Theil, 2009; Hentze *et al.*, 2010.) The regulation of ferritin production is exerted through an intriguing mechanism involving translation of

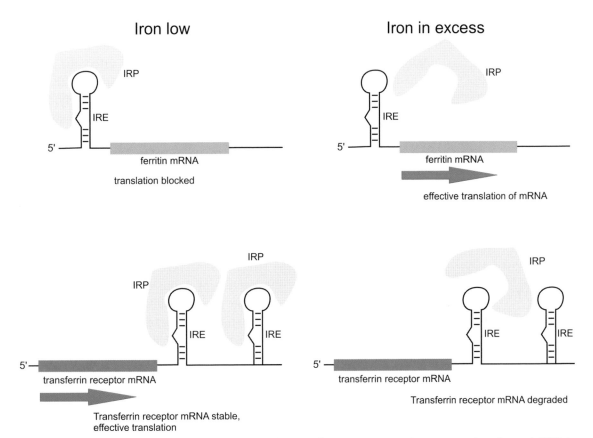

Fig. 6.1 *Translational regulation of iron metabolism.* In ferritin mRNA an iron responsive element (IRE) is located in the 5′ untranslated region (UTR) (upper panel). In the transferrin receptor (Tfr) mRNA there are five copies of IRE located in the 3′ UTR (lower panel). Only two of these IREs are shown here. When the level of iron is low, the IREs are targets of two cytoplasmic iron regulatory proteins, IRP1 and IRP2. The binding of these proteins results in stabilization of Tfr mRNA and inhibition of ferritin translation. Conversely, when iron is abundant IRP proteins tend to have lower affinity to IREs and, as a result, Tfr mRNA degradation and ferritin translation are stimulated.

its mRNA (Fig. 6.1). In the ferritin mRNA a local RNA hairpin structure, the *iron responsive element* (IRE), is located in the 5′ UTR (Fig. 6.2). In cells that are starved of iron, the IREs are targets of two cytoplasmic regulatory proteins, IRP1 and IRP2 (Volz, 2008). The binding of these proteins to IRE results in repression of translation, and hence lowered expression of ferritin. When iron is abundant, the IRP proteins tend to have lower affinity to IREs, and as a result ferritin translation is efficient.

Another iron-binding protein, *transferrin*, has a role in transporting iron ions in the blood. When it delivers iron to a cell it binds to a *transferrin receptor* (Tfr) located at the surface of the cell. The production of the transferrin receptor

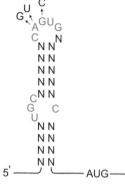

Fig. 6.2 *Structure of the iron responsive element (IRE)*. The loop sequence CAGUG is highly conserved and necessary for binding to the IRE binding proteins, IRP1 and IRP2 (Volz, 2008). Also important for binding are the base-paired regions, although these do not show any sequence conservation. Three mutations leading to hyperferritinaemia cataract syndrome are indicated (Beaumont *et al.*, 1995; Girelli *et al.*, 1995, 1996, 1997; Kato *et al.*, 2001).

is also subject to control that involves the IRE. In the Tfr mRNA there are five copies of IREs located in the 3′ UTR. In this case the IREs are not involved in the translation of the mRNA. Instead, they affect the stability of the Tfr mRNA. Thus, when iron is low, the binding of IRP1 and IRP2 to the IREs results in stabilization of the mRNA. Conversely, when iron is abundant, these proteins do not bind to the UTR. As a result, Tfr mRNA degradation is stimulated and the Tfr protein is produced in lower amounts. IREs are not only present in the mRNAs of ferritin and Tfr; they are also present in other mRNAs whose products are related to iron metabolism.

IREs are not unique as RNA regulatory elements. We know more examples where local RNA structures in untranslated regions of mRNAs play an important role in translational control. Examples include bacterial *riboswitches*, which are RNA structures in mRNA that are part of a regulatory mechanism where the binding of a small molecule to the RNA structure affects the expression of the protein (Winkler and Breaker, 2005). Another example is the selenocysteine insertion sequence (SECIS), which is present in both bacteria and eukaryotes and is responsible for incorporation of selenocysteine in proteins (Walczak *et al.*, 1996; Mix *et al.*, 2007).

Structure of the iron responsive element

As we compare different IREs, a consensus structure emerges (Fig. 6.2). These elements are hairpin-forming sequences 26–30 nucleotides long, with a CAGUGN loop sequence (where N may be any nucleotide), which is conserved in all IREs. Most IREs have a conserved C residue five bases upstream of the CAGUGN sequence, creating a bulge in the hairpin, whereas others, present in ferritin mRNAs, have a conserved bulge/loop UGC/C (Casey *et al.*, 1988; Piccinelli and Samuelsson, 2007).

What were the changes in mRNA sequence in the Italian families discussed above? Analysis of one of the families revealed a single mutation in the highly conserved CAGUG loop sequence. The G in the third position was changed into a C (Fig. 6.2). Interestingly, other families with the same type of ferritinaemia disorder also have mutations in the same loop sequence (Beaumont *et al.*, 1995; Kato *et al.*, 2001). All of these mutations are likely to interfere with the binding of IRPs to the IRE. This will give rise to an overproduction of ferritin because when IRPs are poorly bound to the IRE, translation will be constitutive and poorly regulated.

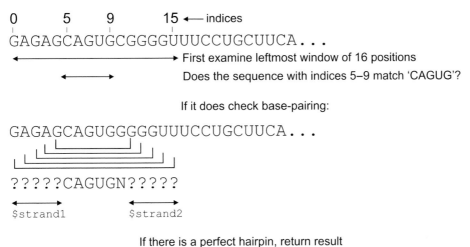

Fig. 6.3 *Outline of procedure used in the Perl script in Code 6.1 to identify IREs.*

BIOINFORMATICS

Identifying the iron responsive element

Tools introduced in this chapter	
Perl	equality operators

IREs are present in the mRNAs of ferritin and transferrin receptors. In fact, they are also present in other mRNAs, most of them coding for proteins related to iron metabolism. We now arrive at a bioinformatics problem. What if we wanted to find IREs in a collection of nucleotide sequences by using computational methods? This would allow us, for example, to identify additional mRNAs which might be regulated in the same way as ferritin and Tfr. What methods could be used to find IREs? In this case, methods based on sequence similarity would not be sensible because the only conserved primary sequence is the CAGUG loop sequence, and such a short sequence would be found by chance all over a genome, without being related to the IRE. What we need is a method that incorporates the CAGUG sequence as well as the conserved base-pairing pattern. This is simple in Perl, as demonstrated in Code 6.1.

The strategy of the script is summarized in Fig. 6.3. We look for hairpin regions with five base pairs where the loop sequence is CAGUGN. The whole structure is 16 nucleotides, so we scan the sequence with a window of this size. The first window to be examined is the sequence in positions 1–16 (with indices 0–15 in Fig. 6.3). This sequence is first checked for the presence of the CAGUG sequence. If it is present, then we check the base-pairing of the flanking sequences. If all bases are paired we have found a potential IRE, and this result is printed. We then go on to analyse the next window, which corresponds to positions 2–17. We keep repeating this until we come to the end of the sequence.

In Code 6.1 we have defined two different functions, `pair` and `findstem`. The function `pair` has two arguments, `$base1` and `$base2`, and tests whether two bases pair or not. In this case we make use of the standard base-pairing rules, where G pairs to C and A pairs to U. However, we also allow the pairing between G and U, as this is common in RNA structures.

In an expression like `$base eq 'G'` the `eq` is the *string equality operator* (`ne` is the 'not equal' operator – for more on comparison operators, see Appendix III). A common beginner mistake in Perl is to use the *numerical equality* symbol, which is `==`. Consider this example:

```
('G' eq 'A')    returns false
('G' == 'A')    returns true because in the context of numerical comparison both
                'G' and 'A' evaluate to 0!
```

A *really bad* thing, but a not too uncommon mistake as a beginner, is to confuse a comparison operator (like `==` and `eq`) with `=`, and so use:

```
if ($base1 = 'G')
```

Here, the equality sign infers an *assignment*, so the content of the variable `$base1` is now 'G'. The expression within parentheses always returns true because the assignment always works. This is a result quite unlike the result you would have got with the correct comparison operator, `eq`.

The `&&` and `||` symbols are the logical AND and OR operators, as discussed earlier. The value 1 (which is equivalent to 'true') is returned from the function `pair` if two of the bases used as input to the function form a pair according to our base-pairing rules.

The other function, `findstem`, is used to find out whether two strands of RNA sequence may form a helix by base-pairing. We test the different pairs of bases one at a time using the `pair` function. For a helix to form we require that all five positions are paired. To keep track of whether we have perfect pairing or not we make use of the variable `$tag`, which is originally set to 1. As soon as we find that a base pair is *not* formed, the `$tag` variable is set to zero. If this happens, the `findstem` function returns zero (false). If base pairs are formed in all five positions, the `$tag` will retain the value of 1 and the `findstem` function will return 1 (true).

Code 6.1 ire.pl

```perl
#!/usr/bin/perl -w

use strict;

my $seq = 'GAGAGCAGUGGGGGUUUCCUGCUUCAACAGUGCUUGGACGGAACCCGGCGCUCGUU';

for ( my $i = 0 ; $i < length($seq) - 15 ; $i++ ) {
    my $test = substr( $seq, $i, 16 );
    if ( substr( $test, 5, 5 ) eq 'CAGUG' ) {
        my $strand1 = substr( $test, 0, 5 );
        my $strand2 = substr( $test, 11, 5 );
        if ( findstem( $strand1, $strand2 ) ) {
            my $pos = $i + 1;
            print "match at position $pos:\n";
            print "$test\n";
            print "<----CAGUGN---->\n";
        }
    }
}

sub findstem {
    my $tag = 1;
    my ( $strand1, $strand2 ) = @_;
    for ( my $j = 0 ; $j < 5 ; $j++ ) {
        my $base1 = substr( $strand1, $j, 1 );
        my $base2 = substr( $strand2, 4 - $j, 1 );
        if ( pair( $base1, $base2 ) == 0 ) { $tag = 0; }
    }
    if ($tag) { return 1; }
}

sub pair {
    my ( $base1, $base2 ) = @_;
    if (   ( ( $base1 eq 'G' ) && ( $base2 eq 'C' ) )
        || ( ( $base1 eq 'G' ) && ( $base2 eq 'U' ) )
        || ( ( $base1 eq 'A' ) && ( $base2 eq 'U' ) )
        || ( ( $base1 eq 'C' ) && ( $base2 eq 'G' ) )
        || ( ( $base1 eq 'U' ) && ( $base2 eq 'A' ) )
        || ( ( $base1 eq 'U' ) && ( $base2 eq 'G' ) ) )
    {
        return 1;
    }
}
```

The output of Code 6.1 is:

```
match at position 23:
UUCAACAGUGCUUGGA
<----CAGUGN---->
```

This shows that one IRE candidate was found in this particular sequence. The script may easily be modified to search a whole database of sequences. If you do this you will find a large number of structures with the criteria outlined in Code 6.1. In fact, many of them are likely to be false positives. In order for this method of identifying IREs to be more specific we need to add more information to our model of the IRE. Hence, we need to add more features of the structure as it is described in Fig. 6.2 (see also Exercise 6.2).

Finally, a word of caution when it comes to the method outlined in this chapter to computationally identify RNA hairpins. The model of the RNA structure is very strict, as it states very precise rules for pairing and for the primary sequence of the loop. Any other RNA structures, including those with only a minor deviation from the consensus structure, are rejected by the model, although some of these RNA structures may be of biological interest. In Chapter 17 we will return to the problem of finding RNA structures, and in that context we will also encounter methods that are much more probabilistic in nature.

EXERCISES

6.1 There are many ways to solve a problem in programming. Consider the construction in Code 6.1 with a number of operations with the `substr` function, such as:

```
if substr( $test, 5, 5) eq 'CAGUG' {}
```

However, we could instead make use of a regular expression:

```
if ($test =~ /(.{5})(CAGUG.)(.{5})/ {}
```

In this expression we may capture not only the loop sequence as $2, but also the $strand1 and $strand2 variables as $1 and $3, respectively. Modify Code 6.1 to use this type of regular expression.

6.2 Design a Perl script to identify the complete structure of the iron responsive element as described in Fig. 6.2. Use your Perl script to search all human mRNAs, the same collection of mRNAs we used in the previous chapter (`refseq_human`). You should find that this search (as compared to that of Code 6.1) is highly specific, with few – if any – false positives.

6.3 A variant of the iron responsive element is present in the EPAS1 (endothelial PAS domain protein 1) mRNA (Sanchez *et al.*, 2007). The characteristic structure is shown below, where the less-than and greater-than symbols

(< >) represent the base-pairing. Write some Perl code to identify this element in the collection of human mRNAs we used in the previous chapter (`refseq_human`).

```
NNNNNNNNNNNNNNCAGUGNNNNNNNNNNNNNNNN
<<<<<  <  <<<<<          >>  >>>>  >>>>>
```

REFERENCES

Beaumont, C., P. Leneuve, I. Devaux, *et al*. (1995). Mutation in the iron responsive element of the L ferritin mRNA in a family with dominant hyperferritinaemia and cataract. *Nat Genet* **11**(4), 444–446.

Casey, J. L., M. W. Hentze, D. M. Koeller, *et al*. (1988). Iron-responsive elements: regulatory RNA sequences that control mRNA levels and translation. *Science* **240**(4854), 924–928.

Girelli, D., R. Corrocher, L. Bisceglia, *et al*. (1995). Molecular basis for the recently described hereditary hyperferritinemia-cataract syndrome: a mutation in the iron-responsive element of ferritin L-subunit gene (the 'Verona mutation'). *Blood* **86**(11), 4050–4053.

Girelli, D., R. Corrocher, L. Bisceglia, *et al*. (1997). Hereditary hyperferritinemia-cataract syndrome caused by a 29-base pair deletion in the iron responsive element of ferritin L-subunit gene. *Blood* **90**(5), 2084–2088.

Girelli, D., O. Olivieri, P. Gasparini and R. Corrocher (1996). Molecular basis for the hereditary hyperferritinemia-cataract syndrome. *Blood* **87**(11), 4912–4913.

Hentze, M. W., M. U. Muckenthaler, B. Galy and C. Camaschella (2010). Two to tango: regulation of mammalian iron metabolism. *Cell* **142**(1), 24–38.

Kato, J., K. Fujikawa, M. Kanda, *et al*. (2001). A mutation, in the iron-responsive element of H ferritin mRNA, causing autosomal dominant iron overload. *Am J Hum Genet* **69**(1), 191–197.

Mix, H., A. V. Lobanov and V. N. Gladyshev (2007). SECIS elements in the coding regions of selenoprotein transcripts are functional in higher eukaryotes. *Nucleic Acids Res* **35**(2), 414–423.

Outten, F. W. and E. C. Theil (2009). Iron-based redox switches in biology. *Antioxid Redox Signal* **11**(5), 1029–1046.

Piccinelli, P. and T. Samuelsson (2007). Evolution of the iron-responsive element. *RNA* **13**(7), 952–966.

Sanchez, M., B. Galy, M. U. Muckenthaler and M. W. Hentze (2007). Iron-regulatory proteins limit hypoxia-inducible factor-2alpha expression in iron deficiency. *Nat Struct Mol Biol* **14**(5), 420–426.

Walczak, R., E. Westhof, P. Carbon and A. Krol (1996). A novel RNA structural motif in the selenocysteine insertion element of eukaryotic selenoprotein mRNAs. *RNA* **2**(4), 367–379.

Winkler, W. C. and R. R. Breaker (2005). Regulation of bacterial gene expression by riboswitches. *Annu Rev Microbiol* **59**, 487–517.

Volz, K. (2008). The functional duality of iron regulatory protein 1. *Curr Opin Struct Biol* **18**(1), 106–111.

7

Human disease
Cancer as a result of aberrant proteins

In the two preceding chapters a number of diseases with a genetic background have been discussed. Many cancer diseases also have a genetic component; this will be the topic of this chapter. We will focus on a chromosomal aberration that gives rise to a specific type of cancer. From a bioinformatics perspective we will approach the problem of sequence alignments and make use of BLAST to study the effect of the chromosomal aberration.

Cancer as a genetic disease

Characteristic of cancer cells is that they divide in an uncontrolled manner (Hanahan and Weinberg, 2000, 2011). This behaviour may be achieved by an overproduction of proteins that stimulate cell growth, or by the inactivation of functions that normally restrict growth. Cancer is mostly a disease developed as a result of environmental and lifestyle factors. For instance, tobacco and obesity are two significant factors. However, there are also important genetic factors. There are changes present in *germline* (reproductive) cells which may result in an inherited predisposition to cancer; for example, there are certain inherited mutations in the BRCA1 and BRCA2 genes that are associated with an elevated risk of breast cancer. However, tumours are more commonly the result of mutations in *somatic* cells, i.e. cells that are not germline cells.

Many cancers have been studied in great detail, and we have a detailed picture of the mutational events involved. One single somatic mutation is typically not enough to develop cancer. For example, several mutations are needed in the well-studied colorectal cancer. This disease originates from epithelial cells of the colon or rectum. The epithelial cells acquire mutations in genes related to the Wnt signalling pathway. This is a pathway initiated through the binding of specific molecules to a class of cell surface receptors, a binding which gives rise to specific intracellular responses. The pathway is important in embryogenesis and cancer. One example of a mutated gene is APC ('adenomatosis polyposis coli'). However, for cancer to develop, additional mutations must occur. One gene often mutated is TP53, a gene encoding the p53 protein. When p53 is not mutated it will initiate cell death (*apoptosis*) when it senses a defect in the

Wnt signalling pathway, but this function may be destroyed through mutation. Additional genes with a function in apoptosis which are often deactivated in colorectal cancer are those that encode the proteins TGF-β and DCC ('deleted in colorectal cancer').

The genes that are mutated and that are implicated in cancer may be classified in either of two groups. *Oncogenes* are genes that, when mutated, will promote cancer development. *Tumour suppressor genes* are genes that normally prevent cells from critical steps towards cancer. However, when these genes are mutated, cells may more easily progress towards cancer.

Cancer and DNA repair

DNA molecules are constantly threatened by chemicals and irradiation, the end result of which is erroneous base sequences or double-strand breaks. Complex biochemical machinery is present in all cells for the repair of DNA damage. DNA repair genes are often mutated in cancer. Indeed, there are cancer diseases that directly result from mutations in DNA repair genes. One example is *Xeroderma pigmentosum*, which results from inherited mutations in genes involved in *nucleotide excision repair* (NER). Individuals with this disease are prone to developing skin cancers. When epidermal (skin) cells are exposed to ultraviolet light, DNA is damaged by the formation of pyrimidine dimers, i.e. a crosslinking of two adjacent pyrimidines in one DNA strand. Normally, these dimers are removed with NER. However, in *Xeroderma pigmentosum*, one component of the NER system is mutated.

Another disease resulting from defective DNA repair is *Lynch syndrome* (also known as *hereditary non-polyposis colorectal cancer*). The genes involved include MSH2 and MLH1, both involved in *DNA mismatch repair*, a mechanism designed to identify and correct errors in a newly synthesized strand of DNA. The repair mechanism will, for example, identify a mismatch between G in the daughter strand and A in the parental strand.

Yet another example of inherited disease with a link to cancer is *Li–Fraumeni syndrome*. Affected individuals are likely to develop a variety of different cancer types, such as breast cancer or brain tumours. The disease is caused by mutations in the gene encoding the p53 protein, TP53. The p53 protein normally takes part in the control of cell division and growth. It will make sure DNA is adequately repaired before proceeding to synthesis of new DNA by inducing growth arrest at a point in the cell cycle referred to as the G1/S checkpoint. Another function of p53 is to initiate apoptosis if DNA damage cannot be repaired sufficiently. When p53 is mutated, however, normal control of cell growth invoked by this protein is disrupted.

Chromosome rearrangements and the Philadelphia chromosome

In the previous chapters we saw examples of how a single nucleotide change or the expansion of a sequence repeat can have dramatic effects on the function of the gene product. Whereas many genetic changes of these types are important in inherited diseases and cancer, there is another important category of changes that involve more extensive rearrangements of genetic material. Such rearrangements are typically the result of *homologous recombination*, a process that, for instance, takes place in germline cells during crossing-over in meiosis. During crossing-over two homologous chromosomes (of parental and maternal origin, respectively[1]) line up and segments of the chromosomes are exchanged; for more details, see Chapter 20.

Another type of chromosomal rearrangements is *chromosomal translocation*, where material is exchanged between *non-homologous* chromosomes. A number of mechanisms may operate to result in such translocations, including homologous recombination. Such translocations are quite frequent. They occur in about 1 in 600 human newborns, but are typically harmless. However, there are also examples of cancer diseases that are the result of such rearrangements. In these cases *fusion proteins* are often produced. A fusion protein is formed whenever gene segments are combined so as to produce an aberrant gene product with components from two different genes. Examples of such protein pairs involved in fusions are (1) c-myc (chr8)/IGH (chr14), resulting in Burkitt's lymphoma; (2) PAX8 (chr 2)/PPARγ1 (chr3), resulting in follicular thyroid cancer; and (3) JAK (chr9)/TEL (chr12), resulting in chronic myelogenous leukaemia (CML) and acute lymphoblastic leukaemia (ALL).

Yet another fusion will be considered for further study in this chapter, BCR–ABL. This fusion protein results from an interchange of material from human chromosomes 9 and 22 (Fig. 7.1). Also this translocation is associated with the leukaemias CML and ALL.

The chromosomal translocation involving chromosomes 9 and 22 was originally identified under the microscope. After the translocation chromosome 22 is much smaller than its normal counterpart. The chromosomal aberration is also referred to as the *Philadelphia* chromosome, named as such because it was discovered in 1960 by Peter Nowell and David Hungerford, both affiliated with

---

[1] The word 'homologous' refers in this case to a pair of chromosomes, where one is inherited from the mother and one from the father. Homologous chromosomes are equivalent in terms of gene sequences and they are paired during meiosis. For instance, in a diploid human cell, chromosome 1 of maternal origin is 'homologous' to chromosome 1 of parental origin.

research facilities in the city of Philadelphia (Nowell and Hungerford, 1960; Nowell, 2007). The Philadelphia chromosome was first detected in patients suffering from CML. This is a rare type of blood cancer that gives rise to a very high production of white blood cells. This was the first time a chromosomal aberration was found to be associated with cancer.

What are the normal functions of BCR and ABL? We still do not fully understand the function of BCR. The protein acts to activate the GTPase activity of the protein named RAC1 (or p21rac) and it has serine–threonine kinase activity. The ABL protein has two variants in humans, ABL1 and ABL2. It is ABL1 that is involved in the BCR–ABL fusion. The ABL1 protein has protein–tyrosine kinase

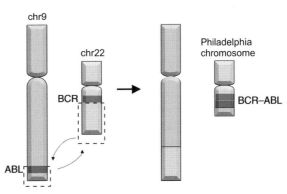

Fig. 7.1 *Philadelphia chromosome*. The chromosomal translocation known as the Philadelphia chromosome involves an interchange of material between chromosomes 9 and 22. This translocation gives rise to the fusion protein BCR–ABL.

activity and is important in processes such as cell differentiation and cell division. Importantly, the fusion of BCR and ABL results in an activation of the tyrosine kinase activity of ABL. This presumably leads to uncontrolled cell growth. Drugs have been developed that are inhibitors to the tyrosine kinase activity of ABL1 and have been exploited as therapeutic agents against CML.

The breakpoint for the BCR–ABL fusion is not always the same. The fusion protein is actually observed in different forms, depending on the breakpoint (Groffen *et al.*, 1984; Fainstein *et al.*, 1987; Saglio *et al.*, 1990). We will now examine one of these fusion proteins more carefully, exploring some widely used bioinformatics tools.

BIOINFORMATICS

Dotplots and alignments

We will eventually compare the BCR–ABL fusion protein with the normal BCR and ABL proteins. We will do so using a procedure of *sequence alignment*, which is to say that

Tools introduced in this chapter	
Software running in Unix environment	BLAST, command-line version

we try to fit the sequences optimally to each other, using some set of rules. Sequence alignment is an extremely common operation in molecular biology.

Fig. 7.2 *Dotplot*. Two nucleotide sequences are compared, one at the horizontal axis and one at the vertical axis. A dot in the plot reflects every instance where there is a match of nucleotides between the two sequences. The diagonal patterns observed reflect local identity between sequences. For instance, the sub-sequence AGCAT in one of the sequences has a perfect match to the other sequence. In this particular plot we may distinguish three different diagonals that may be combined to produce the alignment at the bottom. In this alignment, identity is indicated with a star and we have introduced a gap in one of the sequences.

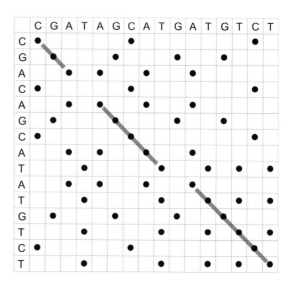

```
CGATAGCATGATGTCT
*** ***** ******
CGACAGCAT-ATGTCT
```

When approaching the problem of sequence alignment a good starting point is a plot referred to as a *dotplot* (Fig. 7.2), a two-dimensional graph in which the two sequences compared are on different axes and all instances of identity have been indicated with a 'dot'. Any local identity between the two sequences is therefore revealed as a diagonal pattern.

We may intuitively infer an alignment from a dotplot. For example, the plot in Fig. 7.2 identifies three different diagonal lines. We could combine these diagonals into an alignment, as shown in the bottom of the figure. In order to accommodate the matches in the alignment we have introduced a *gap* into one of the sequences, represented by a dash (-). In general, a gap in a nucleotide sequence alignment corresponds to a mutational event where one or more nucleotides have been inserted or deleted during evolution.

However, there are more sophisticated methods to produce an optimal alignment, which were developed quite early in the history of bioinformatics. There is a method of *global* alignment, where two sequences in their full length are optimally aligned (Needleman and Wunsch, 1970). In *local* alignment methods (Smith and Waterman, 1981) only regions of significant similarity are reported. The details of one local alignment algorithm, an example of a *dynamic programming* algorithm, is shown in Fig. 7.3. An important element of this algorithm is a scoring scheme where we take into account scores from *matches* and *mismatches*, as well as *gap penalties*. The more general nature and behaviour of local and global alignments is illustrated in Fig. 7.4.

A

$$F(i,j) = \max \begin{cases} 0, \\ F(i\text{-}1, j\text{-}1) + s(x_i,y_j), \\ F(i\text{-}1, j) - d, \\ F(i, j\text{-}1) - d \end{cases}$$

B

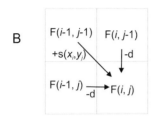

C

		C	A	A	C	A	A
	0	0	0	0	0	0	0
T	0	-1 -2 / -2 0	-1 -2 / -2 0	-1 -2 / -2 0	-1 -2 / -2 0	-1 -2 / -2 0	-1 -2 / -2 0
A	0	-1 -2 / -2 0	2 -2 / -2 2	2 -2 / 0 2	-1 -2 / 0 0	2 -2 / -2 2	2 -2 / 0 2
A	0	-1 -2 / -2 0	2 0 / -2 2	4 0 / 0 4	1 -2 / 2 2	2 0 / 0 2	4 0 / 0 4
A	0	-1 -2 / -2 0	2 0 / -2 2	4 2 / 0 4	3 0 / 2 3	4 0 / 1 4	4 2 / 2 4
A	0	-1 -2 / -2 0	2 0 / -2 2	4 2 / 0 4	3 1 / 2 3	5 2 / 1 5	6 2 / 3 6

D

AA-AA
|| ||
AACAA

Fig. 7.3 *Local alignment of two sequences using dynamic programming.* The two sequences TAAAA and CAACAA are considered here. A two-dimensional matrix F is constructed. This matrix has the indices i and j, for TAAAA and CAACAA, respectively. First the values of F(i,0) and F(0,j) are set to zero (see panel C). The other F(i,j) values are built recursively, using the equation in panel A. In this example we assume that the match score is 2, mismatch is −1 and gap penalty (d) is −2. This means that $s(x_i,y_i)$ in panel A is 2 if x_i is equal to y_i and −1 if they are not equal. For instance, consider the derivation of the F(1,1) value. See panel B as well as the box corresponding to the first pair of nucleotides, C and T. We need to use F(0,0), which is zero. F(0,1) and F(1,0) are both −2 and $s(x_i,y_i)$ is −1. F(0,0) + s = −1. All of these values are negative, but according to the rules in panel A, no negative values are allowed and so we set F(1,1) to zero (box with shaded background in panel C). All the other cells in the matrix are filled out in the same recursive manner. For all the non-zero values we put a traceback arrow to indicate the value from which it was derived. To produce the final alignment we first look up the highest F(i,j) value. We then walk backwards following the traceback arrows (red arrows) until we reach a zero value. In this case the relevant F(i,j) values are those of (5,6), (4,5), (3,4), (3,3) and (2,2). Based on this path we obtain the alignment as indicated in panel D.

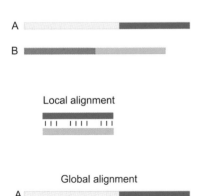

Local alignment

Global alignment

Fig. 7.4 *Local and global alignments.* Two different proteins, A and B, are considered. They are similar (evolutionarily related) in their C-terminal portions, but not at all related in their N-terminal regions. A local alignment method will correctly align the two similar C-terminal regions and only these regions will be included in the alignment. A global alignment method will align both sequences in their full lengths. As a result, the C-terminal regions that are evolutionarily related may be correctly aligned, but the portion of the alignment that is from the N-terminal sequences is not meaningful from a biological point of view and is the result of chance only.

```
A   4
R  -1   5
N  -2   0   6
D  -2  -2   1   6
C   0  -3  -3  -3   9
Q  -1   1   0   0  -3   5
E  -1   0   0   2  -4   2   5
G   0  -2   0  -1  -3  -2  -2   6
H  -2   0   1  -1  -3   0   0  -2   8
I  -1  -3  -3  -3  -1  -3  -3  -4  -3   4
L  -1  -2  -3  -4  -1  -2  -3  -4  -3   2   4
K  -1   2   0  -1  -3   1   1  -2  -1  -3  -2   5
M  -1  -1  -2  -3  -1   0  -2  -3  -2   1   2  -1   5
F  -2  -3  -3  -3  -2  -3  -3  -3  -1   0   0  -3   0   6
P  -1  -2  -2  -1  -3  -1  -1  -2  -2  -3  -3  -1  -2  -4   7
S   1  -1   1   0  -1   0   0   0  -1  -2  -2   0  -1  -2  -1   4
T   0  -1   0  -1  -1  -1  -1  -2  -2  -1  -1  -1  -1  -2  -1   1   5
W  -3  -3  -4  -4  -2  -2  -3  -2  -2  -3  -2  -3  -1   1  -4  -3  -2  11
Y  -2  -2  -2  -3  -2  -1  -2  -3   2  -1  -1  -2  -1   3  -3  -2  -2   2   7
V   0  -3  -3  -3  -1  -2  -2  -3  -3   3   1  -2   1  -1  -2  -2   0  -3  -1   4
    A   R   N   D   C   Q   E   G   H   I   L   K   M   F   P   S   T   W   Y   V
```

Fig. 7.5 *BLOSUM62 amino acid substitution matrix.* The BLOSUM62 matrix may be used in the scoring of an alignment of protein sequences using dynamic programming as described in Fig. 7.3. The numbers in the matrix reflect the probability of observing a particular amino acid substitution in a trusted multiple alignment of evolutionarily related protein sequences. BLOSUM (short for **blo**cks of amino acid **su**bstitution **m**atrix) matrices were constructed from information in the BLOCKS database and were first described in a paper of Henikoff and Henikoff (1992).

The example in Fig. 7.3 is concerned with an alignment of nucleotide sequences. In the case of protein sequences, we also take advantage of a distinct property of protein evolution, which is that not all amino acid changes take place with the same frequency. These frequencies may be observed in trusted alignments of protein sequences. For instance, it is observed that a change from glutamic acid to aspartic acid is much more common than a change from glutamic acid to cysteine. The explanation of these differences is partly related to the physical and chemical differences between amino acids. Thus, glutamic acid and aspartic acid are both negatively charged and replacement of one to the other is not expected to have a major impact on protein structure and function. Replacing glutamic acid with cysteine, on the other hand, will be associated with a more substantial change in protein properties. The differences in the frequencies of amino acid replacements may be exploited in the scoring of alignments. Thus, a likely replacement (like glutamic acid to aspartic acid) receives a higher score than an unlikely one. An *amino acid substitution matrix*, like BLOSUM62 (Fig. 7.5) shows, for each pair of amino acids, a value reflecting the probability of that particular amino acid substitution. This matrix is exploited in the local alignment algorithm. Thus, instead of having one match and one mismatch score as in the example in Fig. 7.3, the score $s(x_i, y_j)$ is obtained from the amino acid substitution matrix.

BLAST

We will use the tool *BLAST* to compare a BCR–ABL fusion protein with the normal BCR and ABL proteins. BLAST is an acronym for 'basic local alignment search tool', and as the name implies it is an example of software implementing a *local* alignment procedure. BLAST is typically used to identify local alignments between a query sequence and a large number of sequences present in databases. The method was first described in 1990 and is probably the most widely used bioinformatics tool ever. The original paper by Stephen Altschul, Warren Gish, Webb Miller, Eugene Myers and David Lipman, published in the *Journal of Molecular Biology* (Altschul *et al.*, 1990) was cited by approximately 29 000 scientific articles during 1990–2010, and an update describing a further development known as PSI-BLAST, published in 1997 (Altschul *et al.*, 1997), has been cited nearly as many times. These papers are a few of the most frequently cited papers in the biomedical literature.

In order to improve on speed, BLAST does not carry out rigorous local alignment methods as described in Fig. 7.3. Instead, the first step of BLAST is to identify word matches between the query sequence and the database sequences. Such matches are then extended in both directions using the dynamic programming algorithm, as in Fig. 7.3. With the procedure based on word matching there is a risk that sensitivity is lost. However, BLAST has a clever way of maintaining sensitivity. During the word matching it considers not only identical words, but also 'near-identical' words, as further explained in Box 7.1.

Box 7.1 Increasing the sensitivity in BLAST by considering word neighbours

BLAST uses a heuristic procedure in which words of the query sequence are matched against database sequences. To obtain a high sensitivity in this step, BLAST uses a method that considers not only the words themselves, but also 'neighbours' to these words. Consider a protein to be used as a query in a BLAST search. Assume it has the sequence MALWGGRFTS...

First, all words of some size are derived from this sequence. Here we consider the size to be three. Hence, we obtain the words MAL, ALW, LWG, WGG, etc. Word neighbours are identified as words that score at least some value T, using information from an amino acid substitution matrix. Consider the word 'WGG' and we set T to 12. Is the word 'WGA' a neighbour? Consider the alignment:

```
WGG
|||
WGA
```

We use the substitution matrix BLOSUM62 in Fig. 7.5 and obtain for the alignment the scores 11 (W:W), 6 (G:G) and 0 (G:A). The sum of these is 17. Therefore, WGA is included in the set of neighbours to WGG. This procedure is carried out for every possible word and we end up with a list of words like this:

```
WGG 11 + 6 + 6 = 23
WGA 11 + 6 + 0 = 17
WEG 11 - 2 + 6 = 15
WGI 11 + 6 - 4 = 13

. . .

. . .

----------- Threshold T -----------

MGG -1 + 6 + 6 = 11
WNS 11 + 0 + 0 = 11

. . .

. . .
```

Here, only some of the words above and below the threshold (*T*) are shown. All words above *T* are used as neighbours to WGG. In the same way, neighbours of all other words of the query sequence are identified. The resulting list of words is then used in the matching against database sequences.

In BLAST a specific DNA or protein sequence is used as the query and local alignments are attempted with every sequence in a database. Sequences that match the query are referred to as 'hits', and the best hits are shown in the output from BLAST.

The object of the BLAST search is to identify sequences that are evolutionarily related to the query sequence. Two proteins or genes that are evolutionarily related and that have a common ancestor are referred to as *homologues*. How do we know if the similarity found by BLAST is reflecting a homology relationship? An important statistical parameter in the output from BLAST is the expectation value, or *E-value*. For the strict mathematical definition of the E-value, see Karlin and Altschul (1990) and http://www.ncbi.nlm.nih.gov/books/NBK21097, but it may be thought of as the number of database hits expected by chance only. For instance, if we observe that the E-value equals 1, this means we should expect one database hit by chance only. Therefore, we cannot reach any conclusions as to the biological relevance of such a hit. If the E-value is sufficiently low – for example, less than 10^{-10} – we are very likely to have identified database sequences evolutionarily related to our query.

In BLAST searches two kinds of evolutionary relationships are detected. A relationship of *orthology* is a result of a speciation event. Orthologous proteins are proteins that carry out the same function but in different species. For example, the human protein ABL is orthologous to the mouse protein ABL. A relationship of *paralogy* is a result of a gene duplication event. Paralogous proteins perform different but related functions within one organism. For instance, gene duplication events gave the two paralogous genes ABL1 and ABL2.

Using BLAST to examine the BCR–ABL fusion protein

We now return to the problem of learning about the BCR–ABL protein. How is the sequence of this protein related to that of the normal BCR and ABL1 proteins? To address this question we carry out a BLAST search using the human BCR–ABL protein as a query, and as a database we use the protein sequence database Swiss-Prot. (For more information on Swiss-Prot and another resource, UniProt, see Box 7.2 and Fig. 7.6.) For the examples below, in which BLAST is run in a Unix environment, we assume you are making use of a computer where the BLAST+ package is installed. If it is not installed, the software is available for download at ftp://ftp.ncbi.nih.gov/blast.

Box 7.2 Swiss-Prot and other protein sequence databases

Swiss-Prot was established in 1986 through the work of Amos Bairoch. Since 2003, Swiss-Prot has been part of UniProt, the Universal Protein Resource (http://www.uniprot.org). UniProt is produced by the UniProt Consortium, formed by the Swiss Institute of Bioinformatics (SIB), the European Bioinformatics Institute (EBI) and the Protein Information Resource (PIR). UniProt provides a comprehensive, high-quality resource of protein sequence and functional information. The centre-piece of the UniProt databases is the UniProt knowledgebase (UniProtKB), which is comprised of two sections: UniProtKB/Swiss-Prot and UniProtKB/TrEMBL.

UniProtKB/Swiss-Prot is a high-quality, manually annotated (reviewed) and non-redundant protein sequence database, which brings together experimental results and computed features. In order to promote minimal redundancy and improve sequence reliability, all protein sequences encoded by the same gene are merged into a single record. In addition to functional information, a special emphasis is placed on the annotation of biological events that generate protein diversity and that cannot be predicted at the genomic level. Alternative products resulting from RNA processing events such as alternative splicing and RNA editing, as well as products of post-translational modifications, are extensively annotated. For additional information, see Boeckmann *et al.* (2005) and Magrane and UniProt Consortium (2011). Swiss-Prot contains numerous cross-references, thus playing the role of a central hub for biological data, linking together relevant resources. Although Swiss-Prot provides annotated entries for all species, it focuses on the annotation of proteins from model organisms of distinct taxonomic groups to ensure the presence of high-quality annotation for representative members of all protein families. Protein families and groups of proteins are continuously reviewed to keep up with current scientific findings.

UniProtKB/TrEMBL is a computer-annotated (unreviewed) supplement to Swiss-Prot, which strives to gather all protein sequences that are not yet represented in Swiss-Prot. The protein sequences are derived from the translation of coding sequences (CDS) submitted to the public nucleic acid databases (EMBL/GenBank/DDBJ) or from other sequence resources, such as Ensembl. Automated annotation of the highest currently available quality is integrated into TrEMBL entries. The usual Swiss-Prot annotation pipeline involves the manual annotation of TrEMBL entries, their integration into Swiss-Prot, with their original accession number, and subsequent deletion from TrEMBL.

Each Swiss-Prot entry contains information about one or more protein sequence(s) derived from one gene in one species. Different sections of the entry report specific biological information.

In January 2012 Swiss-Prot and TrEMBL contained 534 242 and 19 434 245 entries, respectively. An example entry is shown in Fig. 7.6 and explanations of most of the fields are listed below.

ID	Identification; unique but not stable identifier
AC	Primary accession number; stable unique identifier
DT	Entry history
DE	Description
OS	Organism (species)
OC	Organism classification (taxonomy)
RN	Reference number
RX	Reference cross-reference (like PubMed ID)
CC	Comments or notes (often a functional description)
DR	Database cross-references
KW	Keywords
FT	Feature table, description and location of sequence features
SQ	Sequence header. Lines following this header are the actual sequence of the protein, with the amino acids in the one-letter code
//	Termination line

```
ID   CDD_MYCGE               Reviewed;          130 AA.
AC   P47298;
DT   01-FEB-1996, integrated into UniProtKB/Swiss-Prot.
DT   01-FEB-1996, sequence version 1.
DT   30-NOV-2010, entry version 68.
DE   RecName: Full=Cytidine deaminase;
DE            Short=CDA;
DE            EC=3.5.4.5;
DE   AltName: Full=Cytidine aminohydrolase;
GN   Name=cdd; OrderedLocusNames=MG052;
OS   Mycoplasma genitalium.
OC   Bacteria; Tenericutes; Mollicutes; Mycoplasmataceae; Mycoplasma.
OX   NCBI_TaxID=2097;
RN   [1]
RP   NUCLEOTIDE SEQUENCE [LARGE SCALE GENOMIC DNA].
RC   STRAIN=ATCC 33530 / G-37 / NCTC 10195;
RX   MEDLINE=96026346; PubMed=7569993; DOI=10.1126/science.270.5235.397;
RA   Fraser C.M., Gocayne J.D., White O., Adams M.D., Clayton R.A.,
RA   Fleischmann R.D., Bult C.J., Kerlavage A.R., Sutton G.G., Kelley J.M.,
RA   Fritchman J.L., Weidman J.F., Small K.V., Sandusky M., Fuhrmann J.L.,
RA   Nguyen D.T., Utterback T.R., Saudek D.M., Phillips C.A., Merrick J.M.,
RA   Tomb J.-F., Dougherty B.A., Bott K.F., Hu P.-C., Lucier T.S.,
RA   Peterson S.N., Smith H.O., Hutchison C.A. III, Venter J.C.;
RT   "The minimal gene complement of Mycoplasma genitalium.";
RL   Science 270:397-403(1995).
RN   [2]
RP   NUCLEOTIDE SEQUENCE [GENOMIC DNA] OF 108-130.
RC   STRAIN=ATCC 33530 / G-37 / NCTC 10195;
RX   MEDLINE=94075230; PubMed=8253680;
RA   Peterson S.N., Hu P.-C., Bott K.F., Hutchison C.A. III;
RT   "A survey of the Mycoplasma genitalium genome by using random
RT   sequencing.";
RL   J. Bacteriol. 175:7918-7930(1993).
CC   -!- FUNCTION: This enzyme scavenge exogenous and endogenous cytidine
CC       and 2'-deoxycytidine for UMP synthesis (By similarity).
CC   -!- CATALYTIC ACTIVITY: Cytidine + H(2)O = uridine + NH(3).
CC   -!- COFACTOR: Binds 1 zinc ion (By similarity).
CC   -!- SUBUNIT: Homodimer (By similarity).
CC   -!- SIMILARITY: Belongs to the cytidine and deoxycytidylate deaminase
CC       family.
CC   -----------------------------------------------------------------------
CC   Copyrighted by the UniProt Consortium, see http://www.uniprot.org/terms
CC   Distributed under the Creative Commons Attribution-NoDerivs License
CC   -----------------------------------------------------------------------
DR   EMBL; L43967; AAC71268.1; -; Genomic_DNA.
DR   EMBL; U02108; AAD12378.1; -; Genomic_DNA.
DR   PIR; G64205; G64205.
DR   RefSeq; NP_072712.1; NC_000908.2.
DR   ProteinModelPortal; P47298; -.
DR   SMR; P47298; 8-130.
DR   GeneID; 875728; -.
DR   GenomeReviews; L43967_GR; MG052.
DR   KEGG; mge:MG_052; -.
DR   TIGR; MG052; -.
DR   HOGENOM; HBG352939; -.
DR   OMA; YSEYRVG; -.
DR   ProtClustDB; CLSK542120; -.
DR   BioCyc; MGEN243273:MG_052-MONOMER; -.
DR   BRENDA; 3.5.4.5; 110.
DR   GO; GO:0004126; F:cytidine deaminase activity; IEA:EC.
DR   GO; GO:0008270; F:zinc ion binding; IEA:InterPro.
DR   InterPro; IPR016192; APOBEC/CMP_deaminase_Zn-bd.
DR   InterPro; IPR002125; CMP_dCMP_Zn-bd.
DR   InterPro; IPR006262; Cyt_deam_tetra.
DR   InterPro; IPR016193; Cytidine_deaminase-like.
DR   PANTHER; PTHR11644:SF2; Cyt_deam_tetra; 1.
DR   Pfam; PF00383; dCMP_cyt_deam_1; 1.
DR   SUPFAM; SSF53927; Cytidine_deaminase-like; 1.
DR   TIGRFAMs; TIGR01354; cyt_deam_tetra; 1.
DR   PROSITE; PS00903; CYT_DCMP_DEAMINASES; 1.
PE   3: Inferred from homology;
KW   Complete proteome; Hydrolase; Metal-binding; Zinc.
FT   CHAIN         1    130       Cytidine deaminase.
FT                                /FTId=PRO_0000171679.
FT   REGION       43     45       Substrate binding (By similarity).
FT   ACT_SITE     56     56       Proton donor (By similarity).
FT   METAL        54     54       Zinc; catalytic (By similarity).
FT   METAL        88     88       Zinc; catalytic (By similarity).
FT   METAL        91     91       Zinc; catalytic (By similarity).
SQ   SEQUENCE   130 AA;  14974 MW;  1DF02B1718F9495F CRC64;
     MKVNLEWIIK QLQMIVKRAY TPFSNFKVAC MIIANNQTFF GVNIENSSFP VTLCAERSAI
     ASMVTSGHRK IDYVFVYFNT KNKSNSPCGM CRQNLLEFSH QKTKLFCIDN DSSYKQFSID
     ELLMNGFKKS
//
```

Fig. 7.6 *Swiss-Prot database record.* For a description, see Box 7.2. The protein of this record is a cytidine deaminase from *Mycoplasma genitalium* (this protein was also referred to in Fig. 1.5).

First, we need to make sure the Swiss-Prot database is formatted for BLAST searches. If it is not already, this is achieved with the BLAST utility `makeblastdb`.

```
% makeblastdb -in swissprot -dbtype prot -parse_seqids
```

The argument `-in swissprot` specifies that the input sequence file is named `swissprot` and that this file is in the current directory (and if it is not you need to specify the path as well). The `swissprot` file is in the FASTA format. The argument `-dbtype prot` indicates that we are formatting a protein sequence database. As a result of `makeblastdb` a number of binary files with extensions .pnd, .psi, .phr, .pni, .psq, .pin and .psd will be created. The parameter `-parse_seqids` is an instruction to parse sequence identifiers in the FASTA definition line. As a result we can retrieve sequences from the formatted database using the very useful `blastdbcmd` utility. Here is an example to retrieve a Swiss-Prot protein sequence with the identifier ABL1_HUMAN. We direct the output to a file `abl.fa`:

```
% blastdbcmd -entry ABL1_HUMAN -db swissprot > abl.fa
```

It should be noted that here the parameter `-db swissprot` directs `blastdbcmd` to the binary index files, whereas in the `makeblastdb` command the parameter `-in swissprot` refers to the original file of Swiss-Prot sequences in a FASTA format.

We are now ready for the BLAST search. For this example we have stored the BCR–ABL sequence that we want to analyse in a FASTA format in a file named `bcrabl.fa`. This sequence is an entry from the UniProtKB/TrEMBL protein sequence database with accession A9UF02_HUMAN. We type the following to start BLAST in a Unix environment:[2]

```
% blastp -query bcrabl.fa -db swissprot -out bcrabl.blastp
```

`blastp` is the version of BLAST used when using a protein query sequence to search a protein database. As described above for `blastdbcmd`, `-db swissprot` assumes the presence of the different binary database index files created with `makeblastdb` as above. The `-out` parameter is used to specify the name of the output file. (Instead of using `-out` we could have specified the output file by using the Unix redirection symbol '>'.) The output of this BLAST search is shown in Fig. 7.7.

Let's examine the BLAST result to learn about its general structure, as well as about the biological implications of this specific search. There are two major

---

[2] This command-line syntax assumes we are using the BLAST+ version of NCBI BLAST. See Appendix II for information on syntax for older versions.

```
BLASTP 2.2.23+

Database: swissprot
           434,882 sequences; 161,124,461 total letters

Query= tr|A9UF02|A9UF02_HUMAN BCR/ABL fusion protein isoform X9 OS=Homo
sapiens GN=BCR/ABL fusion PE=2 SV=1
Length=1644
                                                               Score      E
Sequences producing significant alignments:                   (Bits)   Value

  gi|85681908|sp|P00519.4|ABL1_HUMAN RecName: Full=Tyrosine-prote...  2268    0.0
  gi|59802613|sp|P00520.3|ABL1_MOUSE RecName: Full=Tyrosine-prote...  1885    0.0
  gi|143811366|sp|P11274.2|BCR_HUMAN RecName: Full=Breakpoint clu...  1038    0.0
  gi|1168268|sp|P42684.1|ABL2_HUMAN RecName: Full=Tyrosine-protei...   991    0.0
  gi|118582158|sp|Q4JIM5.1|ABL2_MOUSE RecName: Full=Tyrosine-prot...   963    0.0
  gi|125134|sp|P10447.1|ABL_FSVHY RecName: Full=Tyrosine-protein ...   932    0.0
  gi|125136|sp|P00521.1|ABL_MLVAB RecName: Full=Tyrosine-protein ...   895    0.0
  gi|124007119|sp|Q6PAJ1.2|BCR_MOUSE RecName: Full=Breakpoint clu...   867    0.0
  gi|62512130|sp|P00522.3|ABL_DROME RecName: Full=Tyrosine-protei...   780    0.0
  gi|27808642|sp|P03949.4|ABL1_CAEEL RecName: Full=Tyrosine-prote...   619  8e-176
  gi|1174436|sp|P42686.1|SRK1_SPOLA RecName: Full=Tyrosine-protei...   419  1e-115
  ....
```

HSP for ABL1_HUMAN

```
> gi|85681908|sp|P00519.4|ABL1_HUMAN RecName: Full=Tyrosine-protein
kinase ABL1; AltName: Full=Abelson murine leukemia viral
oncogene homolog 1; AltName: Full=Proto-oncogene c-Abl; AltName:
Full=p150
Length=1130

 Score = 2268 bits (5876),  Expect = 0.0, Method: Compositional matrix adjust.
 Identities = 1105/1105 (100%), Positives = 1105/1105 (100%), Gaps = 0/1105 (0%)

Query  540   EEALQRPVASDFEPQGLSEAARWNSKENLLAGPSENDPNLFVALYDFVASGDNTLSITKG   599
             EEALQRPVASDFEPQGLSEAARWNSKENLLAGPSENDPNLFVALYDFVASGDNTLSITKG
Sbjct  26    EEALQRPVASDFEPQGLSEAARWNSKENLLAGPSENDPNLFVALYDFVASGDNTLSITKG   85

Query  600   EKLRVLGYNHNGEWCEAQTKNGQGWVPSNYITPVNSLEKHSWYHGPVSRNAAEYLLSSGI   659
             EKLRVLGYNHNGEWCEAQTKNGQGWVPSNYITPVNSLEKHSWYHGPVSRNAAEYLLSSGI
Sbjct  86    EKLRVLGYNHNGEWCEAQTKNGQGWVPSNYITPVNSLEKHSWYHGPVSRNAAEYLLSSGI   145
.....
Query  1560  NSEQMASHSAVLEAGKNLYTFCVSYVDSIQQMRNKFAFREAINKLENNLRELQICPATAG  1619
             NSEQMASHSAVLEAGKNLYTFCVSYVDSIQQMRNKFAFREAINKLENNLRELQICPATAG
Sbjct  1046  NSEQMASHSAVLEAGKNLYTFCVSYVDSIQQMRNKFAFREAINKLENNLRELQICPATAG  1105

Query  1620  SGPAATQDFSKLLSSVKEISDIVQR  1644
             SGPAATQDFSKLLSSVKEISDIVQR
Sbjct  1106  SGPAATQDFSKLLSSVKEISDIVQR  1130
```

Fig. 7.7 *Output from BLAST search*. Portions of the output from a BLAST search where the fusion protein BCR–ABL was used as a query to search the Swiss-Prot protein sequence database. In the first section of a BLAST report, the hits from the search are listed and ordered by score or E-value (circled). In a second section all alignments (HSPs) are listed. The HSPs for ABL1_HUMAN and BCR_HUMAN are shown here. The line indicated with 'Query' is the amino acid sequence of the query; 'Sbjct' refers to the database sequence. The line between is to indicate positions in the alignment of identity or similarity. In the case of a match the relevant amino acid is shown. In this case the two sequences in the alignments are all identical. The position numbers to indicate start and end positions in the alignment are circled.

HSP for BCR_HUMAN

```
> gi|143811366|sp|P11274.2|BCR_HUMAN RecName: Full=Breakpoint cluster
region protein; AltName: Full=Renal carcinoma antigen
NY-REN-26
Length=1271

 Score = 1038 bits (2684),  Expect = 0.0, Method: Compositional matrix adjust.
 Identities = 518/520 (99%), Positives = 519/520 (99%), Gaps = 0/520 (0%)

Query  1    MVDPVGFAEAWKAQFPDSEPPRMELRSVGDIEQELERCKASIRRLEQEVNQERFRMIYLQ  60
            MVDPVGFAEAWKAQFPDSEPPRMELRSVGDIEQELERCKASIRRLEQEVNQERFRMIYLQ
Sbjct  1    MVDPVGFAEAWKAQFPDSEPPRMELRSVGDIEQELERCKASIRRLEQEVNQERFRMIYLQ  60

Query  61   TLLAKEKKSYDRQRWGFRRAAQAPDGASEPRASASRPQPAPADGADPPPAEEPEARPDGE  120
            TLLAKEKKSYDRQRWGFRRAAQAPDGASEPRASASRPQPAPADGADPPPAEEPEARPDGE
Sbjct  61   TLLAKEKKSYDRQRWGFRRAAQAPDGASEPRASASRPQPAPADGADPPPAEEPEARPDGE  120

...


Query  421  AFHGDADGSFGTPPGYGCAADRAEEQRRHQDGLPYIDDSPSSSPHLSSKGRGSRDALVSG  480
            AFHGDADGSFGTPPGYGCAADRAEEQRRHQDGLPYIDDSPSSSPHLSSKGRGSRDALVSG
Sbjct  421  AFHGDADGSFGTPPGYGCAADRAEEQRRHQDGLPYIDDSPSSSPHLSSKGRGSRDALVSG  480

Query  481  ALESTKASELDLEKGLEMRKWVLSGILASEETYLSHLQML  520
            ALESTKASELDLEKGLEMRKWVLSGILASEETYLSHL+ L
Sbjct  481  ALESTKASELDLEKGLEMRKWVLSGILASEETYLSHLEAL  520
```

Fig. 7.7 (*cont.*)

parts to the output. The first is a list of the best matching hits; the second is a section with details on the local alignments that were identified. In the first section we note that the hits are ordered according to Score/E-value. All of the top hits of this search have very low E-values, indicating that they are biologically significant and not the result of chance only. The very best hit is against the human ABL1 protein. The third hit is against the human BCR protein, and there is also a hit against human ABL2, the human paralogue to the ABL1 protein. There are also hits to database entries listed as ABL2_MOUSE and BCR_MOUSE, suggesting to us that these are orthologues to the human ABL2 and BCR proteins. Therefore, we here see examples of how BLAST is useful to identify both orthologues and paralogues.

Looking at the section where local alignments are described, we can gather additional information. In BLAST terminology, the alignments shown are referred to as High Scoring Pairs, or HSPs. Looking at the HSP for the top hit, there are lines like:

```
Query 540 EEALQRPVASDFEPQGLSEAARWNSKENLLAGPSENDPNLFVALYDFVASGDNTLSITKG 599
          EEALQRPVASDFEPQGLSEAARWNSKENLLAGPSENDPNLFVALYDFVASGDNTLSITKG
Sbjct 26  EEALQRPVASDFEPQGLSEAARWNSKENLLAGPSENDPNLFVALYDFVASGDNTLSITKG 85
```

where Query as expected refers to the query sequence, and the sequence labelled Sbjct refers to the database sequence aligned to the query. The sequence in

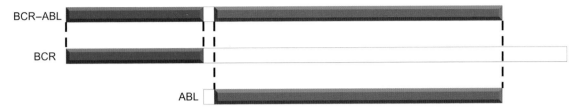

Fig. 7.8 *Fusion protein BCR–ABL compared to the normal BCR and ABL proteins.* The figure is based on information obtained from the BLAST report in Fig. 7.7. The fusion protein contains a major portion of ABL, but only a minor part of BCR.

between is to indicate where there is sequence identity between the query and database sequence. In this case the two sequences are completely identical. For an example where the sequences are not identical, consider these lines:

```
Query 1320  PGSSPPNLTPKPLRRQVTVAPASGLPHKEEAGKGSA--LGTPAAAEPVTPTSKAGSGAPG 1377
            PGSSPP+LTPK LRRQVT +P+SGL HKEEA KGSA  +GTPA AEP  P++K
Sbjct 804   PGSSPPSLTPKLLRRQVTASPSSGLSHKEEATKGSASGMGTPATAEPAPPSNKV------ 857
```

Here, the symbol + means that the amino acid pair in this position receives a relatively high score from the amino acid substitution matrix, whereas a space indicates that it does not.

A closer examination of the HSPs of BCR_HUMAN and ABL1_HUMAN can inform us about the point of fusion between these proteins. The numbers of the alignment lines tell us about positions of query and database sequence (Fig. 7.7). Thus, according to the ABL1_HUMAN HSP, the fusion protein region between positions 540 and 1644 matches the ABL1_HUMAN protein region between positions 26 and 1130. The total length of the fusion protein is 1644 amino acids, whereas the length of the ABL1_HUMAN protein is 1130. Therefore, it may be concluded from this HSP that a large C-terminal portion of the fusion protein corresponds to ABL, and that more or less the complete ABL sequence is present in the fusion protein. As to BCR_HUMAN (Fig. 7.7), with a total length of 1271 amino acids we can learn from the alignment that the N-terminal part of this protein (positions 1–520) is present in the fusion protein. The relationship between the fusion protein and the normal ABL and BCR proteins is further illustrated in Fig. 7.8.

This chapter has shown how BLAST may be used to identify evolutionarily related sequences; orthologues as well as paralogues. Furthermore, the local alignment procedure of BLAST allowed a detailed comparison of the BCR–ABL fusion protein to the normal BCR and ABL components. In Chapter 11 we will encounter BLAST again, then illustrating how it is useful when it comes to assigning a function to a sequence.

EXERCISES

7.1 A typical application of BLAST is to identify sequences that are homologous to a query sequence. Modify the BLAST search of this chapter so you will be able to identify homologues of the human BCR protein. This protein has the Swiss-Prot identifier BCR_HUMAN. You can use the `blastdbcmd` utility to retrieve it from the Swiss-Prot database. In the output from BLAST, examine the HSP(s) of the match to ABR_HUMAN. Which portion (amino acid positions) of BCR_HUMAN is similar to ABR_HUMAN?

7.2 Use the dynamic programming method as outlined in Fig. 7.3 to obtain a local alignment of the two nucleotide sequences GCACGC and GGACGTTT. As in Fig. 7.3, assume a match score of 2, a mismatch score of −1 and a gap penalty of −2.

7.3 Write a Perl program that will make a dotplot type of output (Fig. 7.2) based on the two nucleotide sequences in Exercise 7.2.

REFERENCES

Altschul, S. F., W. Gish, W. Miller, E. W. Myers and D. J. Lipman (1990). Basic local alignment search tool. *J Mol Biol* **215**(3), 403–410.

Altschul, S. F., T. L. Madden, A. A. Schaffer, *et al*. (1997). Gapped BLAST and PSI-BLAST: a new generation of protein database search programs. *Nucleic Acids Res* **25**(17), 3389–3402.

Boeckmann, B., M. C. Blatter, L. Famiglietti, *et al*. (2005). Protein variety and functional diversity: Swiss-Prot annotation in its biological context. *C R Biol* **328**(10–11), 882–899.

Fainstein, E., C. Marcelle, A. Rosner, *et al*. (1987). A new fused transcript in Philadelphia chromosome positive acute lymphocytic leukaemia. *Nature* **330**(6146), 386–388.

Groffen, J., J. R. Stephenson, N. Heisterkamp, *et al*. (1984). Philadelphia chromosomal breakpoints are clustered within a limited region, bcr, on chromosome 22. *Cell* **36**(1), 93–99.

Hanahan, D. and R. A. Weinberg (2000). The hallmarks of cancer. *Cell* **100**(1), 57–70.

Hanahan, D. and R. A. Weinberg (2011). Hallmarks of cancer: the next generation. *Cell* **144**(5), 646–674.

Karlin, S. and S. F. Altschul (1990). Methods for assessing the statistical significance of molecular sequence features by using general scoring schemes. *Proc Natl Acad Sci U S A* **87**(6), 2264–2268.

Magrane, M. and Uniprot Consortium (2011). UniProt Knowledgebase: a hub of integrated protein data. Database.

Needleman, S. B. and C. D. Wunsch (1970). A general method applicable to the search for similarities in the amino acid sequence of two proteins. *J Mol Biol* **48**(3), 443–453.

Nowell, P. C. (2007). Discovery of the Philadelphia chromosome: a personal perspective. *J Clin Invest* **117**(8), 2033–2035.

Nowell, P. C. and D. A. Hungerford (1960). Chromosome studies on normal and leukemic human leukocytes. *J Natl Cancer Inst* **25**, 85–109.

Saglio, G., A. Guerrasio, C. Rosso, *et al.* (1990). New type of Bcr/Abl junction in Philadelphia chromosome-positive chronic myelogenous leukemia. *Blood* **76**(9), 1819–1824.

Smith, T. F. and M. S. Waterman (1981). Identification of common molecular subsequences. *J Mol Biol* **147**(1), 195–197.

8

Evolution
What makes us human?

The concept of evolution is of fundamental importance in genomics and bioinformatics, as already emphasized in Chapter 1. Over the next three chapters we will be looking at questions related to the evolution of species, as well as the evolution of individual genes and proteins. First, we will specifically address the question of how humans are genetically different to other animals, such as our close relative the chimpanzee. At the level of bioinformatics and computational tools we will introduce *multiple sequence alignments* and show how to analyse such alignments with programming tools.

Genetic differences between humans and chimpanzees

Now that the complete genome sequences of humans and many other animals are known, we are in a position to address interesting questions about similarities and differences between animals in terms of genetic information. Humans clearly have functions that other animals seem to be lacking, such as advanced spoken and written languages and abstract-level thinking. Many cultural activities such as music and painting also seem to be specific to humans. We are definitely distinct from other animals in these respects, even when considering our close primate relatives. With the genome sequences now available, there is an obvious question to answer: what are the significant genetic differences between us and other primates such as the chimpanzee? It would be particularly interesting to identify the genetic elements that are likely to affect brain function. First of all, it should be kept in mind that the identity between the human and chimpanzee genomes is quite extensive. Thus, if we consider the portions of the human and chimpanzee genomes that may be aligned to each other, there are approximately 35 million single nucleotide differences (Chimpanzee Sequencing and Analysis Consortium, 2005). This is to say that in this respect the genomes are in fact 99% identical. In addition, there are about five million insertion/deletion events, accounting for another 3% difference between humans and chimpanzees.

The genomic differences that we observe between humans and other animals may have very different functional implications. For example, there are a number of genomic regions that are conserved among mammals, including chimpanzees, that have been deleted in humans. These deletions are expected to affect the regulation of gene expression (McLean *et al.*, 2011). Furthermore, gene duplication events in humans may have introduced paralogous genes with new functions. Local changes like point mutations could affect regulation of transcription or processes such as splicing. Point mutations could also directly affect a protein-coding sequence. Such mutations may also be of interest when learning about functions that are specific to humans. We could examine the evolution of different proteins and try to identify proteins that have changed specifically in the human lineage. We may consider genes or protein sequences that are conserved during vertebrate or mammalian evolution, but where the human sequences have evolved at an unexpectedly high rate. One such example is the non-coding RNA HAR1F (HAR for 'human accelerated region'), which turns out to be expressed specifically in the developing human neocortex (Pollard *et al.*, 2006). This is a part of the brain involved in functions such as conscious thought and language.

Another example of a gene implicated as being developed specifically in the human lineage is that encoding the protein *alpha tectorin* (Clark *et al.*, 2003). This protein is located in the tectorial membrane of the inner ear. Human families with congenital deafness have been identified as having a mutated alpha tectorin-encoding gene (Naz *et al.*, 2003). The mutations in this gene are associated with poor frequency response in the ear, and make it difficult to understand speech.[1] It has been suggested that the changes in the alpha tectorin protein in humans adjusted our hearing to adapt it to human-specific development of speech.

A protein related to human speech

A third example of a gene that has been implicated as being of particular interest in terms of human evolution is the FOXP2 gene (Fisher and Scharff, 2009). The FOXP2 protein is a transcription factor, and the acronym FOX refers to the 'forkhead box' DNA-binding domain contained within the protein. It is expressed in the central nervous system (CNS) during development and is likely to regulate a number of genes of importance for brain function. The FOXP2 protein has been identified in all vertebrates and its amino acid sequence is fairly well conserved.

The human FOXP2 gene is strongly associated with human speech and language. Inherited diseases which give rise to speech deficiency have been

---

[1] 'It's something like replacing the soundboard of a Stradivarius violin with a piece of plywood', Andrew G. Clark, Cornell Professor of Molecular Biology and Genetics (quotation from http://www.news.cornell.edu, describing the effect of mutations in the alpha tectorin gene).

identified with mutations in the FOXP2 gene. For example, in one specific extended family, known as KE, the affected individuals have difficulties with grammar and expressive language; in addition they are characterized by defective articulation (verbal dyspraxia), such that their speech is quite incomprehensible to the inexperienced listener (Lai *et al.*, 2000, 2001; Vargha-Khadem *et al.*, 2005). It would seem that a critical issue of the disease is a deficiency in the ability to carry out the fine movements of face and mouth muscles required for effective speech.

The cause of the disease is a mutation R553H in the FOXP2 gene. This is an amino acid substitution in the DNA-binding domain of the FOXP2 protein. Analysis of a number of other individuals with verbal dyspraxia identified yet another mutation in the FOXP2 gene, an R328X nonsense mutation which is predicted to truncate the protein product. The inherited defects related to FOXP2 are dominant, which is to say that both copies of a healthy FOXP2 gene are required for normal speech and language development. The FOXP2 gene is located on chromosome 7 and has at least three isoforms.

FOXP2 is likely to regulate the expression of more than one gene. One of its known targets is CNTNAP2 (contactin-associated protein-like 2), a protein associated with language defects. For instance, changes in the CNTNAP2 gene were found to be predominant in families in which affected individuals had difficulties repeating nonsense words (Vernes *et al.*, 2008).

FOXP2 in other animals

Other animals do not have the advanced speech of humans. However, other vertebrates do present cases in which mutations in the FOXP2 gene homologue have effects that are reminiscent of those in humans. One example is mice. Normal mice use different kinds of sounds to communicate, both sonic and ultrasonic. For instance, young mouse pups emit ultrasounds whenever they are aroused, such as when they are removed from their parents. This may be thought of as analogous to young human babies crying to attract the attention of their parents. Mice that do not have a functional copy of FOXP2 are not able to produce these isolation calls (Shu *et al.*, 2005; Fujita *et al.*, 2008; Groszer *et al.*, 2008). However, other studies show that mouse vocalization may be produced using other stress conditions and it would seem that the relationship between mouse

Fig. 8.1 *Zebra finch*. Birds learn to sing from a tutor, such as an adult male bird. In one experiment with zebra finches the expression of the protein FOXP2 was reduced to about 50% of its normal value; the birds imitated tutor songs incompletely. Thus, some notes were omitted and the notes copied were less accurately reproduced as compared to normal birds (Haesler *et al.*, 2007). Photograph with permission from Dreamstime.

vocalization and FOXP2 is not that simple (Gaub *et al.*, 2010). Rather than vocalization as such, many investigations point to a role of FOXP2 in *synaptic plasticity*, i.e. the ability of the connection between two neurons to be modulated (Groszer *et al.*, 2008; Enard *et al.*, 2009). Such plasticity would in turn be important for motor-skill learning.

FOXP2 has been shown to be important for neural development in birds. Bird-song has similarities and differences to human speech. In humans the speech learning is dependent on hearing other individuals, something also found in birds. Experiments have been carried out with zebra finches in which the expression of FOXP2 was reduced to 50% of its normal value. Interestingly, these birds were not as efficient as normal birds when it came to imitating the songs they were taught (Haesler *et al.*, 2007).

Now that we know a few things about the role of FOXP2 in neural development in animals, we might ask: during the evolution of humans, did anything happen to the amino acid sequence of the FOXP2 protein that could be related to the development of human speech? To answer this question we would like to compare FOXP2 sequences from a range of animals, and find out whether there are mutations that are unique to humans.

BIOINFORMATICS

Comparing FOXP2 in different animals

The first step in comparing FOXP2 sequences from a number of species is to produce a *multiple alignment*. For this example we will make use of a number of sequences collected from the Swiss-Prot/UniProtKB database. The sequence identifiers and species names are:

Tools introduced in this chapter	
Perl	breaking out of a loop with `last`
Software running in a Unix environment	ClustalW

FOXP2_HUMAN	*Homo sapiens*
FOXP2_GORGO	*Gorilla gorilla*
FOXP2_MACMU	*Macaca mulatta* (Rhesus macaque)
FOXP2_PANTR	*Pan troglodytes* (Chimpanzee)
FOXP2_HYLLA	*Hylobates lar* (Common gibbon)
FOXP2_PONPY	*Pongo pygmaeus* (Bornean orangutan)
FOXP2_MOUSE	*Mus musculus* (Mouse)
FOXP2_XENLA	*Xenopus laevis* (African clawed frog)

For this example we will assume that the sequences are collected in a file named `foxp2.fa`, with all sequences in FASTA format. These sequences can be downloaded using the NCBI tool Entrez, further described in Chapter 14; or they may be retrieved from the command line using the `blastdbcmd` tool of the previous chapter. A command like

```
% blastdbcmd -entry FOXP2_HYLLA -db swissprot
```

would retrieve a single sequence, but you could also do the following to retrieve all of them in one command:

```
% blastdbcmd -entry_batch seqids.txt -db swissprot
```

Here, the file `seqids.txt` contains a list of the identifiers, one in each line of the file.

With the sequences collected in the file `foxp2.fa`, we now want to produce a multiple sequence alignment. We have previously seen how two different sequences may be aligned using a local alignment algorithm, such as in BLAST. In a multiple sequence alignment, however, more than two sequences are aligned and typically we make use of a *global* alignment. Multiple sequence alignments are computationally more demanding than pairwise alignments when a strict dynamic programming method is employed. As a result, the most widely used programs such as ClustalW (Thompson *et al.*, 2002), Muscle (Edgar, 2004) and T-Coffee (Notredame *et al.*, 2000) use heuristic methods. ClustalW, a further development of Clustal, which was introduced in the late 1980s, has been widely used and will be used for this example, although many other multiple alignment programs may be used to produce virtually identical results with the specific sequences used here. ClustalW uses a *progressive alignment* procedure, meaning that it uses a stepwise procedure to produce the final alignment. Thus, in each step a single sequence or a subalignment is merged (aligned) with a previous sequence or subalignment. The order in which sequences are aligned is determined by a *guide tree*. This guide tree is deduced in a first step of the program and is the result of pairwise comparison of all sequences (see also Fig. 8.2).

The ClustalW method is available through a graphical user interface named ClustalX. For our example, however, we will be using the command-line version of ClustalW (see also Appendix II). We provide as an argument to ClustalW (version 2) the name of the file containing the sequences that are to be aligned:

```
% clustalw2 foxp2.fa
```

As you execute this command you will see as output the progression of the multiple alignment procedure, containing both the initial analysis leading to the guide tree and the final stage in which the multiple alignment is created. The actual results, however, will be stored in the files `foxp2.dnd` and `foxp2.aln`. The file with extension '.dnd' (for 'dendrogram') is the guide tree and may be viewed with software designed to draw trees, such as `njplot` (Perriere and Goultz, 1996).

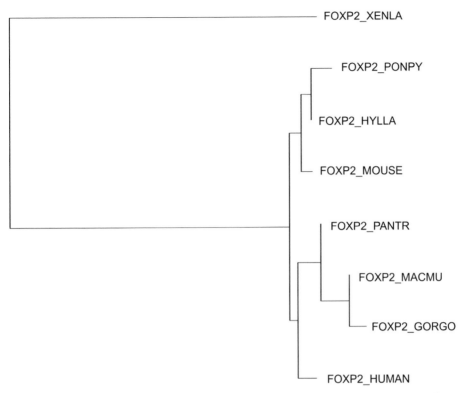

Fig. 8.2 *ClustalW guide tree*. A tree is deduced in a first step of the program and is the result of pairwise comparison of all sequences. This tree is then used to guide the multiple alignment. The tree depicted here was obtained during alignment of a collection of FOXP2 sequences.

The contents of the `foxp2.dnd` file are:

```
((((FOXP2_HUMAN:0.00122,((FOXP2_GORGO:0.00117,
FOXP2_MACMU:-0.00117):0.00196,FOXP2_PANTR:-0.00196)
:0.00157):0.00059,FOXP2_XENLA:0.04051)
:0.00083,FOXP2_MOUSE:0.00074)
:0.00066,FOXP2_HYLLA:0.00000,
FOXP2_PONPY:0.00140);
```

This is the guide tree shown with a syntax known as the *Newick* format. This tree is shown graphically in Fig. 8.2. The tree may be thought of as a phylogenetic tree, but we are now a bit ahead of ourselves because such trees are the subject of the next chapter. What we are mainly interested in for the topic of the present chapter is the actual alignment. If you examine the output file `foxp2.aln`, you will note that there is a difference compared to the input file in that the sequences have now been aligned and gaps have been introduced at some locations.

To address the question of how the human FOXP2 is related to its orthologues in the other species we need to construct a Perl script to analyse the alignment. The alignment in the file foxp2.aln has a somewhat awkward format as each individual sequence is divided up into several sections of the alignment (an 'interleaved' format). Although we could handle this through some scripting, we will avoid that problem by producing the output alignment from ClustalW in a different format. When running the ClustalW alignment we can specify that we want to have the output in a FASTA format:

```
% clustalw2 foxp2.fa -output=fasta
```

The output file from ClustalW will now be foxp2.fasta, and the first part of the file, with two sequences, is:

```
>FOXP2_GORGO
MMQESATETISNSSSMNQNGMSTLSSQLDAGSRDGRSSGDTSSEVSTVELLHLQQQQALQA
ARQLLLQQQTSGLKSPKSSDKQRPLQVPVSVAMMTPQVITPQQMQQILQQQVLSPQQLQA
LLQQQQAVMLQQQQLQEFYKKQQEQLHLQLLQQQQQQQQQQQQQQQQQQQQQQQQQQQQQQ
QQQQQQQQQ---HPGKQAKEQQQQQQQQQQLAAQQLVFQQQLLQMQQLQQQQHLLSLQRQ
GLISIPPGQAALPVQSLPQAGLSPAEIQQLWKEVTGVHSMEDNGIKHGGLDLTTNNSSST
TSSTTSKASPPITHHSIVNGQSSVLNARRDSSSHEETGASHTLYGHGVCKWPGCESICED
FGQFLKHLNNEHALDDRSTAQCRVQMQVVQQLEIQLSKERERLQAMMTHLHMRPSEPKPS
PKPLNLVSSVTMSKNMLETSPQSLPQTPTTPTAPVTPITQGPSVITPASVPNVGAIRRRH
SDKYNIPMSSEIAPNYEFYKNADVRPPFTYATLIRQAIMESSDRQLTLNEIYSWFTRTFA
YFRRNAATWKNAVRHNLSLHKCFVRVENVKGAVWTVDEVEYQKRRSQKITGSPTLVKNIP
TSLGYGAALNASLQAALAESSLPLLSNPGLINNASSGLLQAVHEDLNGSLDHIDSNGNSS
PGCSPQPHIHSIHVKEEPVIAEDEDCPMSLVTTANHSPELEDDREIEEEPLSEDLE
>FOXP2_MACMU
MMQESATETISNSSSMNQNGMSTLSSQLDAGSRDGRSSGDTSSEVSTVELLHLQQQQALQA
ARQLLLQQQTSGLKSPKSSDKQRPLQVPVSVAMMTPQVITPQQMQQILQQQVLSPQQLQA
LLQQQQAVMLQQQQLQEFYKKQQEQLHLQLLQQQQQQQQQQQQQQQQQQQQQQQQQQQQQQ
QQQQQQQQQ--HPGKQAKEQQQQQQQQQQQLAAQQLVFQQQLLQMQQLQQQQHLLSLQRQ
GLISIPPGQAALPVQSLPQAGLSPAEIQQLWKEVTGVHSMEDNGIKHGGLDLTTNNSSST
TSSTTSKASPPITHHSIVNGQSSVLNARRDSSSHEETGASHTLYGHGVCKWPGCESICED
FGQFLKHLNNEHALDDRSTAQCRVQMQVVQQLEIQLSKERERLQAMMTHLHMRPSEPKPS
PKPLNLVSSVTMSKNMLETSPQSLPQTPTTPTAPVTPITQGPSVITPASVPNVGAIRRRH
SDKYNIPMSSEIAPNYEFYKNADVRPPFTYATLIRQAIMESSDRQLTLNEIYSWFTRTFA
YFRRNAATWKNAVRHNLSLHKCFVRVENVKGAVWTVDEVEYQKRRSQKITGSPTLVKNIP
TSLGYGAALNASLQAALAESSLPLLSNPGLINNASSGLLQAVHEDLNGSLDHIDSNGNSS
PGCSPQPHIHSIHVKEEPVIAEDEDCPMSLVTTANHSPELEDDREIEEEPLSEDLE
```

The FASTA format is such that all sequences are separated (sequentially arranged and not in an 'interleaved' format), and therefore easier to handle with a Perl script. We will make use of such a script to analyse this alignment columnwise. An important property of all multiple alignments is that the number of columns is the same for all sequences. We want the script to identify all positions (columns) where the human sequence is different from all the other species.

Our script to analyse the alignment is shown in Code 8.1. In the first part (labelled # 1 #) we read the sequence using a similar technique as in previous chapters. One difference is that we are here collecting all the non-human sequences as a hash. The human sequence is stored in a separate variable as it has a special role. Thus, we have as a result of the first part of the script the following variables, each containing a sequence with the same length:

```
$humanseq
$nonhuman{'FOXP2_GORGO'}
$nonhuman{'FOXP2_MACMU'}
$nonhuman{'FOXP2_PANTR'}
$nonhuman{'FOXP2_HYLLA'}
$nonhuman{'FOXP2_PONPY'}
$nonhuman{'FOXP2_MOUSE'}
$nonhuman{'FOXP2_XENLA'}
```

In the second part (labelled # 2 #) we analyse these variables and go through each of the different columns with a loop:

```
for (my $i = 0; $i < length($humanseq); $i++) {}
```

The amino acid in the human sequence at position $i is obtained with the substr function:

```
substr ($humanseq, $i, 1)
```

and a similar expression is used to extract the corresponding position in a non-human sequence.

We use the loop

```
foreach my $key (keys %nonhuman) {}
```

to go through all the non-human sequences and compare each of them to the human sequence.

The variable $unique is used to keep track of whether the human sequence is different from all the non-human sequences. It is first set to the value of 1 (true) but changes to 0 (false) as soon as we observe that the human position is different from the non-human position.

The observant reader may note that the foreach loop is not so well designed. Thus, we go through all keys of %nonhuman even after we have reached a zero value for $unique. We can live with this, as in this particular case the script produces the desired results very quickly anyway. However, more generally speaking such use of a loop is a waste of time and is computationally inefficient. We can avoid it in Perl by using `last` to exit the `foreach` loop. This command is used whenever you want to break out of a loop once a certain condition is met:

```perl
foreach my $key (keys %nonhuman) {
    if (substr($humanseq, $i, 1) eq substr($nonhuman{$key}, $i, 1)) {
        $unique=0;
        last;
    }
}
```

Code 8.1 foxp2.pl

```perl
#!/usr/bin/perl -w

use strict;

# 1 # Read sequences and store them in a hash

open(IN, 'foxp2.fasta') or die "Could not open file\n";
my %nonhuman;
my $id = undef;
my $seq;
my $id_human;
my $humanseq;
while (<IN>) {
    chomp;
    if (/^>/) {
        if ($id) {
            if ( $id !~ /HUMAN/ ) {
                $nonhuman{$id} = $seq;
            }
            else {
                $humanseq = $seq;
                $id_human = $id;
            }
        }
        $id = $_;
        $id =~ s/>//;
        $seq = '';
    }
    else {
        $seq .= $_;
    }
}
```

```
$nonhuman{$id} = $seq;      # this assumes that we know that the
                            # final sequence read is non-human

close IN;

# 2 # Analyse the multiple alignment

for ( my $i = 0 ; $i < length($humanseq) ; $i++ ) {
    my $unique = 1;
    foreach my $key ( keys %nonhuman ) {

        # check if the amino acid in the human sequence
        # is the same as in the non-human sequence
        if ( substr( $humanseq, $i, 1 ) eq
             substr( $nonhuman{$key}, $i, 1 ) ) {
            $unique = 0;
        }
    }
    if ($unique) {
        my $pos = $i + 1;
        print "At position $pos\n";
        my $aa = substr( $humanseq, $i, 1 );
        print "$id_human\t$aa\n";
        foreach my $key ( keys %nonhuman ) {
            my $aa = substr( $nonhuman{$key}, $i, 1 );
            print "$key\t$aa\n";
        }
    }
}
```

The output from Code 8.1 is:

```
At position 304
FOXP2_HUMAN      N
FOXP2_XENLA      T
FOXP2_PONPY      T
FOXP2_PANTR      T
FOXP2_MOUSE      T
FOXP2_GORGO      T
FOXP2_HYLLA      T
FOXP2_MACMU      T
```

Changes in FOXP2 specific to humans

From the results of our Perl script we see that there is only one position, 304, where the human sequence is different from all other species. The human sequence has asparagine (N), whereas all the non-human sequences have threonine (T). It should be noted that the position numbers in the multiple alignment are not directly comparable to the position numbers in the individual proteins. For instance, position 304 in the alignment corresponds to position 303 in the human protein, and to position 302 in mouse protein. The reason for this difference is of course that the sequences are not of equal length and gaps have been inserted in the alignment.

The human-specific T303N mutation is not a finding which first came about as this chapter was conceived. It was noted already in publications of Enard *et al.* (2002) and Zhang *et al.* (2002). These authors noted yet another mutation, N325S, and hypothesized that both T303N and N325S were subject to positive selection because of the effects on language and speech. The N325S mutation did not show up in our analysis here, because the amino acid serine is found also in the frog *Xenopus laevis*. The T303N and N325S mutations have been further explored experimentally. Thus, mice were constructed that, instead of their normal FOXP2, carried a 'humanized' version with both of these mutations (Enard *et al.*, 2009). These mice appeared rather normal and did not develop into Mickey Mouse-like animals with human speech and language. However, basal ganglia, a group of nuclei in the brain associated with motor control and learning, seemed to be affected. The striatum is a part of the basal ganglia and the medium spiny neurons, a type of inhibitory cell within the striatum, showed increased dendritic length and increased synaptic plasticity. These studies would therefore support the idea that the specific changes in human FOXP2 contributed to the development of human speech.

EXERCISES

8.1 Modify Code 8.1 such that you may identify (1) positions where the human sequence is different from all other primates; and (2) positions where the human and primate sequences are identical but different to both mouse and frog.

8.2 Construct a multiple alignment of FOXP2 mRNA sequences (make use of the unaligned sequences in file `foxp2_mrna.fa`). Use Perl code to answer the question: in what positions is the human mRNA different from all other primates?

8.3 In the default output from the ClustalW program (a file with the extension `.aln`) there are asterisks (*) that indicate positions where the nucleotide is the same in all sequences. Make a Perl script to count the number of such positions in an alignment (like `foxp2.aln`). Also count the number of positions where asterisks are absent.

REFERENCES

Chimpanzee Sequencing and Analysis Consortium (2005). Initial sequence of the chimpanzee genome and comparison with the human genome. *Nature* **437**(7055), 69–87.

Clark, A. G., S. Glanowski, R. Nielsen, *et al.* (2003). Inferring nonneutral evolution from human–chimp–mouse orthologous gene trios. *Science* **302**(5652), 1960–1963.

Edgar, R. C. (2004). MUSCLE: a multiple sequence alignment method with reduced time and space complexity. *BMC Bioinformatics* **5**, 113.

Enard, W., S. Gehre, K. Hammerschmidt, *et al.* (2009). A humanized version of Foxp2 affects cortico-basal ganglia circuits in mice. *Cell* **137**(5), 961–971.

Enard, W., M. Przeworski, S. E. Fisher, *et al.* (2002). Molecular evolution of FOXP2, a gene involved in speech and language. *Nature* **418**(6900), 869–872.

Fisher, S. E. and C. Scharff (2009). FOXP2 as a molecular window into speech and language. *Trends Genet* **25**(4), 166–177.

Fujita, E., Y. Tanabe, A. Shiota, *et al.* (2008). Ultrasonic vocalization impairment of Foxp2 (R552H) knockin mice related to speech–language disorder and abnormality of Purkinje cells. *Proc Natl Acad Sci U S A* **105**(8), 3117–3122.

Gaub, S., M. Groszer, S. E. Fisher and G. Ehret (2010). The structure of innate vocalizations in Foxp2-deficient mouse pups. *Genes Brain Behav* **9**(4), 390–401.

Groszer, M., D. A. Keays, R. M. Deacon, *et al.* (2008). Impaired synaptic plasticity and motor learning in mice with a point mutation implicated in human speech deficits. *Curr Biol* **18**(5), 354–362.

Haesler, S., C. Rochefort, B. Georgi, *et al.* (2007). Incomplete and inaccurate vocal imitation after knockdown of FoxP2 in songbird basal ganglia nucleus Area X. *PLoS Biol* **5**(12), e321.

Lai, C. S., S. E. Fisher, J. A. Hurst, *et al.* (2000). The SPCH1 region on human 7q31: genomic characterization of the critical interval and localization of translocations associated with speech and language disorder. *Am J Hum Genet* **67**(2), 357–368.

Lai, C. S., S. E. Fisher, J. A. Hurst, F. Vargha-Khadem and A. P. Monaco (2001). A forkhead-domain gene is mutated in a severe speech and language disorder. *Nature* **413**(6855), 519–523.

McLean, C. Y., P. L. Reno, A. A. Pollen, *et al.* (2011). Human-specific loss of regulatory DNA and the evolution of human-specific traits. *Nature* **471**(7337), 216–219.

Naz, S., F. Alasti, A. Mowjoodi, *et al.* (2003). Distinctive audiometric profile associated with DFNB21 alleles of TECTA. *J Med Genet* **40**(5), 360–363.

Notredame, C., D. G. Higgins and J. Heringa (2000). T-Coffee: a novel method for fast and accurate multiple sequence alignment. *J Mol Biol* **302**(1), 205–217.

Perriere, G. and M. Gouy (1996). WWW-query: an on-line retrieval system for biological sequence banks. *Biochimie* **78**(5), 364–369.

Pollard, K. S., S. R. Salama, N. Lambert, *et al.* (2006). An RNA gene expressed during cortical development evolved rapidly in humans. *Nature* **443**(7108), 167–172.

Shu, W., J. Y. Cho, Y. Jiang, *et al.* (2005). Altered ultrasonic vocalization in mice with a disruption in the Foxp2 gene. *Proc Natl Acad Sci U S A* **102**(27), 9643–9648.

Thompson, J. D., T. J. Gibson and D. G. Higgins (2002). Multiple sequence alignment using ClustalW and ClustalX. *Curr Protoc Bioinformatics*, Chapter 2, Unit 2.3.

Vargha-Khadem, F., D. G. Gadian, A. Copp and M. Mishkin (2005). FOXP2 and the neuroanatomy of speech and language. *Nat Rev Neurosci* **6**(2), 131–138.

Vernes, S. C., D. F. Newbury, B. S. Abrahams, *et al.* (2008). A functional genetic link between distinct developmental language disorders. *N Engl J Med* **359**(22), 2337–2345.

Zhang, J., D. M. Webb and O. Podlaha (2002). Accelerated protein evolution and origins of human-specific features: Foxp2 as an example. *Genetics* **162**(4), 1825–1835.

Evolution
Resolving a criminal case

9

This chapter will introduce *molecular phylogeny*, a science where DNA, RNA or protein sequences are used to deduce relationships between organisms. Such relationships are typically shown in the form of a tree. The entities under study are often species, but phylogenetic methods could be used to examine other types of evolutionary relationships, such as how individuals in a population are related or how different members of a protein family are related by orthology and paralogy. For the example in this chapter a phylogenetic tree will show the relationship between different HIV isolates. But we will start out with a somewhat ghastly criminal story.

A fatal injection

This story takes place in Lafayette in Louisiana. Maria Jones, a married nurse of age 20, met a gastroenterologist named Robert White.[1] Robert, aged 34, was also married and he had three children. As Robert and Maria entered a relationship, Maria divorced her husband. In return Robert promised to divorce his wife, but never followed through with it. Still, the relationship between Maria and Robert continued. After five years Maria had become pregnant three times, but Robert convinced her every time to have an abortion. Maria did give birth to a child; Robert was the father. Robert eventually became exceedingly jealous and controlling. When Maria saw other men, Robert would sometimes threaten to kill them. He also threatened Maria. After ten turbulent years, Maria finally decided to leave Robert in July 1994.

Before the lovers had broken up, Robert had regularly given Maria vitamin B-12 injections. Late one fatal evening in August 1994 Maria was in bed with her son beside her. Robert appeared and said he wanted to give her yet another injection. Maria protested as she was tired and it was late in the evening. However, Robert proceeded with the injection. Maria felt something was not right this time; she felt a strong pain that she did not recognize from her previous injections.

Two weeks later Maria consulted a physician. Her lymph nodes were swollen, indicating that she had a viral infection. In December Maria was subject to an

---

[1] This is a true story, but assumed names are used for the individuals involved.

annual check-up, which included a test for HIV. She was found HIV positive. Not only that, she was also infected with hepatitis C.

Maria suspected Robert was responsible and that she received the HIV infection through the injection in August. She turned to the police in January 1995. At first the police did not take Maria's accusations seriously. However, it turned out that all seven men Maria reported as sexual contacts for the period 1984–1995 were tested and found to be HIV negative. Furthermore, investigations showed that Robert lied when asked questions about Maria. When the police examined Robert's patient records there were two blood samples withdrawn in early August that lacked the proper laboratory references. One of the blood samples was from a patient with AIDS and the other from a patient with hepatitis C. These two patients had been asked by Robert to leave blood samples; he told them he needed the samples for a research project.

From this evidence it was strongly suspected that Robert had given Maria injections containing blood or blood products from the AIDS patient he had in his care. However, there was a forensic challenge in this case, because the prosecution wanted firm evidence that the HIV carried by Maria was the same as – or very closely related to – the AIDS patient's HIV.[2] In other words, it was important to exclude the possibility that Maria received the infection from some other source. There was a need for sequence data to clarify the relationship of the different HIV isolates. HIV was thus isolated and selected genes were sequenced. The virus isolates that were analysed were from Maria, the AIDS patient, as well as from other individuals from the same geographical area.

In order to determine the relationship of sequences we commonly reconstruct a phylogenetic tree. In this tree it is easy to identify the closest neighbours and to estimate relative distances between nodes in the tree. Molecular phylogeny is the science of constructing phylogenetic trees using molecular sequence data. The criminal case discussed here was the first time molecular phylogeny was used in court in the United States. For this chapter we will try and reproduce the procedure used to obtain a phylogenetic tree.

BIOINFORMATICS

Tools introduced in this chapter	
Software running under Unix	More on ClustalW

Methods of molecular phylogeny

Before we proceed with the phylogenetic analysis of the HIV sequences for the

---

[2] HIV mutates very rapidly and therefore the sequence of one isolate is not exactly the same as another, although they are closely related and derived from the same original strain.

criminal case, we will more closely examine some of the methods of molecular phylogeny. The majority of tree construction methods rely on analysis of a multiple alignment, like the one we examined in the previous chapter for the FOXP2 proteins.

One can distinguish two major categories of tree construction: *distance based* and *character based*. The distance-based methods take into consideration distances between sequences as measured by the number of nucleotide or amino acid differences, as discussed below. The character-based methods rely on analysis of columns in a multiple sequence alignment. They interpret molecular changes in the context of some model of evolution and are often of a probabilistic nature. We will see examples of character-based methods in the following chapter.

In general, character-based methods are more reliable as they make use of much more biological information than is used in distance-based methods. However, distance-based methods are computationally simple and fast, and may therefore be applied to large datasets. They are often surprisingly accurate compared to character-based methods. They will be used for the example in this chapter.

The first step in a distance-based method is to generate a distance matrix, showing all pairwise distances between sequences. In the example in Fig. 9.1, four sequences are analysed; as the distance between two sequences we simply use the number of nucleotide differences between them. The distance matrix obtained is symmetrical and it is often shown in a triangular form (Fig. 9.1).

The most popular distance-based method is one referred to as *neighbour-joining* (Saitou and Nei, 1987). The neighbour-joining method is an example of a *star decomposition* method, as it starts out with the tree in a star-like configuration with a single node connecting to all external nodes. Then, in a stepwise manner, it 'decomposes' the star by identifying neighbours and appropriately introducing new nodes. An important principle of the neighbour-joining method is that it uses a principle of *minimal evolution* such that in each step of neighbour-joining the tree is preferred with the smallest total branch length. A detailed worked example of the neighbour-joining method is shown in Box 9.1.

>A
GGACCACTACGAGCGCCTACGACGTA
>B
GGACCCCTACGAGCCCCTACGACGTA
>C
GGACCGCTGCGAGCTTCTACGACGTA
>D
GGACCTCTCCGGGCAGCTAGGACGTA

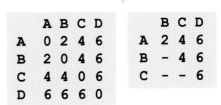

	A	B	C	D			B	C	D
A	0	2	4	6	A		2	4	6
B	2	0	4	6	B		-	4	6
C	4	4	0	6	C		-	-	6
D	6	6	6	0					

Fig. 9.1 *Construction of distance matrix.* A multiple alignment is analysed such that the number of substitutions between all pairs of sequences are counted and introduced in a two-dimensional matrix. The resulting matrix is shown in two forms, a symmetric matrix and an upper triangular matrix.

Box 9.1 A worked example of the neighbour-joining method of phylogenetic tree reconstruction

Consider the species as listed below for analysis with the neighbour-joining method. We may be interested in identifying the primate which is most closely related to humans.

A *Gorilla gorilla* (gorilla)
B *Pan troglodytes* (chimpanzee)
C *Homo sapiens* (humans)
D *Pongo pygmaeus* (orangutang)
E *Macaca fascicularis* (macaque)

As a starting point we use the following distance matrix:

	B	C	D	E
A	11	12	17	24
B		9	16	24
C			16	24
D				24

Step 1 ($n = 5$)

We start by calculating the S_X values defined by $S_X = \sum_{i=1}^{N} d_{Xi}$, where N is the number of OTUs (operational taxonomic units) – in this case five.
For instance, $S_A = d_{AB} + d_{AC} + d_{AD} + d_{AE}$.

$S_A = 11 + 12 + 17 + 24 = 64$
$S_B = 11 + 9 + 16 + 24 = 60$
$S_C = 12 + 9 + 16 + 24 = 61$
$S_D = 17 + 16 + 16 + 24 = 73$
$S_E = 24 + 24 + 24 + 24 = 96$

We then calculate a δ matrix where $\delta_{ij} = d_{ij} - (S_i + S_j) / 2$:

$\delta_{AB} = 11 - (64 + 60) / 3 = -30.3$
$\delta_{AC} = 12 - (64 + 61) / 3 = -29.7$
$\delta_{AD} = 17 - (64 + 73) / 3 = -28.7$

etc. for all δ_{ij}.

The new matrix becomes

	B	C	D	E
A	−30.3	−29.7	−28.7	−29.3
B		−29.7	−28.3	−28
C			−28.7	−28.3
D				(−32.3)

The numbers in this matrix reflect the relative total branch lengths of trees where the nodes i and j have been joined as neighbours. As we prefer the tree with the smallest total branch length we identify the minimum value, which in this case is δ_{DE}. Thus, the nodes D and E are the first neighbours to be joined. They are connected to a new node, X. The distances d_{DX} and d_{EX} are calculated by

$$d_{DX} = (d_{DE} + (S_D - S_E)/(N - 2)) / 2 = (24 + (73 - 96) / 3) / 2 = 8.2$$
$$d_{EX} = d_{DE} - d_{DX} = 24 - 8.2 = 15.8$$

We have now arrived at the following tree:

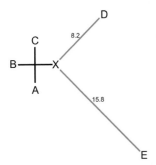

Step 2 ($n = 4$)

In the next step we proceed to identify the next pair of nodes to be joined as neighbours. We set up a matrix with X instead of D and E. Thus, this matrix has four OTUs, A, B, C and X. We need to calculate the distances d_{XA}, d_{XB} and d_{XC}.

$$d_{XA} = (d_{DA} + d_{EA} - d_{DE}) / 2 = (17 + 24 - 24) / 2 = 8.5$$
$$d_{XB} = (d_{DB} + d_{EB} - d_{DE}) / 2 = (16 + 24 - 24) / 2 = 8$$
$$d_{XC} = (d_{DC} + d_{EC} - d_{DE}) / 2 = (16 + 24 - 24) / 2 = 8$$

We then arrive at the matrix

	B	C	X
A	11	12	8.5
B		9	8
C			8

Again, we construct a δ matrix, using the same principles as above.

$$S_A = 11 + 12 + 8.5 = 31.5$$
$$S_B = 11 + 9 + 8 = 28$$
$$S_C = 12 + 9 + 8 = 29$$
$$S_X = 8.5 + 8 + 8 = 24.5$$

	B	C	X
A	−18.75	−18.25	−19.5
B		−19.5	−18.25
C			−18.75

This matrix has two minimum values. Here, we select nodes B and C to be joined (the final tree would be the same if we had selected nodes A and X at this time). B and C are connected to the new node Y.

$$d_{BY} = (d_{BC} + (S_B - S_C) / (N - 2)) / 2 = (9 + (60 - 61) / 2) / 2 = 4.25$$
$$d_{CY} = d_{BC} - d_{BY} = 9 - 4.25 = 4.75$$

The tree is now

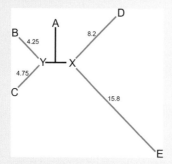

Step 3 ($n = 3$)

We set up a matrix with Y instead of B and C. This matrix has three OTUs: A, Y and X. The distances d_{YA} and d_{YX} are calculated:

$$d_{YA} = (d_{BA} + d_{CA} - d_{BC}) / 2 = (11 + 12 - 9) / 2 = 7$$
$$d_{YX} = (d_{BX} + d_{CX} - d_{BC}) / 2 = (8 + 8 - 9) / 2 = 3.5$$

The new matrix is

	Y	X
A	7	8.5
Y		3.5

The δ matrix becomes:

	Y	X
A	-19	-19
Y		-19

In this matrix all values are equal. We select A and Y to be joined, and they are connected to the new node Z.

$$d_{YZ} = 1$$
$$d_{AZ} = 6$$

The tree is now:

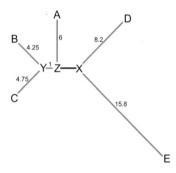

Step 4 ($n = 2$)

The final matrix is very simple, with only:

	X
Z	2.5

We use the distance d_{ZX} to obtain the final unrooted *nj* tree. According to this tree, the chimpanzee is the closest relative of humans.

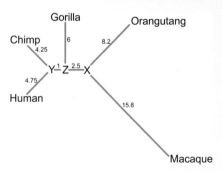

Neighbour-joining is available through programs such as ClustalW, which will be used for our example here. The method is also implemented in the program Neighbor, which is part of PHYLIP, a program package with numerous molecular phylogeny methods (Felsenstein, 1989, 2005) (Dnapars of this package will be used in the following chapter).

In order to test the reliability of a tree, a common procedure is *bootstrapping*. In this procedure a large number of new alignments, typically between 100 and 1000, are produced from the original alignment. Each of these alignments is constructed by randomly selecting columns from the original alignment and will have the same number of columns as the original alignment (Box 9.2).

Box 9.2 Bootstrap test of phylogeny

Bootstrapping is a commonly used technique to test how reliable your dataset (alignment) is when it comes to reconstructing a phylogeny. Consider a multiple alignment of five sequences, each ten nucleotides in length:

```
A: T C C G A G G T T T
B: T C T G A T G T T T
C: T T T G A A G T T T
D: T T C G A A A T T T
E: T T T G A C A A T T
```

This alignment gives rise to the following distance matrix:

```
0 2 3 3 5
2 0 2 4 4
3 2 0 2 3
3 4 2 0 3
5 4 3 3 0
```

This matrix may in turn be used to construct the following unrooted neighbour-joining tree, in which we ignore branch lengths:

Next we sample columns from the alignment above. We produce new alignments with the same number of columns (ten) as the original alignment. Below are two alignments in which the numbers on top of the columns refer to positions in the original alignment. For each of the two alignments are shown the derived distance matrices and neighbour-joining trees.

Alignment	Distance matrix	*Nj* tree

4 4 6 6 1 8 5 9 7 3
G G G G T T A T G C
G G T T T T A T G T
G G A A T T A T G T
G G A A T T A T A C
G G C C T A A T A T

```
0 3 3 3 5
3 0 2 4 4
3 2 0 2 4
3 4 2 0 4
5 4 4 4 0
```

3 6 3 9 9 1 10 3 5 8
C G C T T T T C A T
T T T T T T T T A T
T A T T T T T T A T
C A C T T T T C A T
T C T T T T T T A A

```
0 4 4 1 5
4 0 1 4 2
4 1 0 3 2
1 4 3 0 5
5 2 2 5 0
```

In the same manner we may produce a large number of randomly generated alignments. By analysing the trees derived from these alignments we obtain information about how frequently specific branches are observed. For instance, below is a rooted version of a neighbour-joining tree with 1000 bootstrapping replicates of the same original alignment as above. Numbers at the branches refer to the number of times that branching is observed. Thus, A and B group together 635 times and D and E group together 514 times. In general, branches need at least 70% support to be considered reliable. Therefore, in this particular case the tree should be treated with caution.

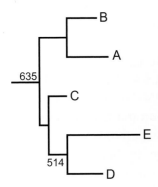

Examination of the criminal case

We are going to apply the method of neighbour-joining to a set of HIV sequences obtained in the criminal case (Metzker *et al.*, 2002). We need to know a few basic facts about HIV. The HIV genome is very small; it is an RNA molecule

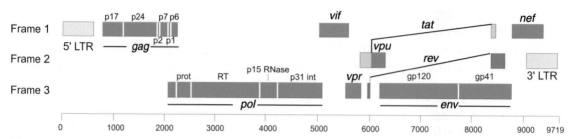

Fig. 9.2 *HIV genome*. Genes of the HIV-1 genome are shown schematically. Each of the genes will be in one of three possible reading frames. The *gag*, *pol* and *env* regions all produce polyproteins that are proteolytically cleaved to produce fully mature proteins. The *tat* and *rev* gene products are translated from spliced transcripts. For more explanations of the genes and the LTR regions, see Table 9.1.

with only 9749 nucleotides. Less than 20 proteins are encoded by this genome. Two genes frequently used for phylogenetic studies are the *env* and *reverse transcriptase* (RT) genes. The *env* gene encodes gp160, a precursor to gp120 and gp41, two proteins that are on the surface of the HIV virus particle. The reverse transcriptase is an enzyme that is used to make a DNA copy of the viral RNA molecule once the virus has entered a host cell. The result is a double-stranded DNA version of the HIV genome that may be incorporated into the chromosomal DNA of the host organism. This viral DNA may then be replicated in the same way as any other host DNA. For more information on the HIV genome and its genes, the reader is referred to Fig. 9.2 and Table 9.1.

For the forensic investigation, blood samples were collected from the HIV-infected patient of the doctor, as well as from the nurse that received the injection (we will in the following refer to her as the victim). As controls in the study, blood samples were obtained from a number of other HIV-infected individuals in Lafayette. These were sampled from the same metropolitan area where the patient and victim resided. Genomic DNA was then prepared from the blood samples and PCR was used to amplify an 858-bp *env* gene fragment and a 1147-bp reverse transcriptase (RT) gene fragment (for more on PCR, see Chapter 4). Finally, the resulting PCR products were subject to sequencing.

As the first step in our computational analysis we need to obtain the sequences that were used in the forensic studies. They may easily be retrieved in this case because they are linked to the publication describing this analysis (Metzker *et al.*, 2002). There are 174 sequence records with accessions in the range AY156734–AY156907 and they may be retrieved using NCBI Entrez, as will be discussed in Chapter 14. When the sequences are retrieved in a FASTA format, each identifier line will have sequence database identifiers as well as information about what HIV gene is represented in the sequence. The nomenclature of the clones of HIV that have been sequenced is such that the first letter(s) in an expression like

Table 9.1 *HIV genes* (information from http://www.hiv.lanl.gov).

GAG   The genomic region encoding the capsid proteins (group-specific antigens). The precursor is the p55 protein, which is processed to p17 (MAtrix), p24 (CApsid), p7 (NucleoCapsid) and p6 proteins, by the viral protease. *Gag* p55 associates with the plasma membrane, where virus assembly takes place.

POL   The genomic region encoding the viral enzymes protease (prot), reverse transcriptase (RT) and integrase (int). These enzymes are produced as a *gag–pol* precursor polyprotein, which is processed by the viral protease. The *gag–pol* precursor is produced by ribosome frameshifting near the 3′ end of the *Gag* gene.

ENV   Viral glycoproteins produced as a precursor (gp160), which is processed to give a non-covalent complex of the external glycoprotein gp120 and the transmembrane glycoprotein gp41. The mature gp120–gp41 proteins are bound by non-covalent interactions and are associated as a trimer on the cell surface. A substantial amount of gp120 can be found released in the medium. gp120 contains the binding site for the CD4 receptor, and the seven transmembrane domain chemokine receptors that serve as co-receptors for HIV-1.

TAT   Transactivator of HIV gene expression. One of two essential viral regulatory factors (*tat* and *rev*) for HIV gene expression. Two forms are known, *tat-1* exon (minor form) of 72 amino acids and *tat-2* exon (major form) of 86 amino acids. Low levels of both proteins are found in persistently infected cells. Tat has been localized primarily in the nucleolus/nucleus by immunofluorescence. It acts by binding to the TAR RNA element and activating transcription initiation and elongation from the LTR promoter, preventing the 5′ LTR AATAAA polyadenylation signal from causing premature termination of transcription and polyadenylation. It is the first eukaryotic transcription factor known to interact with RNA rather than DNA and may have similarities with prokaryotic anti-termination factors. Extracellular *tat* can be found and can be taken up by cells in culture.

REV   The second necessary regulatory factor for HIV expression. A 19-kD phosphoprotein, localized primarily in the nucleolus/nucleus, *rev* acts by binding to RRE and promoting the nuclear export, stabilization and utilization of the viral mRNAs containing RRE. *Rev* is considered the most functionally conserved regulatory protein of lentiviruses. *Rev* cycles rapidly between the nucleus and the cytoplasm.

VIF   Viral infectivity factor, a basic protein typically 23 kD. Promotes the infectivity but not the production of viral particles. In the absence of *vif*, the produced viral particles are defective, while the cell-to-cell transmission of virus is not affected significantly. Found in almost all lentiviruses, *vif* is a cytoplasmic protein, existing in both a soluble cytosolic form and a membrane-associated form. The latter form of *vif* is a peripheral membrane protein that is tightly associated with the cytoplasmic side of cellular membranes. *Vif* prevents the action of the cellular APOBEC-3G protein, which deaminates DNA–RNA heteroduplexes in the cytoplasm.

(cont.)

Table 9.1 (*cont.*)

VPR	*Vpr* (viral protein R) is a 96-amino acid (14-kD) protein, which is incorporated into the virion. It interacts with the p6 *gag* part of the Pr55 *gag* precursor. *Vpr* detected in the cell is localized to the nucleus. Proposed functions for *vpr* include targeting the nuclear import of preintegration complexes, cell-growth arrest, transactivation of cellular genes and induction of cellular differentiation. In HIV-2, SIV-SMM, SIV-RCM, SIV-MND-2 and SIV-DRL the *vpx* gene is apparently the result of a *vpr* gene duplication event, possibly by recombination.
VPU	*Vpu* (viral protein U) is unique to HIV-1, SIVcpz (the closest SIV relative of HIV-1), SIV-GSN, SIV-MUS, SIV-MON and SIV-DEN. There is no similar gene in HIV-2, SIV-SMM or other SIVs. *Vpu* is a 16-kD (81-amino acid) type I integral membrane protein with at least two different biological functions: (1) degradation of CD4 in the endoplasmic reticulum; and (2) enhancement of virion release from the plasma membrane of HIV-1-infected cells. *Env* and *vpu* are expressed from a bicistronic mRNA. *Vpu* probably possesses an N-terminal hydrophobic membrane anchor and a hydrophilic moiety. It is phosphorylated by casein kinase II at positions Ser52 and Ser56. *Vpu* is involved in *env* maturation and is not found in the virion. *Vpu* has been found to increase susceptibility of HIV-1 infected cells to Fas killing.
NEF	A multifunctional 27-kD myristoylated protein produced by an ORF located at the 3′ end of the primate lentiviruses. Other forms of *nef* are known, including non-myristoylated variants. *Nef* is predominantly cytoplasmic and associated with the plasma membrane via the myristoyl residue linked to the conserved second amino acid (Gly). *Nef* has also been identified in the nucleus and found associated with the cytoskeleton in some experiments. One of the first HIV proteins to be produced in infected cells, it is the most immunogenic of the accessory proteins. The *nef* genes of HIV and SIV are dispensable in vitro, but are essential for efficient viral spread and disease progression in vivo. *Nef* is necessary for the maintenance of high viral loads and for the development of AIDS in macaques. Viruses with defective *nef* have been detected in some HIV-1-infected long-term survivors. *Nef* downregulates CD4, the primary viral receptor, and MHC class I molecules, and these functions map to different parts of the protein. *Nef* interacts with components of host cell signal transduction and clathrin-dependent protein-sorting pathways. It increases viral infectivity. *Nef* contains PxxP motifs that bind to SH3 domains of a subset of Src kinases and are required for the enhanced growth of HIV, but not for the downregulation of CD4.
LTR	Long terminal repeat, the DNA sequence flanking the genome of integrated proviruses. It contains important regulatory regions, especially those for transcription initiation and polyadenylation.

'P1.BCM.RT' reflect the origin of the clone. Thus, P = patient, V = victim and LA = Lafayette area:

```
>gi|24209939|gb|AY156734.1| HIV-1 clone P1.BCM.RT from USA reverse
transcriptase (pol) gene, partial cds
```

For the analysis in this chapter these sequences were divided up into the *env* and RT categories, with 132 *env* sequences in the file `env.fa` and 42 RT sequences in the file `rt.fa`. In the following we will be using only the RT sequences for alignment and tree construction. The *env* sequences may be handled in exactly the same way, except that the processing of these is somewhat more time-consuming, as they form a larger collection of sequences.

The ClustalW alignment of the RT sequences is initiated with:

```
% clustalw2 rt.fa
```

The output files will be `rt.aln`, the multiple alignment in the ClustalW format, and `rt.dnd`, the guide tree produced in the first step of the ClustalW algorithm, as explained in the previous chapter. This guide tree may be regarded as a phylogenetic tree, but we may also exploit ClustalW to produce a tree based on the multiple alignment produced above:

```
% clustalw2 rt.aln -tree
```

The multiple alignment in `rt.aln` is now used to construct a neighbour-joining tree, and that tree will be stored in the file `rt.ph`.

A bootstrapping procedure is also available as an option using ClustalW. The number of replicates may be specified with the `-bootstrap` parameter, like `-bootstrap=100`. The default value is 1000:

```
% clustalw2 rt.aln -bootstrap
```

The resulting tree is now stored in the file `rt.phb`. The tree in this file is the same as in `rt.ph`, but in addition will have the bootstrap information. The bootstrap values are in the range 1–1000, reflecting the number of replicates consistent with a specific branch.

The trees may be visualized using a tree viewer, such as the simple viewer `njplot` (Perriere and Gouy, 1996). As we examine these trees, we note a problem with the identifiers associated with the different sequences displayed in the tree. The FASTA definition line of the sequences has been truncated to leave only the identifiers. As a consequence, there is no information on the origin of a sequence, i.e. whether the sequence is from the victim, the patient or any of the independent HIV isolates.

As explained above, the nomenclature is such that in an expression like 'P1.BCM.RT', the first letter(s) reflect the origin. It would be best to have these descriptions in the final phylogenetic tree. To fix this, we could edit the tree file, or we could edit the original sequence file and start over with the ClustalW alignment. Here we try the second method. This is again an occasion where Perl comes in handy, and we can edit the definitions of the sequence file with a very simple script.

Code 9.1 reformat_giline.pl

```perl
#!/usr/bin/perl -w

use strict;

open(IN, 'rt.fa') or die "Could not open file\n";
while (<IN>) {
    if (/>.*clone (\S+) /) {
        print ">$1\n";
    }
    else { print; }
}
close IN;
```

Consider the script in Code 9.1. We open the file rt.fa and read one line at a time. When we encounter a definition line we capture the word which follows immediately after 'clone'. In the regular expression \S refers to any non-whitespace character.[3] The match to \S+ is captured in $1; instead of printing the original definition line we are printing this expression only. Therefore, a line like:

```
>gi|24209939|gb|AY156734.1| HIV-1 clone P1.BCM.RT from USA reverse
transcriptase (pol) gene, partial cds
```

is transformed into:

```
>P1.BCM.RT
```

We can save the results of this editing with the command:

```
% perl reformat_giline.pl > rt_reformat.fa
```

We may now apply ClustalW to the file rt_reformat.fa.[4] With bootstrapping we arrive at the tree in file rt_reformat.phb, shown in Fig. 9.3. This tree clearly

---

[3] A whitespace is any character that does not give rise to anything visible when printed, such as when you use the space bar, tab key or enter key on your keyboard. The whitespace symbol in Perl is \s. If we are convinced that spaces are the only whitespace characters in this example, we could, as an alternative to \S, make use of [ˆ] to represent everything which is not space.

[4] Careful examination of the multiple alignment reveals yet another problem. The sequences are not of equal length as the victim and patient sequences are a bit longer both on the 5′ and 3′ sides as compared to the other sequences. Ideally, we should edit the alignment to remove this extra material. We can do that with a suitable multiple sequence editor, such as Jalview.

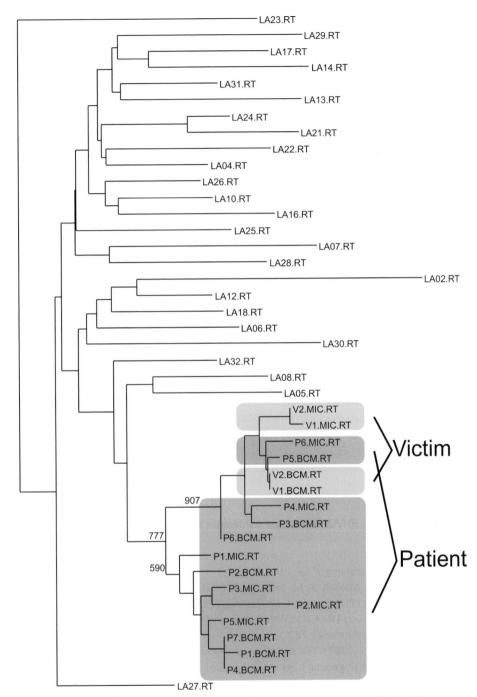

Fig. 9.3 *HIV phylogeny in a criminal case.* The tree was constructed using the neighbour-joining method. It is based on a ClustalW alignment of HIV reverse transcriptase sequences, using 1000 bootstrap replicates. Sequences are from the patient (P followed by a number, orange background), victim (V with a number, yellow background) and various isolates from the Lafayette area (LA with a number). Bootstrapping values for the victim and patient branches are shown. For instance, in 777 out of 1000 replicates all victim and patient sequences are grouped together.

shows that the victim sequences are grouped with the patient sequences. Further analysis by other methods such as those demonstrated in the next chapter will give further support to this notion. Analyses of this type were considered strong evidence in court that Maria was infected as a result of the injection by Dr Robert White in August 1994.

In 1998 Robert White was convicted of attempted second degree murder and is currently serving a 50-year prison sentence.

EXERCISES

9.1 The tree in Fig. 9.3 was obtained using sequences of the HIV reverse transcriptase gene. Construct a corresponding tree based on *env* sequences in the file `env.fa`.

9.2 Make a Perl script to derive one of the distance matrices shown in Fig. 9.1. The starting point for the script is the multiple alignment in the same figure.

9.3 The tree file `rt.ph` was produced from the ClustalW program with the command `clustalw2 rt.aln -tree`. Design a Perl script to modify that file by replacing the GenBank identifiers with identifiers like those used in Fig. 9.3. For instance, `gi|24209975|gb|AY156780.1|` is to be replaced with `LA14.RT`. You can read identifier information from the file `rt.fa` and make a hash table to be used when converting between the two categories of identifiers.

REFERENCES

Felsenstein, J. (1989). PHYLIP – Phylogeny Inference Package (Version 3.2). *Cladistics* **5**, 164–166.

Felsenstein, J. (2005). PHYLIP (Phylogeny Inference Package) version 3.6. Distributed by the author. Department of Genetics, University of Washington, Seattle.

Metzker, M. L., D. P. Mindell, X. M. Liu, *et al.* (2002). Molecular evidence of HIV-1 transmission in a criminal case. *Proc Natl Acad Sci U S A* **99**(22), 14292–14297.

Perrière, G. and M. Gouy (1996). WWW-query: an on-line retrieval system for biological sequence banks. *Biochimie* **78**(5), 364–369.

Saitou, N. and M. Nei (1987). The neighbor-joining method: a new method for reconstructing phylogenetic trees. *Mol Biol Evol* **4**(4), 406–425.

Evolution
The sad case of the Tasmanian tiger

10

From 1878 to 1896, 3482 Tiger skins were despatched from [a tannery] to London where they were made into waistcoats.

> (Norman Laird, article in *The Mercury*, 7 October 1968; cited in (Owen, 2003)

This chapter will deal further with phylogenetic analysis. We will introduce methods in addition to those of neighbour-joining and we will use a Perl script to examine taxonomy data. For these topics we will take a closer look at an extinct animal, the Tasmanian tiger.

Extinction

The Tasmanian tiger was not, in fact, a tiger; it was a dog-like marsupial animal. *Thylacine* is the more adequate scientific name. In the early twentieth century it existed only in Tasmania, and even there it was very scarce. A farmer named Wilf Batty lived in the Mawbanna district of northeastern Tasmania. On 13 May 1930 he spotted a thylacine attempting to break into his chicken coop. Batty had observed the thylacine around his house for weeks, and this day he took his rifle and shot the animal. As it happened, this was the last wild thylacine to be killed. Another specimen, most likely a female, was captured in 1933 and kept at the Hobart Zoo in Tasmania. She died on 7 September 1936, apparently as a result of neglect. The animal was kept outdoors and was not allowed access to her den, despite difficult temperatures. Ironically, the death took place only two months after the thylacine species was given full legal protection by the Tasmanian government. There are sightings of the thylacine reported after 1936, but none of these are well documented and we unfortunately need to regard the thylacine as being extinct.

Fig. 10.1 *Thylacines in National Zoo, Washington DC*. A female thylacine with three cubs arrived at the National Zoo in 1902. One of the cubs died only a few days after arrival. This photograph depicts the two surviving offspring a few years later, around 1903 to 1905. One of these animals, presumably the one in front, was a male that died in 1905. The skin of this animal was preserved and was used as a source of DNA for sequencing the mitochondrial DNA by Miller *et al.* (2009). The sequence is being used in this chapter for phylogenetic analysis. Photograph used with permission from the Smithsonian Institution Archives (SIA).

The thylacine

What kind of animal was the thylacine and why did it become extinct? It was a marsupial, and as with any species of this group of animals, the female thylacine carried its young in a pouch. The thylacine has been known by a variety of names: it was named the Tasmanian tiger because of its striped back; it has been called the Tasmanian or marsupial wolf because of its resemblance to a wolf. It was first described scientifically in the early nineteenth century. The scientific name *Thylacinus cynocephalus* means literally 'the pouched thing with a dog head'.[1] The thylacine was similar to members of the dog family in the northern hemisphere (Figs 10.1 and 10.2). Indeed, although it was a marsupial, the thylacine resembled a large dog, and a dog skull is very similar to that of a

---

[1] The word 'Thylacinus' is derived from a Greek word meaning 'pouch' and 'cynocephalus' means 'dog-headed'.

Fig. 10.2 *Thylacines at Beaumaris Zoo, Hobart, Tasmania.* This photograph is from 1911 and has often been used to illustrate the difference in size between a male and female thylacine. It is unfortunate for that discussion that both animals in this photograph are most likely males (Paddle, 2008). Photograph by permission from the Tasmanian Museum & Art Gallery.

thylacine. However, this resemblance does not imply an evolutionary relationship. Instead, the similarity between the thylacine and dogs is an example of *convergent evolution*, i.e. when independent paths during evolutionary history lead to species with similar properties.

The thylacine was a predator and it was the largest known carnivorous marsupial in modern times. Like a wolf, it hunted in family packs and would typically pursue its target animal prey, such as a wallaby or a smaller kangaroo, for a long period of time until that animal became exhausted. It had very strong jaws that could open to almost 100 degrees and there are stories that it was able to crush the skull of its prey.

Tiger history

What about the early history of the thylacine? The modern thylacine appeared about four million years ago. It was widespread in Australia and New Guinea. Over time its numbers were reduced significantly. About 4000 years ago it became near-extinct in these regions, but survived on the island of Tasmania. It has been proposed that the thylacine was out-competed by dingoes (wild dogs) that had been introduced on the Australian mainland. These dogs went feral and

Fig. 10.3 *Thylacine with a chicken*. This photograph by Harry Burrell appeared in 1921 in *The Australian Museum Magazine*. The photograph was widely spread and presumably contributed to the reputation of the thylacine as a poultry thief.

competed effectively with the thylacine for prey. However, it is also possible that human activities played a critical role in the extinction of thylacines from Australia and New Guinea.

Unfortunately, thylacines were not safe in Tasmania either. In the early nineteenth century Europeans settled there and many of them became sheep farmers. The early years of Tasmanian sheep farming were characterized by a severe lack of competence, giving rise to great losses in stock. Farmers were upset to see their sheep killed by predators. Most of these killings were presumably carried out by wild dogs and Aborigines, but thylacines became scapegoats. They had an unjustified bad reputation as a result of superstition and misinformation. They were considered bloodthirsty and were described as 'a sort of nightmare wolf' in a 1940s edition of Arthur Mee's *Children's Encyclopedia*. As late as the 1980s thylacines were incorrectly described in the scientific literature as 'blood-feeding', suggesting that they killed their prey mainly for the blood. A photograph taken in 1921 by Henry Burrell, showing a thylacine with a chicken has been widely used in publications and it is believed that this picture significantly contributed to the thylacine's reputation as a poultry thief (Freeman, 2005; Paddle, 2008) (Fig. 10.3).

In 1830 a private bounty scheme was introduced in Tasmania, which offered 'rewards for the destruction of noxious animals'. By the end of the nineteenth century sheep farmers put pressure on the Tasmanian government to compensate them for their loss of sheep, and between 1888 and 1909 the government paid trappers £1 for each dead thylacine. This was a very large amount of money at the time. The effects of this governmental bounty were fatal for the thylacine; more than 2000 animals were killed during the period.

Our planet is dominated by human activities and it is often feared that wildlife is threatened as a result of these activities. Evolution of life on our planet has been going on for 3.5 billion years and numerous species during this period have become extinct. In fact, species becoming extinct may be thought of as a natural component of evolution, and it certainly occurred long before humans entered the scene. For instance, dinosaurs surely went extinct without the help of man as humans were not around at the time. At the same time there is justified fear that human activities are seriously influencing wildlife today. The 2008 IUCN (International Union for Conservation of Nature) *Red List of Threatened Species* shows that at least 1141 of the 5487 mammals on Earth are threatened with extinction. Incidentally, one of them is the Tasmanian devil, a close relative of the thylacine. As we have seen, there is strong evidence that the thylacine itself went extinct as a result of bounty hunting, although it may not be ruled out that other factors contributed to its decline. For an exhaustive account of the extinction, see Paddle (2000).

Recent DNA analysis

We know that the thylacine was a marsupial, but how was it related to the marsupials still present today? In the more recent days of gene technology the thylacine has been somewhat 'revived' in the sense that it has been possible to amplify DNA sequences from remains of thylacines. PCR was used in 1989 to amplify DNA from two different thylacine animals (Thomas *et al.*, 1989) (for more information on PCR, see Chapter 4). One sample was comprised of two small pieces of untanned hide with attached hair that were collected in the early twentieth century. The other sample was a piece of dried muscle adhering to a bone collected before 1893. Additional animals have been analysed (Krajewski *et al.*, 1992, 1997, 2000); more recently, the complete mitochondrial genome sequences of two different thylacine individuals were reconstructed (Miller *et al.*, 2009). One of these animals was the offspring of a female animal shipped to the National Zoo in Washington, DC in 1902. The sequenced offspring, an adult male, died at this zoo in 1905 and is depicted in Fig. 10.1. DNA was obtained from the dry skin of this animal. The other animal sequenced died at London Zoo in 1893, and the almost-complete animal has since been stored in ethanol.

The complete thylacine mitochondrial sequences offer opportunities to examine the phylogeny of the thylacine. Below, we will see explicit examples of how these mitochondrial sequences may be used together with sequences of other

Table 10.1 *Marsupial mitochondrial genomes*. The table lists mitochondrial sequences of marsupial species as well as that of a wolf (*Canis lupus*) that were used for phylogenetic analysis, as shown in Fig. 10.4. In addition to the thylacine mitochondrial genome shown in this table, the mitochondrial genome of yet another thylacine animal has been sequenced (Miller *et al.*, 2009). However, in the multiple alignment used in this chapter for phylogenetic analysis, in which terminal gapped regions were removed, the two thylacine sequences are identical.

Species	GenBank accession
Monodelphis domestica	NC_006299.1
Isoodon macrourus	AF358864.1
Echymipera rufescens australis	NC_007632.1
Trichosurus vulpecula	AF357238.1
Phalanger interpositus	NC_008137.1
Macropus robustus	NC_001794.1
Dactylopsila trivirgata	NC_008134.1
Potorous tridactylus	NC_006524.1
Vombatus ursinus	NC_003322.1
Dasyurus hallucatus	AY795973.1
Phascogale tapoatafa	NC_006523.1
Sminthopsis crassicaudata	NC_007631.1
Myrmecobius fasciatus	NC_011949.1
Thylacinus cynocephalus	FJ515781.1
Notoryctes typhlops	NC_006522.1
Canis lupus	AB499825.1

animals to reconstruct a phylogenetic tree. We will also examine available taxonomy data of the species involved.

BIOINFORMATICS

Tools introduced in this chapter	
Perl	arrays and split
	formatting output
Software running in a Unix environment	Dnapars
	MrBayes

Inferring the phylogeny of marsupials

We will reconstruct phylogenetic trees on the basis of mitochondrial genome sequences. A number of aligned mitochondrial genomes are in the file `mito.fa`. The species involved are listed in Table 10.1 – scientific names are provided along with their database accession codes. The sequences were originally retrieved using the NCBI Entrez system that is discussed in Chapter 14. They have been aligned with ClustalW (see Chapters 8 and 9) and edited to remove columns with numerous gaps. The multiple alignment is in the FASTA format, in which the original content of the FASTA identifier line has been replaced with the species name.

The multiple alignment of mitochondrial sequences may now be subjected to a number of phylogenetic methods. Here, we will try out three different methods: (1) the neighbour-joining method of ClustalW; (2) maximum parsimony analysis with the PHYLIP program Dnapars; and (3) MrBayes.

Neighbour-joining was introduced in the previous chapter. A neighbour-joining tree may be created from a multiple alignment using ClustalW. At the command line, as discussed in the previous chapter, we can generate a bootstrapped tree with:

```
% clustalw2 mito.fa -bootstrap
```

The output file is `mito.phb`. This tree is depicted in Fig. 10.4.

Dnapars is a parsimony method and part of the PHYLIP package developed by Joe Felsenstein (Felsenstein, 1989, 2005; http://evolution.genetics.washington. edu/phylip). 'Parsimony' in this context means that the tree construction makes use of the principle that the tree to be preferred is the one associated with the smallest number of evolutionary changes (Felsenstein, 2004). The MrBayes program performs Bayesian inference of phylogeny using a variant of Markov chain Monte Carlo (Ronquist and Huelsenbeck, 2003). The theory of these methods will not be described here, but Appendix II offers detailed information about the practical steps involved in generating the trees in Fig. 10.4.

Returning to our thylacine story, how is the thylacine related to other marsupials? What are the conclusions of the trees that were obtained with the three different methods: neighbour-joining, Dnapars and MrBayes? First of all, we note that all three trees have more or less the same topology, although they were obtained with methods that are algorithmically very different. Therefore, there is very strong support for the observed branching. According to all trees, the thylacine is a member of the genus *Dasyuromorphia* (Fig. 10.4). This is consistent with previous studies that used only smaller pieces of sequence from mitochondrial genomes (Thomas *et al.*, 1989). Furthermore, the thylacine is seen to have a basal position in this group. The numbat (*Myrmecobius fasciatus*) seems to be its closest relative.[2]

In earlier studies of thylacine taxonomy based on morphology it was suggested that the thylacine was most closely related to an extinct group of South American carnivorous marsupials (Sinclair, 1906). Most importantly, there are

---

[2] Numbats are now only found in Western Australia. Very much unlike the thylacine, its major source of food is termites. The numbat is listed as an endangered species.

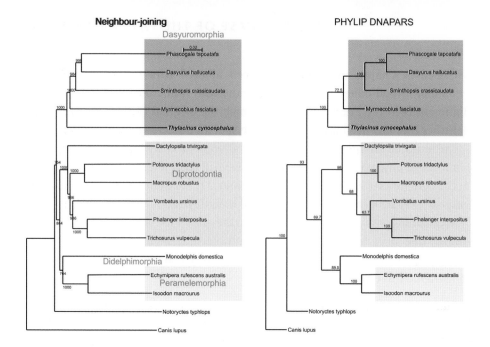

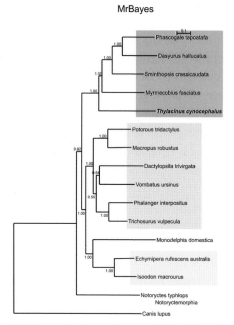

Fig. 10.4 *Phylogenetic tree of marsupials*. Three different tree construction methods were used; neighbour-joining as part of ClustalW, parsimony analysis using Dnapars of the PHYLIP package and Bayesian analysis using MrBayes. See Appendix II for more details on how these programs were used. For the neighbour-joining method, bootstrapping with 1000 replicates was used and bootstrap values are indicated at each node. In the case of Dnapars, 100 replicates were used in the bootstrap analysis. The MrBayes method generates for each branch a probability that the branch is correct. For each of the three trees, the major genus groups are shown with a shaded background. The names of these groups are shown in red in the neighbour-joining tree.

dental and pelvic features shared by these animals. However, such a relationship seems very unlikely based on more recent phylogenetic analysis in which
extant South American marsupial sequences are included (Thomas *et al.*, 1989).
Again, convergent evolution is most likely the explanation of the morphological
features shared by these animals. There is no mitochondrial genome sequence
of the Tasmanian devil (*Sarcophilus harrisii*) available at the time of writing,
but analyses using shorter mitochondrial sequences reveal that this animal is
also positioned in the *Dasyuromorphia* group (Thomas *et al.*, 1989; Krajewski
et al., 1992, 1997).

Examining taxonomy

A number of different animals, as listed in Table 10.1, are included in the
phylogenetic analysis in this chapter. What if you wanted to get an idea about
the different species and how they are classified? Of course, our phylogenetic
analysis shows how these different species are related, but what if we wanted to
know about what is *already* known about their classification? Or, as another kind
of practical problem, if you encounter a scientific name such as *Vulpes vulpes*,
how do you know what kind of species that is? We typically want to find out
about the common name. One useful database to turn to for these problems is the
NCBI Taxonomy database. This database attempts to incorporate phylogenetic
and taxonomic knowledge from a variety of sources, such as literature, other
databases and the advice of scientists. Although the NCBI Taxonomy database
should not be considered as a phylogenetic or taxonomic authority, it is useful
for many applications. We will use it to examine the taxonomy of the species
listed in Table 10.1. We could have done that by using the web interface at
NCBI, but as this is a book with a command-line and Unix-based approach
(*serious* bioinformatics) we will make use of a local text version of the database
and explore that resource with a Perl script.

In February 2012 the Taxonomy database contained 861 824 entries. A file
to be used here, `tax.txt`, contains extracts from each entry of the database
where each entry is described by values in four columns.[3] The first column
is the Taxonomy database identifier (a number). The second is the scientific
name of the species. The third is the 'GenBank common name' (note that not
all species have a common name) along with alternative names (separated by a
pipe symbol |). The fourth column has information on the *lineage*, i.e. the full

---

[3] This information may be obtained in different ways, but in this case NCBI Entrez was used. This query
system is further described in Chapter 14. All entries in a specific database may be retrieved with the
query `all[filter]`. This query was applied to the Taxonomy database and the downloaded data
was further processed with a Perl script to produce the `tax.txt` file with the contents as described.

path from the root of the taxonomic tree to the current species. This is how the species *Vulpes vulpes* is presented:

Column 1: `9627`
Column 2: `Vulpes vulpes`
Column 3: `silver fox | red fox`
Column 4: `Eukaryota; Metazoa; Chordata; Craniata; Gnathostomata;`
   `Mammalia; Laurasiatheria; Carnivora; Caniformia; Canidae; Vulpes`

In case we previously had no idea about *Vulpes vulpes*, we learn from this data that one common name is 'red fox', and we learn from its lineage that it is a mammalian carnivore. We will now make use of the information in the file `tax.txt` to learn about the species in Table 10.1. The species scientific names are in the file `mito.fa`.

The script in Code 10.1 will be used to analyse the taxonomy data. We first define two different hashes, `%lineage` and `%common`. Both of these have as keys the scientific species name, and the values are lineage and common name, respectively. After reading the species name from the file `mito.fa`, we look up the lineage and common name using these hashes. We encounter a new function, `split`, when reading the taxonomy file.

The line

```
my @columns = split(/\t/,$_);
```

has the effect that we put all the tab-separated values into an array named `@columns`. The Perl function `split` takes a string as input and splits it up into substrings, using a specific separator that may be a distinct character or a regular expression. It returns these substrings as an array. In this case the separator is defined by the pattern `/\t/`, where the `\t` represents a tab. The string to be processed is `$_`, which in this case is the line read from the file. In fact, as the string is `$_`, there is no need to specify it as an argument to split. Thus, we could have said:

```
my @columns = split(/\t/);
```

Incidentally, the default separator to `split` is the blank space character, so if that was our separator we could have used an even simpler expression:

```
my @columns = split;
```

The different elements of the `@columns` array are referred to as `$columns[0]`, `$columns[1]`, etc. In this case the taxonomy identifier will be stored in `$columns[0]`, the scientific name in `$columns[1]`, etc.

A new command shown in Code 10.1 is `format` and `write` to obtain the output in a specific format. Before discussing the `format` example in the script, consider first a simple example in which we use the following lines of code to produce an output of a hash:

```
my %species = (
    'Vulpes vulpes'                  => 'red fox',
    'Procavia capensis habessinica' => 'Abyssinian hyrax',
    'Sus scrofa'                     => 'pig'
);
foreach my $key ( keys %species ) {
    print "$key\t$species{$key}\n";
}
```

The output from this code is:

```
Procavia capensis habessinica    Abyssinian hyrax
Vulpes vulpes    red fox
Sus scrofa       pig
```

The formatting is not ideal as we may want to have the common names vertically aligned. A solution is to make use of format. The format is defined in a separate section and then printing is achieved with the write function:

```
my $sciname;
my $common;
format =
@<<<<<<<<<<<<<<<<<<<<<<<<<   @<<<<<<<<<<<<<<<<<<<<<<<<<<<<
$sciname, $common

.
my %species = (
    'Vulpes vulpes'                  => 'red fox',
    'Procavia capensis habessinica' => 'Abyssinian hyrax',
    'Sus scrofa'                     => 'pig'
);
foreach my $key ( keys %species ) {
    $sciname = $key;
    $common = $species{$key};
    write;
}
```

The output of this script is

```
Procavia capensis habessini    Abyssinian hyrax
Vulpes vulpes                  red fox
Sus scrofa                     pig
```

The actual report format is in this case specified with the section:

```
format =
@<<<<<<<<<<<<<<<<<<<<<<<<<<    @<<<<<<<<<<<<<<<<<<<<<<<<<<<<<
$sciname, $common
```

The second line is a 'picture' line specifying the format of one output line. The two *fieldholders* @<<<<<<<<<<<<<<<<<<<<<<<<<< specify that exactly 27 characters are be used in those two fields. The '<' symbol indicates that the text fields should be left-justified. The line following the fieldholders has information about what variables are to be in the fieldholders as specified in the previous line. Finally, the format definition is terminated by a line containing only a dot (.). The format may be specified anywhere in the script, but is invoked with the write function.

Another example of formatting is shown in Code 10.1. Here we want the following type of format, in which fields COMMON and LINEAGE may be divided into several lines:

```
SPECIES: Trichosurus vulpecula
COMMON:  common brush-tailed possum | silver-gray brushtail
            possum
LINEAGE: Eukaryota; Metazoa; Chordata; Craniata; Vertebrata;
            Euteleostomi; Mammalia; Metatheria; Diprotodontia;
            Phalangeridae; Trichosurus
```

A line like this:

```
^<<<<<<<<<<<<<<<< ~
```

specifies as before the number of characters, but the caret (^) in front specifies a 'filled field' fieldholder. This allows the breaking up of the text into conveniently sized lines at word boundaries, wrapping the lines as needed. The tilde symbol (~) is there to prevent blank lines from being printed.

Code 10.1 tax.pl

```perl
#!/usr/bin/perl -w

use strict;

my $taxfile = 'tax.txt';
my %lineage;
my %common;

open(TAX, $taxfile) or die "Could not open file\n";
while (<TAX>) {
    chomp;
    my @columns = split(/\t/, $_);
```

```
        $lineage{ $columns[1] } = $columns[3];
        $common{ $columns[1] } = $columns[2];
}
close TAX;

my @species = ();
my $infile = 'mito.fa';
my $species;
my $common;
my $lineage;

format =
---------------------------------------------------------------------
SPECIES: @<<<<<<<<<<<<<<<<<<<<<<<<<<<<<<<<<<<<<<<<<
COMMON:  ^<<<<<<<<<<<<<<<<<<<<<<<<<<<<<<<<<<<<<<<<< ~
         ^<<<<<<<<<<<<<<<<<<<<<<<<<<<<<<<<<<<<<<<<< ~
LINEAGE: ^<<<<<<<<<<<<<<<<<<<<<<<<<<<<<<<<<<<<<<<<< ~
         ^<<<<<<<<<<<<<<<<<<<<<<<<<<<<<<<<<<<<<<<<< ~
         ^<<<<<<<<<<<<<<<<<<<<<<<<<<<<<<<<<<<<<<<<< ~

.

open(FASTA, $infile) or die "Could not open file\n";
while (<FASTA>) {
    if (/^>(.*)/) {
        $species = $1;
        $species =~ s/_/ /g;    # replace underscore with blank space
        $common = $common{$species};
        $lineage = $lineage{$species};
        write;
    }
}
close FASTA;
```

The output from Code 10.1 is truncated here to show only the first four species:

```
---------------------------------------------------------------------
SPECIES: Dasyurus hallucatus
COMMON:  northern quoll
LINEAGE: Eukaryota; Metazoa; Chordata; Craniata; Vertebrata;
         Euteleostomi; Mammalia; Metatheria; Dasyuromorphia;
         Dasyuridae; Dasyurus
```

Table 10.2 *Marsupial taxonomy.* Information is displayed about marsupial species as learned from the script shown in Code 10.1, which analyses taxonomy information.

Species	Common name	Group
Monodelphis domestica	Grey short-tailed opossum	Didelphimorphia
Isoodon macrourus	Northern brown bandicoot	Peramelemorphia
Echymipera rufescens australis	Long-nosed echymipera	Peramelemorphia
Trichosurus vulpecula	Common brushtail possum	Diprotodontia
Phalanger interpositus	Stein's cuscus	Diprotodontia
Macropus robustus	Eastern wallaroo	Diprotodontia
Dactylopsila trivirgata	Striped possum	Diprotodontia
Potorous tridactylus	Long-nosed potoroo	Diprotodontia
Vombatus ursinus	Common wombat	Diprotodontia
Dasyurus hallucatus	Northern quoll	Dasyuromorphia
Phascogale tapoatafa	Brush-tailed phascogale	Dasyuromorphia
Sminthopsis crassicaudata	Fat-tailed dunnart	Dasyuromorphia
Myrmecobius fasciatus	Numbat (banded ant eater)	Dasyuromorphia
Thylacinus cynocephalus	Tasmanian tiger	Dasyuromorphia
Notoryctes typhlops	Southern marsupial mole	Notoryctemorphia
Canis lupus	Wolf	(Eutheria)

```
-----------------------------------------------------------------
SPECIES: Phascogale tapoatafa
COMMON:  brush-tailed phascogale
LINEAGE: Eukaryota; Metazoa; Chordata; Craniata; Vertebrata;
         Euteleostomi; Mammalia; Metatheria; Dasyuromorphia;
         Dasyuridae; Phascogale
-----------------------------------------------------------------
SPECIES: Sminthopsis crassicaudata
COMMON:  fat-tailed dunnart
LINEAGE: Eukaryota; Metazoa; Chordata; Craniata; Vertebrata;
         Euteleostomi; Mammalia; Metatheria; Dasyuromorphia;
         Dasyuridae; Sminthopsis
-----------------------------------------------------------------
SPECIES: Myrmecobius fasciatus
COMMON:  numbat
LINEAGE: Eukaryota; Metazoa; Chordata; Craniata; Vertebrata;
         Euteleostomi; Mammalia; Metatheria; Dasyuromorphia;
         Myrmecobiidae; Myrmecobius
```

Based on the information from this script we have compiled a new table on marsupials (Table 10.2). What do we learn from the output of the script and this table? First, all species except the wolf are members of the group Metatheria, meaning they are marsupials. The wolf belongs to the group Eutheria, i.e. placental mammals. We also learn the common names for all of the animals. Finally, we see that the NCBI taxonomy is consistent with the phylogenetic trees as shown in Fig. 10.4. For instance, all members of Diprotodontia are in the same branch in all three trees of this figure.

EXERCISES

10.1 Design a Perl script to examine the taxonomy information in the file `tax.txt`. Consider the following questions:

(a) How many species are from Eukaryota, Archaea and Bacteria, respectively?

(b) How many marsupials are listed?

(c) The Tasmanian devil is a species closely related to the thylacine. What is the scientific name of this animal?

(d) How many species have a common name?

How may the same questions be addressed using Unix commands such as `grep`, `cut`, `sort` and `uniq`? See Appendix I and Chapter 14 for brief information on these commands.

10.2 Use the alignment of FOXP2 mRNA sequences from Exercise 8.2 and MrBayes and Dnapars to infer phylogenetic trees. See Appendix II for more details on these programs.

10.3 The sequence of the cytochrome b gene from the Tasmanian devil, a GenBank nucleotide database entry with accession M99465.1 (or SCHMTCYTB), is available as the file `tasmanian_devil.fa` (see the web resources for this book). Construct a multiple sequence alignment containing this sequence as well as the mitochondrial sequences in the file `mito.fa`, used in this chapter. To construct the alignment you may use ClustalW with the options `-profile1=mito.fa` and `-profile2=tasmanian_devil.fa`, which are used to align two profiles (in this way you will reuse the existing alignment `mito.fa` and save time). Edit the resulting alignment with a multiple sequence editor like Jalview to remove all pieces of sequence that are not part of the cytochrome b gene. Construct a phylogenetic tree on the basis of the final alignment. What species are most closely related to the Tasmanian devil according to your tree?

REFERENCES

Felsenstein, J. (1989). PHYLIP – Phylogeny Inference Package (Version 3.2). *Cladistics* **5**, 164–166.

Felsenstein, J. (2004). *Inferring Phylogenies*. Sunderland, MA, Sinauer Associates.

Felsenstein, J. (2005). PHYLIP (Phylogeny Inference Package) version 3.6. Distributed by the author. Department of Genetics, University of Washington, Seattle.

Freeman, C. (2005). Is this picture worth a thousand words? An analysis of Harry Burrell's photograph of a thylacine with a chicken. *Aust Zoo* **33**, 1–16.

Krajewski, C., M. J. Blacket and M. Westerman (2000). DNA sequence analysis of familial relationships among Dasyuromorphian marsupials. *J Mamm Evol* **7**(2), 95–108.

Krajewski, C., L. Buckley and M. Westerman (1997). DNA phylogeny of the marsupial wolf resolved. *Proc Biol Sci* **264**(1383), 911–917.

Krajewski, C., A. C. Driskell, P. R. Baverstock and M. J. Braun (1992). Phylogenetic relationships of the thylacine (Mammalia: Thylacinidae) among dasyuroid marsupials: evidence from cytochrome b DNA sequences. *Proc Biol Sci* **250**(1327), 19–27.

Miller, W., D. I. Drautz, J. E. Janecka, *et al.* (2009). The mitochondrial genome sequence of the Tasmanian tiger (*Thylacinus cynocephalus*). *Genome Res* **19**(2), 213–220.

Owen, D. (2003). *Thylacine: The Tragic Tale of the Tasmanian Tiger*. Sydney, Allen & Unwin.

Paddle, R. (2000). *The Last Tasmanian Tiger: The History and Extinction of the Thylacine*. Cambridge and New York, Cambridge University Press.

Paddle, R. (2008). The most photographed of thylacines: Mary Robers' Tyenna male – including a response to Freeman (2005) and a farewell to Laird (1968). *Aust Zoo* **34**(4), 459–470.

Ronquist, F. and J. P. Huelsenbeck (2003). MrBayes 3: Bayesian phylogenetic inference under mixed models. *Bioinformatics* **19**(12), 1572–1574.

Sinclair, W. J. (1906). *Reports of the Princeton University Expeditions to Patagonia, 1896–1899. Volume 4. Palaeontology. Part 3. Marsupialia of the Santa Cruz Beds*. Princeton, NJ, Princeton University Press.

Thomas, R. H., W. Schaffner, A. C. Wilson and S. Paabo (1989). DNA phylogeny of the extinct marsupial wolf. *Nature* **340**(6233), 465–467.

A function to every gene
Termites, metagenomics and learning about the function of a sequence

11

Great fleas have little fleas upon their backs to bite 'em,
And little fleas have lesser fleas, and so ad infinitum.

(Augustus De Morgan, 1806–1871)

For this chapter, as well as Chapters 12 and 13, we turn to the important genomics and bioinformatics problem of identifying biological function based on nucleotide and amino acid sequences.

Assigning function based on sequence similarity

A common problem in molecular biology is that you are faced with a gene or a gene product and you have no clue from experimental studies as to its function. In this context a critical contribution of bioinformatics is to attribute the sequence of a gene or a gene product a function. As one example, a genome sequencing project may give rise to tens of thousands of predicted protein sequences. In such a case we want to assign as many of these as possible a biological function using computational tools. In this manner we avoid many laborious wetlab experiments. In addition to genome sequencing projects, there are other more specialized situations where we want to find functions of genes. For instance, we could identify genes as being related to a specific genetic trait or disease, or a set of genes as being expressed under certain conditions.

A number of computational tools are available to predict a biological function associated with a protein sequence. In this chapter we will see an example in which we assign a function to a protein based on sequence similarity. Consider the human gene encoding the protein BRCA1, originally sequenced in 1994 (Miki *et al.*, 1994). It was found to be related in sequence to a yeast protein RAD9. This yeast protein is involved in cell cycle control. This observation gave scientists a hint about possible roles of the BRCA1 gene. We see here an example of inferring a function based on a homology relationship to a protein that has already been functionally characterized. We will see yet another example of this situation in this chapter, where we will make use of BLAST to identify a homology relationship. We already encountered BLAST in the context of the BCR–ABL fusion protein in Chapter 7.

The example selected here is concerned with the assignment of function to DNA sequences of a group of insects, the termites. These sequences were obtained in a *metagenomics* project. Before we come to our actual investigation of function, the concept of metagenomics will be introduced, and I will explain why termites are very interesting creatures.

Metagenomics

In the early days of DNA sequencing, scientists would try and isolate a smaller portion of DNA and have that region sequenced. Typically, this region corresponded to a single gene or very few genes. As sequencing technology improved, larger and larger pieces of DNA were subject to analysis. Eventually, entire genomes were sequenced, even very large genomes such as that of humans (see also Chapters 18–20). Still, the technology of conventional DNA sequencing projects is such that the DNA under study needs to be purified to contain material as homogeneous as possible. For example, the sample DNA typically represents one species only. This type of sequencing assumes that the organism under study may be isolated in pure form in the laboratory.[1] However, there are many instances where this is not possible. For instance, most bacteria on our planet have so far not been cultivated in the laboratory. However, as current sequencing technology is extremely powerful, excessive DNA sequence information may be obtained in only one experiment. This means that in many cases we may also subject a mixture of DNA molecules from different sources to sequencing. Such an approach may be used to learn about what species are present in a specific environment. *Metagenomics* or *environmental sequencing* refers to a procedure in which a sample is taken from some habitat, and then all DNA from this sample is isolated and sequenced. One is thereby able to reach a conclusion on the distribution of species in that environment, as well as on the properties of the species based on identification of individual genes.

Craig Venter was one of the first scientists to pursue metagenomics and has sampled ocean seawater from a large number of locations around the world (Venter *et al.*, 2004; Nealson and Venter, 2007; Rusch *et al.*, 2007). These studies have revealed thousands of bacterial species, many of which had not previously

---

[1] This discussion applies mainly to microorganisms. If we are considering large animals like humans, there is no need to isolate them in a laboratory.

been observed. The studies of Venter were concerned with near-surface seawater, where sunlight reaches and photosynthesis takes place. However, investigations of deep seawater in the North Atlantic identify an even more extensive biodiversity in these ecological niches. Tens or even hundreds of thousands of different bacterial species may be found in a single litre of deep seawater (Sogin *et al.*, 2006; Huber *et al.*, 2007; Agogue *et al.*, 2011).

In another metagenomics project scientists went 2.8 km below the Earth's surface in a South African gold mine and collected microorganisms (Chivian *et al.*, 2008). This particular environment is rather hostile to humans as there is no sunlight, the temperature is about 60 °C and there is radioactive radiation from uranium. In a metagenomics project we often observe a wealth of different species in the habitat being studied. However, in the case of the gold mine samples, the DNA was surprisingly homogeneous. Nearly all of the isolated DNA was derived from a previously unknown bacterium named *Candidatus Desulforudis audaxviator*. Its complete genome could be assembled from the sequences obtained in the metagenomics project. It is remarkable as it has a biochemical capacity to produce all necessary compounds without the aid of other organisms in its environment. The source of energy and nutrients to the system is ultimately chemical compounds generated by radioactive irradiation. A whole ecosystem is in this case apparently encoded within a single genome.

The other genome

Not only in the sea and underground are there bacterial communities that are interesting to explore. Many animals harbour bacteria. In fact, bacteria are, in many animals, quantitatively and functionally important. For instance, in the human body there is in the order of 100 trillion (10^{14}) bacterial cells, which is ten times the number of 'human' cells. There are at least 500–1000 different bacterial species present in the human body. The bacterial genomes present in the human body have sometimes been referred to as the 'other' genome, to highlight the fact that not only the 'normal' human genome is important in human genomics.

There are bacteria on all surfaces in the human body that are exposed to the environment, but the vast majority of bacteria are found in the large intestine of the gut. We benefit a lot from gut bacteria. For example, bacteria carry enzymes for carbohydrate digestion that human cells are not able to produce. For this reason, the bacteria are able to digest carbohydrates that are left undigested by human cells. Gut bacteria also protect against pathogens – known as the 'barrier effect'. Non-pathogenic bacteria that are normally present in the gut effectively out-compete pathogenic bacteria when it comes to nutrition and attachment sites to the epithelium of the intestine. There is also a role for the gut flora in the context of bowel diseases and obesity (Guarner and Malagelada, 2003; Turnbaugh *et al.*, 2009).

The human gut microbial flora has been explored by metagenomics sequencing. In one study, faecal samples of 124 European individuals were analysed, thus revealing a total of around three million different bacterial genes (Qin *et al.*, 2010). This number is impressive, considering that the human genome contains only around 21 000 protein-coding genes.

Other animals than man have an extensive bacterial flora, but the bacterial population is different depending on what the animal feeds on. For instance, herbivores such as cattle and sheep make use of bacteria with the capacity to digest cellulose. This digestion takes place in the rumen. However, for this chapter we are concerned with the bacterial flora of an entirely different group of organisms. It is time to introduce termites, herbivore insects that feed on wood.

Termites and cellulose digestion

Termites are insects that typically feed on dead plant material. One important reason why termites are highly successful is that they are able to digest wood. Each termite is able to digest in the order of 0.1 mg of material every day. This may not sound like much, but remember that there might be 500 000 individuals in a single termite colony.

From an evolutionary perspective the termites are most closely related to wood-eating cockroaches. Termites and cockroaches are the only animals that are able to digest wood. A very primitive termite is the Australian *Mastotermes darwiniensis*, also referred to as the giant northern termite or the Darwin termite[2] (Fig. 11.1). A colony of these termites may have more than 100 000 individuals, and it is one of the most destructive termite species in the world. Like many other termites, it may cause severe damage to buildings.

Fig. 11.1 *A destructive termite species.* The Australian termite *Mastotermes darwiniensis* is also referred to as the giant northern termite or Darwin termite and is depicted here in a drawing from Desneux (1904).

How are termites able to digest cellulose? Enzymes referred to as *cellulases* are needed for breakdown of cellulose. They are able to hydrolyse beta-1,4 glycosidic bonds formed between the glucose units of cellulose (Fig. 11.2).

We have seen that herbivorous mammals like cows depend on bacteria in their guts that are able to digest cellulose. In the case of termites that feed on wood, they are known to produce at least one cellulase encoded by the termite genome. However, cellulose digestion in these species is also aided by a wealth of

---

[2] Named after the city of Darwin in Australia, which is in turn named after Charles Darwin.

bacteria and flagellate protozoa (unicellular eukaryotes) in the termite gut. There is a remarkable flagellate protozoon in *M. darwiniensis* named *Mixotricha paradoxa*. This protozoon is entirely restricted to that termite species. When first examined under the microscope it seemed to have both cilia and flagellates, with the cilia responsible for the motility of the *Mixotricha*. Surprisingly, the cilia of the protozoon were eventually shown to be bacteria that exist in close symbiosis with the *Mixotricha*! Richard Dawkins is reminded in this context of a verse of the mathematician Augustus De Morgan, cited in the introduction to this chapter (Dawkins, 2004).

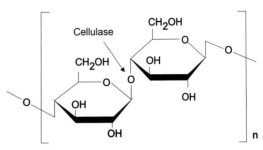

Fig. 11.2 *Cellulase action*. In cellulose, glucose units are joined with 1,4-beta-glycosidic bonds. Breakdown of cellulose is mediated by cellulase enzymes such as endo-1,4-beta-glucanase that catalyse hydrolysis of the glycosidic bond (arrow).

The role of the flagellate protozoa with respect to the digestion of cellulose has not been clarified in detail so far, but metagenomics has now made it possible to explore in great detail the genes that reside within termites, independent of whether they originate from the termite itself, or from protozoa and bacteria residing within the insect. For the discussion in this chapter we will be examining one such specific analysis.

Apart from the fact that it is crucial from a basic scientific point of view to understand how cellulose is digested, there are also important biotechnology applications of cellulases. Cellulose is one interesting potential energy source as an alternative to fossil fuels. For this reason there is an interest in identifying novel cellulases that may be exploited at an industrial scale to digest plant cellulose to sugars. To aid in this identification, many projects have been initiated to sequence the genomes of many organisms that encode cellulase enzymes (Rubin, 2008). Either individual microbial species have been analysed or, as in the example explored in this chapter, metagenomics has been used to characterize a whole population of gut bacteria residing within a herbivorous animal.

BIOINFORMATICS

Assigning function to termite sequences

For the bioinformatics part of this chapter we will attempt to assign function to a set of sequences that originate from termites. We will make use of the results of one

Tools introduced in this chapter	
Software running in a Unix environment	More on BLAST
Perl	Analysing output from BLAST

Table 11.1 *Versions of BLAST.* Descriptions are from NCBI documentation.

BLAST version	Description
blastp	Compares an amino acid query sequence against a protein sequence database
blastn	Compares a nucleotide query sequence against a nucleotide sequence database
blastx	Compares a nucleotide query sequence translated in all reading frames against a protein sequence database
tblastn	Compares a protein query sequence against a nucleotide sequence database dynamically translated in all reading frames
tblastx	Compares the six-frame translations of a nucleotide query sequence against the six-frame translations of a nucleotide sequence database

specific investigation of the termite hindgut (Warnecke *et al.*, 2007). In this case, 'higher' termites were studied. These are termites that do not possess cellulase-producing flagellates, but to a large extent rely on bacteria for the digestion of wood. The termites under study were collected in Costa Rica, and they are most closely related to previously characterized termites named *Nasutitermes ephratae* and *N. corniger*.

A total of 56 447 sequences were obtained in this project; for the coming practical exercise these are in the file `termite_metagenomics.fa`.[3] These sequences could not be effectively merged into longer sequences. We could analyse one sequence at a time and try to figure out what gene is associated with it. In particular, we want to identify a probable function of the gene. How do we do this? We will use BLAST. Remember, we used BLAST in Chapter 7 to examine the fusion protein BCR–ABL and showed how BLAST may be used to reveal orthologues and paralogues. Here, we will use it to find a function of a previously anonymous sequence. As a demonstration, we will select one of the 56 447 metagenomics sequences, a sequence stored in the file `sample.fa`.

When BLAST was introduced earlier (Chapter 7) we made use of *blastp*, which is used when a protein query sequence is used to search a protein database. There are a few other flavours of BLAST, however (see Table 11.1), and the one of interest right now is *blastx*, which uses a nucleotide sequence as the query but searches a protein database. This variant of BLAST will consider all possible translation products of the query sequence and test all of those against the different protein sequences in the database. Remember, there are six possible reading frames to each nucleotide sequence, as discussed in Chapter 1. Why

---

[3] These sequences may be downloaded using NCBI Entrez (Chapter 14). The majority of sequences are a collection of nucleotide sequences referred to in the master record with accession ABDH00000000.1.

do we carry out the search in this manner and why do we not search against a nucleotide database? The most important reason is that homologous protein sequences are more similar to each other than the corresponding nucleotide sequences. This has the consequence that we are more likely to capture an evolutionary relationship when considering proteins as compared to DNA. And why are protein sequences more similar than the corresponding DNA sequences? The most important reason is that the genetic code is degenerate. Thus, one amino acid may be represented by more than one codon. For instance, two proteins may be identical at the level of amino acid sequence but clearly different at the level of nucleotide sequence.

We carry out the BLAST search by typing a command similar to the one we used in Chapter 7.

```
% blastx -db swissprot -query sample.fa -out sample.blastx
```

In this command line it is assumed that the files referred to by the parameters, including the Swiss-Prot binary database files, are in the current directory. This BLAST search is somewhat extensive and you should not expect it to complete immediately. Once it is finished we can examine the output file `sample.blastx`:

```
. . .
Query= gi|161051957|gb|ABDH01022858.1| Termite gut metagenome
tgut2b_Contig23490, whole genome shotgun sequence
Length=954
                                                 Score    E
Sequences producing significant alignments:     (Bits)  Value

sp|P54937.1|GUNA_CLOLO RecName: Full=Endoglucanase A; ...  265  2e-070
sp|Q12647.1|GUNB_NEOPA RecName: Full=Endoglucanase B; ...  243  1e-063
sp|P10477.1|GUNE_CLOTM RecName: Full=Endoglucanase E; ...  241  6e-063
sp|P23661.1|GUNB_RUMAL RecName: Full=Endoglucanase B; ...  233  1e-060

. . .
```

The best hits in the Swiss-Prot database are proteins described as 'Endo-1,4-beta-glucanase'. This is in fact nothing but cellulase proteins, being able to hydrolyse 1,4 glycosidic bonds (Fig. 11.2). One important conclusion from this BLAST search is therefore that *our sequence most likely encodes a cellulase*. Hence, we have been able to assign a function to a sequence, which is a major theme of this chapter.

In the BLAST search above we analysed only one sequence of the metagenomics project. We could go further in our studies and carry out a BLAST search of the whole collection of sequences from the metagenomics project (in the file named `termite_metagenomics.fa`), although it will of course take a while for this analysis to complete. Enter (in one line):

```
% blastx -db swissprot -query termite_metagenomics.fa
-out all2swissprot.blastx
```

If you try this BLAST search you will note that the output file will be fairly large. One way of avoiding that is to restrict the number of descriptions and alignments to include in the output file. We might decide that we are satisfied with the two best hits – if so, we can add the parameters `-num_descriptions 2` and `-num_alignments 2` to indicate that we want to include only the best two hits in the output from the BLAST descriptions and alignments.

We return to our result of the BLAST search in the file `sample.blastx`. What are the species encoding the top hits from the BLAST search? The last five characters of any Swiss-Prot entry represent the species in an abbreviated form. In the case of a name like 'HUMAN' it is easy to guess the identity of the species, but in many other cases it is more difficult. In this case we can look up the entries in Swiss-Prot to see that most of the species are eubacteria (*Clostridium longisporum* (CLOLO), *Clostridium thermocellum* (CLOTM), *Ruminococcus albus* (RUMAL)), and only one is a fungus (*Neocallimastix patriciarum*, a rumen fungus (NEOPA)). However, it is important to note that we cannot reach a conclusion as to the species from which our query sequence has originated.

What if we wanted to know if there are more potential cellulases in the collection of sequences from the termite project? Again, we take advantage of BLAST, but now we use a query sequence which is the best hit from our previous search (the sequence from *Clostridium longisporum)* and we use it to search against a database of the termite sequences. Here, we need to use *tblastn*, which uses a protein sequence as the query in searches against a nucleotide database. It works like blastx, but the other way round, i.e. the protein sequence in the query is compared to all possible translation products of the nucleotide database.

First, the database of termite sequences needs to be formatted for BLAST searches (enter in one line):

```
% makeblastdb -in termite_metagenomics.fa -dbtype nucl
-parse_seqids
```

The option `-dbtype nucl` indicates that we are formatting a nucleotide sequence database.

As the next step we retrieve the query sequence from Swiss-Prot with the BLAST utility `blastdbcmd`:

```
% blastdbcmd -db swissprot -entry GUNA_CLOLO > guna_clolo.fa
```

Then we carry out the tblastn search. In this case we want to use a different output format to what we did before. We want to have the output in a tabular

format that is more easily parsed by a Perl script. Thus, the parameter `-outfmt 6`
is an instruction to produce the output in such a format (enter in one line):

```
% tblastn -query guna_clolo.fa -db termite_metagenomics.fa
-outfmt 6 -out guna_clolo_against_termite_metagenomics.tblastn
```

We could now exploit the resulting output to address several questions. For
instance, how many sequences have a significant match to the query sequence?
How many of the hits cover a significant part of the query sequence? In a case
like this, a script is useful to extract the features we want. Such an example
is shown in Code 11.1. It is worth noting that this simple type of script may
be used to analyse almost any BLAST report, and as BLAST is one of the most
widely used bioinformatics applications, such scripts are highly useful.

The Perl script in Code 11.1 is designed to print all hits with an E-value of less
than 10^{-10} (1E−10). It reads one line at a time from the BLAST output.[4] Each
line is a set of tab-separated values that are the (1) name of query sequence; (2)
name of the database sequence; (3) percentage identity; (4) alignment length;
(5) number of mismatches; (6) number of gap openings; (7) query start position;
(8) query end position; (9) subject start position; (10) subject end position; and
(11) E-value.[5]

As in the previous chapter, we use the function `split` to create an array of
the tab-separated values:

```
my @columns = split (/\t/);
```

The different elements of the `@columns` array are referred to as `$columns[0]`,
`$columns[1]`, etc. More specifically, the database identifiers are in `$columns[1]`
and the query begin and end positions in `$columns[6]` and `$columns[7]`, respectively. The E-value will be stored in `$columns[10]`.

Code 11.1 parse_blast.pl

```
#!/usr/bin/perl -w

use strict;

# print header information
print "id\teval\tbegin\tend\n";

open(IN, 'guna_clolo_against_termite_metagenomics.tblastn')
```

---

[4] In Appendix III there is information on an alternative method to parse the output from BLAST
searches, a method exploiting the Perl module SearchIO. In this case is it assumed that the BLAST
report is in the default output format.
[5] The output with '`-outfmt 6`' may actually be customized; see `blastn -help` for more
information.

```
    or die "Could not open file\n";
while (<IN>) {
    chomp;
    my @columns = split(/\t/);
    if ( $columns[10] < 1E-10 ) {
        print "$columns[1]\t$columns[10]\t$columns[6]\t$columns[7]\n";
    }
}
close IN;
```

The first lines of the output from Code 11.1 are:

id	eval	begin	end
gi|158451944|gb|EF428063.1|	1e-082	38	381
gi|158451958|gb|EF428070.1|	1e-082	38	381
gi|158451956|gb|EF428069.1|	2e-081	25	379
gi|158451964|gb|EF428073.1|	3e-077	43	380

The data is shown graphically in Fig. 11.3, in which the green line represents the extent of the query sequence and the red lines show the matching to the database sequence based on the begin and end positions, as in the output above.

We learn from this data that there are a number of sequences in the metagenomics project that match the *Clostridium* query sequence. To be more precise, there are 44 sequences that are below the E-value threshold 10^{-10} (1E–10). Another thing to note from the output in Fig. 11.3 is that the C-terminal end of the query protein does not have a match to any of the metagenomics sequences. This type of result could be artefactual if the sensitivity of BLAST was not enough to reveal the similarity in this part.[6] However, it could also indicate that this particular region of the query is lacking in the metagenomics sequences. In this case the C-terminal region in the *C. longisporum* cellulase protein is a *cellulose-binding domain* which is characteristic of a group of cellulases. It would seem from this result that it is present in the sequence from *C. longisporum*, but absent in all the termite cellulase sequences from the metagenomics project. We therefore see here an example of how BLAST may be used to analyse the *domain structure* of proteins. In the next chapter we will see even more effective methods of doing so.

---

[6] Alternatively, it could be because a *low-complexity sequence* was filtered out in the query sequence in this region. Regions of low-complexity sequence have an unusual composition and may often be identified by visual inspection. For example, the protein sequence PPPPPPPKDKKKKDDKK has low complexity. Filters are used in BLAST by default to remove low-complexity sequence because it can cause artefactual hits. If you want to avoid filtering of such sequences, see the BLAST documentation for more information.

BLAST hits

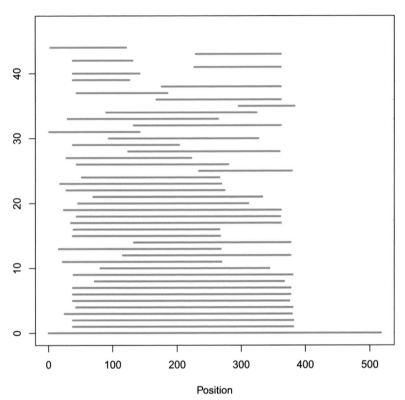

Position

Fig. 11.3 *Length distribution of BLAST hits.* BLAST was used to search a database of metagenomics nucleotide sequences, using a *Clostridium longisporium* cellulase sequence as the query. The green line at the bottom corresponds to the query sequence, with a length of 517 amino acids. Red lines, from the bottom and up, are hits against the database to the query sequence. Note that none of the hits match the C-terminal region of the query sequence, which is known to contain a cellulose-binding domain. The plot was obtained using an R script (see the web resources for this book).

EXERCISES

11.1 Perform the blastx search as referred to in this chapter with `sample.fa` as the query and `swissprot` as the database, but use the option to provide tabular output. Use the type of script in Code 11.1 to produce information about what hits are found with an E-value less than 1E–20. For each of these hits print the Swiss-Prot identifier (e.g. 'GUNA_CLOLO') as well as the begin and end positions for the query and database ('subject') sequences.

11.2 Consider the tblastn search carried out in this chapter:

```
% tblastn -query guna_clolo.fa -db termite_metagenomics.fa
-outfmt 6 -out guna_clolo_against_termite_metagenomics.tblastn
```

Modify the output format of this search so the aligned part of the subject (database) *sequences* are printed. For example, you may use the option `outfmt '6 qseqid sseqid evalue sseq'` (see `tblastn -help` for more information). Using the output from BLAST, extract the sequences with a Perl script to form a file with the sequences in a FASTA format. Include only sequences with an E-value less than 10^{-10}. Finally, produce a ClustalW alignment of these sequences together with GUNA_CLOLO. Compare the results to the output in Fig. 11.3.

11.3 Consider a more extensive BLAST search. Extract the first 200 sequences of the file `termite_metagenomics.fa` and save them in a file `200.fa`. Then perform a blastx search (enter in one line):

```
% blastx -query 200.fa -db swissprot -out swissprot.blastx
-num_descriptions 2 -num_alignments 2 -outfmt 6
```

This search will take a while to complete. Use Perl code to analyse the results. How many of the query sequences have a hit to Swiss-Prot with an E-value less than 1E–20?

REFERENCES

Agogue, H., D. Lamy, P. R. Neal, M. L. Sogin and G. J. Herndl (2011). Water mass-specificity of bacterial communities in the North Atlantic revealed by massively parallel sequencing. *Mol Ecol* **20**(2), 258–274.

Chivian, D., E. L. Brodie, E. J. Alm, *et al.* (2008). Environmental genomics reveals a single-species ecosystem deep within Earth. *Science* **322**(5899), 275–278.

Dawkins, R. (2004). *The Ancestor's Tale: A Pilgrimage to the Dawn of Evolution*. Boston, MA, Houghton Mifflin.

Desneux, J. (1904). *Genera Insectorum. 25, Isoptera: Fam. Termitidae/Jules Desneux*. Bruxelles, Verteneuil & Desmet.

Guarner, F. and J. R. Malagelada (2003). Gut flora in health and disease. *Lancet* **361**(9356), 512–519.

Huber, J. A., D. B. M. Welch, H. G. Morrison, *et al.* (2007). Microbial population structures in the deep marine biosphere. *Science* **318**(5847), 97–100.

Miki, Y., J. Swensen, D. Shattuck-Eidens, *et al.* (1994). A strong candidate for the breast and ovarian cancer susceptibility gene BRCA1. *Science* **266**(5182), 66–71.

Nealson, K. H. and J. C. Venter (2007). Metagenomics and the global ocean survey: what's in it for us, and why should we care? *ISME J* **1**(3), 185–187.

Qin, J., R. Li, J. Raes, *et al.* (2010). A human gut microbial gene catalogue established by metagenomic sequencing. *Nature* **464**(7285), 59–65.

Rubin, E. M. (2008). Genomics of cellulosic biofuels. *Nature* **454**(7206), 841–845.

Rusch, D. B., A. L. Halpern, G. Sutton, *et al.* (2007). The Sorcerer II Global Ocean Sampling expedition: northwest Atlantic through eastern tropical Pacific. *PLoS Biol* **5**(3), e77.

Sogin, M. L., H. G. Morrison, J. A. Huber, *et al*. (2006). Microbial diversity in the deep sea and the underexplored 'rare biosphere'. *Proc Natl Acad Sci U S A* **103**(32), 12115–12120.

Turnbaugh, P. J., M. Hamady, T. Yatsunenko, *et al*. (2009). A core gut microbiome in obese and lean twins. *Nature* **457**(7228), 480–484.

Venter, J. C., K. Remington, J. F. Heidelberg, *et al*. (2004). Environmental genome shotgun sequencing of the Sargasso Sea. *Science* **304**(5667), 66–74.

Warnecke, F., P. Luginbuhl, N. Ivanova, *et al*. (2007). Metagenomic and functional analysis of hindgut microbiota of a wood-feeding higher termite. *Nature* **450**(7169), 560–565.

12 A function to every gene
Royal blood and order in the sequence universe

The boy is in such indescribable pain day and night that no one from among his closest relatives, though they do not spare themselves, has the strength to bear looking after him too long, not to mention his mother, with her chronically ill heart. . . . His attendant . . . after a few sleepless nights filled with agony, becomes totally worn out and wouldn't be able to take it at all.

(Steinberg and Khrustalöv, 1995)

These words are from a letter written by Yevgeny Botkin, the family doctor of the Russian Tsar Nicholas II and his wife Alexandra (Steinberg and Khrustalëv, 1995). The suffering boy is Alexei, the son of the couple, born in 1904 and the successor to the throne of all Russians. He suffered from a congenital bleeding disease known as *haemophilia*, in which the defect is a missing or malfunctioning blood-clotting factor. The particular haemophilia that affected Alexei is a genetic disease linked to the X chromosome, and for this reason it is more prevalent in males. Females rarely show symptoms of the disease but may carry the gene on to the next generation.

Royal disease

Alexei inherited his disease gene from his great grandmother, Queen Victoria of England (Fig. 12.1). In fact, the queen transmitted her gene to a number of members of European royalty and thus decimated the thrones of Britain, Germany, Russia and Spain. Typically the affected males are exceptionally sensitive to injuries – for example, both Infante Gonzalo of Spain and Prince Rupert of Teck died young as a result of bleeding following car accidents. The last carrier of the disease in the royal family was Prince Waldemar of Prussia, who died in 1945.

Alexei's parents feared that he would not survive the first month of his life. He did survive childhood, but he was never to become a tsar. He was assassinated in 1918, along with his whole family, in the aftermath

of the Russian revolution. The assassination is an interesting story of its own, but here we will focus on the
molecular basis for Alexei's disease. We will examine
the process of blood coagulation and eventually use a
number of tools to examine different components of the
blood coagulation system in vertebrates.

The bleeding disorders referred to as haemophilia
come in different variants. The most important include
haemophilia A, which is caused by a deficiency in the
activity of coagulation factor VIII; *haemophilia B*, or
Christmas disease,[1] which is a factor IX deficiency and
the *von Willebrand disease*, a deficiency of the protein
von Willebrand factor.[2] What type of haemophilia did
Alexei have? Both haemophilia A and B are X-linked
disorders, and for this reason it was realized early that
Alexei was likely to have had either of those. But we
now know more, thanks to gene technology and PCR.
In 2009, scientists carried out a DNA analysis of historical specimens from the Nicholas II branch of the royal
family. This was possible because remains of all family
members that were assassinated in 1918 had been identified (Rogaev *et al.*, 2009b). Therefore, it was possible to
include Alexei's whole family in the analysis. The results
showed that Alexei suffered from haemophilia B; a fatal
point mutation in the factor IX gene was identified in Alexei and in his mother,
Alexandra, and his sister, Tatiana (Fig. 12.1). The mutation is predicted to alter
RNA splicing and lead to production of a truncated form of factor IX (Rogaev
et al., 2009a).

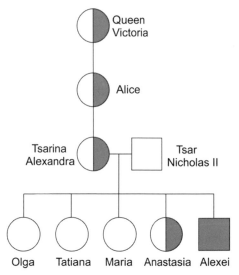

Fig. **12.1** *Inheritance of haemophilia B*.
The transmission of haemophilia B from
Queen Victoria to Alexei, son of Tsar
Nicholas II is shown in a simplified family
tree. The symbols are: open rectangle,
normal male; red rectangle, affected male;
open circle, normal female (non-carrier)
and white/red circles, carrier female.
Based on information in Rogaev *et al.*
(2009).

Blood-clotting pathways

In an injury in which a blood vessel is damaged, blood will eventually form a
clot. The formation of the clot proceeds through a multi-step process. There are
two major pathways; the intrinsic and extrinsic pathways (Fig. 12.2).

The most important pathway for initiating clot formation is the extrinsic
pathway. Here, damage to a blood vessel exposes *tissue factor*, a membrane

---

[1] It is named as such because the first patient to be described with this disease was Stephen Christmas.
    Also, it was first described in a Christmas issue of the *British Biomedical Journal*.
[2] The von Willebrand disease is named after the Finnish paediatrician Erik Adolf von Willebrand, who
    first described the disease.

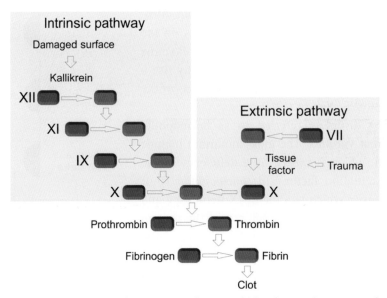

Fig. 12.2 *Intrinsic and extrinsic pathways of blood coagulation.* For the initiation of clot formation the extrinsic pathway is the most important. It is caused by damage to a blood vessel which exposes *tissue factor*, a membrane protein. Tissue factor activates factor X, which in turn converts prothrombin to thrombin through proteolytic cleavage. Thrombin converts fibrinogen to fibrin, which is the molecule forming the actual clot. The intrinsic pathway (also known as contact activation pathway) is initiated when prekallikrein, high molecular weight kininogen (HK or HMWK), factor XI and factor XII are exposed to a negatively charged surface. The conversion of prekallikrein to kallikrein activates factor XII, which in turn activates factor XI. Factor XI then activates factor IX, which in turn activates factor X. Finally, factor X finally acts on prothrombin as described for the extrinsic pathway.

protein. Tissue factor activates factor X, which in turn converts prothrombin to thrombin through proteolytic cleavage. Thrombin converts fibrinogen to fibrin, which is the molecule forming the actual clot. Blood-clotting pathways are dominated by reactions in which an enzyme is activated by proteolysis. This activated form of the enzyme catalyses the next reaction of the pathway, in which another enzyme is activated. Blood coagulation is one example of an *enzymatic cascade*, in which at each step a molecular signal is amplified. The purpose is to achieve an efficient and rapid response to an early trauma, the initial molecular signal of which may be rather weak.

A closer look at the enzymes that are part of the intrinsic and extrinsic pathways reveals that they are biochemically related. Thus, they are all enzymes with proteolytic activity. But not only that, they also seem to have other properties in common, and analysis of their sequences and protein domain structure indicates that they are related by evolution and came about by gene duplication events.

(a)

(b)

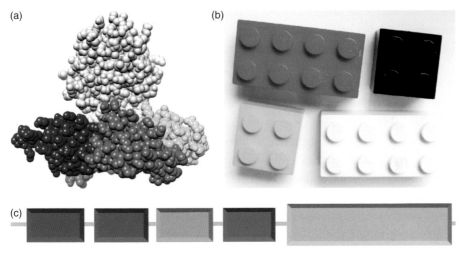

(c)

Fig. 12.3 *Structure of blood coagulation factor XI.* (a) Three-dimensional structure of factor XI in its inactive precursor (zymogen) form. The protein is composed of four PAN domains (coloured) as well as one serine–protease domain (grey). Note the PAN and protease domains in a 'cup and saucer' arrangement. It is the protease domain that is responsible for the important catalytic activity of factor XI, which is to cleave and activate factor IX in the intrinsic blood-coagulation pathway. Structure is from the PDB entry 2F83 and is represented with the UCSF Chimera software (Pettersen *et al.*, 2004). (b) Lego® pieces. (c) Linear representation of factor XI, showing the PAN (coloured) and serine–protease (grey) Pfam domains.

Protein domain architecture

We will examine the blood-clotting proteins in some detail and see how they are related by sequence. In particular, we will focus on the *domain architecture* of the proteins. To become familiar with the concepts of protein domains, consider coagulation factor XI.[3] The three-dimensional structure as elucidated with X-ray crystallography is shown in Fig. 12.3a (Papagrigoriou *et al.*, 2006). A number of distinct structural domains may be distinguished: four PAN domains[4] (Tordai *et al.*, 1999) as well as one serine–protease domain. These are structural units where each unit contributes a distinct function. The protease domain is responsible for the important catalytic activity of factor XI, which is to cleave and activate factor IX in the intrinsic blood-coagulation pathway. The PAN domains mediate a number of interactions with other proteins in the coagulation system. As shown in Fig. 12.3, the PAN and protease domains assemble into

---

[3] This factor is also associated with a clotting disease as it causes haemophilia C, which is a rare disease that mainly occurs among Ashkenazi Jews.

[4] The acronym PAN stems from the occurrence of this domain in (1) the *P*lasminogen/hepatocyte growth factor family; (2) the *A*pple domains of the plasma prekallikrein/coagulation factor XI family; and (3) domains of various *N*ematode proteins.

a 'cup and saucer' arrangement. All of the domains are encoded within the same polypeptide chain and we may alternatively represent the domain structure of factor XI in a linear manner, as shown in Fig. 12.3c.

A majority of proteins are built from more than one domain. We can think of the domain *architecture* of a protein as the sequence of protein domains. From an evolutionary point of view, protein domains are nature's own set of molecular Lego pieces, used to build the protein universe (Fig. 12.3b). The expression 'order in the sequence universe' in the title of this chapter refers to the fact that a huge number of different protein sequences may be classified into a highly restricted number of different protein architectures. How do new architectures come about during evolution? An important mechanism of protein gene evolution is chromosomal rearrangement, which gives rise to new genes with protein domain modules in new combinations. We will see examples of this below when examining the evolution of coagulation proteins.

BIOINFORMATICS

Tools introduced in this chapter	
Software running in a Unix environment	HMMER
Perl	Analysing output from HMMER

Bioinformatics of protein domains

What are the computational methods used when elucidating a linear domain structure like that of Fig. 12.3c? In most cases we do not have access to the three-dimensional structure of the protein. What methods are available to elucidate the domain structure on the basis of amino acid sequence? We saw in the previous chapter that BLAST (Altschul *et al.*, 1990, 1997) could be useful in this respect. Thus, we could in principle use different well-defined protein domains as queries in BLAST searches to identify homologous domains in a set of sequences present in databases. However, a problem with this approach is that BLAST is often not sensitive enough to identify the evolutionary relationship between two homologous sequences. This is because proteins are allowed to alter their amino acid sequences quite extensively because the same three-dimensional structure and biological function of a protein may be obtained with very different amino acid sequences. Therefore, protein sequences may have as little as 5–10% of sequence identity, although they are evolutionarily related.

BLAST is often not an adequate method to identify all members of a family of homologous proteins, but fortunately there are other methods at our disposal. One important category of methods are those based on *profiles* (or *position-specific scoring matrices*, PSSMs). In such methods a statistical model of a family of evolutionarily related protein sequences is created. This model is based on analysis of a multiple alignment of sequences. In essence, the profile is a two-dimensional matrix in which the columns represent the different columns (or positions) of the multiple alignment and the rows represent the characters characteristic of the sequences, i.e. the 20 amino acids for proteins and the four bases for nucleotide sequences. The values in each cell of the matrix represent the probability of seeing a particular character in a specific position. For more details, see Chapter 16, where all the steps in constructing a nucleotide sequence profile are shown. The important application of a protein profile is that we can use it to identify new members of the family being described by the profile. Therefore, whenever we have the sequence of an experimentally uncharacterized protein, we can search it against a library of profiles and thereby learn about the structural and functional properties of the protein.

What software is available for profile-based searches? PSI-BLAST (for position-specific iterated BLAST) is a further development of BLAST that makes use of profiles (Altschul *et al.*, 1997). The first step in PSI-BLAST is a 'regular' BLAST search using a query protein sequence to search a protein sequence database. The resulting hits of that search are then used to construct a profile. In the next step of PSI-BLAST, that profile is used to search the same database. This profile construction and database search may then be iterated any number of times (Fig. 12.4).

Another important piece of software making use of profiles is the HMMER package, developed by Sean Eddy (Eddy, 1998; http://hmmer.org). This package implements *profile hidden Markov models* (*profile HMMs*). The profile HMMs are reminiscent of the PSSMs used by PSI-BLAST, but have a formal probabilistic basis. For instance, HMMs have a consistent theory for setting gap and insertion scores. A detailed description of profile HMMs – or any HMM – is beyond the scope of this book. Instead, the reader is referred to other textbooks (e.g. Durbin *et al.*, 2007). At any rate, we will be using HMMER programs later in this chapter and we will also be analysing the output from these programs.

As with PSI-BLAST, an important application of HMMER is the identification of protein homology. The most important search possibilities within HMMER are listed in Table 12.1. For instance, *hmmscan* is used to search a sequence against a profile HMM database.

The database *Pfam* is a large collection of protein families. It collects protein multiple alignments and their corresponding profile HMMs (Finn *et al.*, 2010). There are two components of Pfam, Pfam-A and Pfam-B. The Pfam-A entries are high-quality, manually curated protein families – this is the most useful component of Pfam. Each of the Pfam-A families corresponds to a structural and functional domain of a protein. A very important application of Pfam is that any query protein sequence may be searched against the whole library of families and the resulting hits may give important clues as to the function of

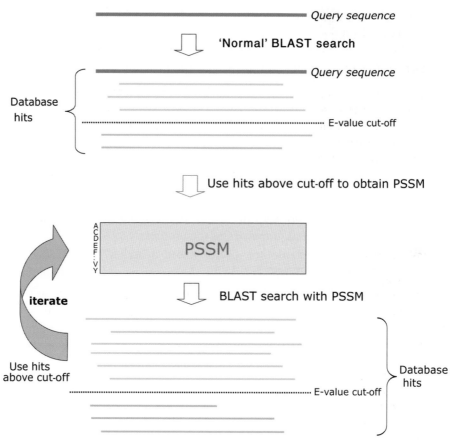

Fig. 12.4 *Principle of PSI-BLAST (position-specific iterative BLAST)*. PSI-BLAST first takes as an input a single protein sequence and compares it to a protein database, using the 'standard' BLAST program. All significant alignments resulting from this search, i.e. all alignments below a specific E-value, are used to construct a multiple alignment where the query sequence is used as a template. A profile is then constructed on the basis of this alignment and the profile is used to search the same database. Because the profile search is more sensitive, more significant hits against the database are identified in this step compared to the first BLAST search. The process of profile construction and database search may be iterated any number of times, until no more novel sequences are found.

the query protein. For instance, for the factor XI discussed above and shown in Fig. 12.3, the structure with four PAN domains and one protease domain is easily identifiable using Pfam, as we will see below.

Bioinformatic analysis of blood-clotting proteins

We will examine a set of 14 proteins related to coagulation. We will use hmmscan of the HMMER package. The proteins are from Swiss-Prot/UniProt and are

Table 12.1 *Programs in HMMER.* From the *HMMER User's Guide,* http://hmmer.org, Version 3, March 2010.

Program	Description
phmmer	Search a sequence against a sequence database (BLASTP-like)
jackhmmer	Iteratively search a sequence against a sequence database (PSI-BLAST-like)
hmmbuild	Build a profile HMM from an input multiple alignment
hmmsearch	Search a profile HMM against a sequence database
hmmscan	Search a sequence against a profile HMM database
hmmalign	Make a multiple alignment of many sequences to a common profile HMM

Table 12.2 *Human Swiss-Prot proteins related to blood coagulation.*

Swiss-Prot/UniProt identifier	Protein description
FA7_HUMAN	Coagulation factor VII
FA8_HUMAN	Coagulation factor VIII (haemophilia A)
FA9_HUMAN	Coagulation factor IX (haemophilia B)
FA10_HUMAN	Coagulation factor X
FA11_HUMAN	Coagulation factor XI
FA12_HUMAN	Coagulation factor XII
TF_HUMAN	Tissue factor (thromboplastin)
PLMN_HUMAN	Plasminogen
TPA_HUMAN	Tissue-type plasminogen activator (t-PA)
UROK_HUMAN	Urokinase-type plasminogen activator (u-PA)
THRB_HUMAN	Prothrombin
KLKB1_HUMAN	Plasma kallikrein
HGF_HUMAN	Hepatocyte growth factor
HGFA_HUMAN	Hepatocyte growth factor activator

listed in Table 12.2. The command line with `hmmscan` below assumes that these sequences are in a FASTA format and collected in a file named `clotting.fa`.

As a database of profile HMMs, we will be using Pfam. This dataset may be downloaded in a text-file format from the Pfam website (http://pfam.sanger.ac. uk) as a file named `Pfam-A.hmm`. It has to be formatted for HMMER searches with a utility of HMMER, `hmmpress`:

```
% hmmpress Pfam-A.hmm
```

A set of files with extensions h3m, h3i, h3f and h3p are produced as a result of `hmmpress`. We are then ready for an examination of our clotting proteins. The program `hmmscan` is used with an option to produce tabular output (`--domtblout [filename]`); in this example this output goes to the file `clotting.tab`:

```
% hmmscan --domtblout clotting.tab Pfam-A.hmm clotting.fa
```

The file `clotting.tab` will have contents in a tabular, space-delimited form in which each line has a large number of different values. The columns and values that will be of interest here are:

(1) Pfam family name;
(4) query sequence name;
(6) query sequence length;
(13) domain i-E-value ('independent' E-value);
(18) query sequence start position in alignment;
(19) query sequence end position in alignment;

We will extract these pieces of information from the file `clotting.tab` with the Perl script shown in Code 12.1. This is actually very similar to the script used in the previous chapter to extract data from a BLAST report in tabular form. One small difference is that there is no tab separator in the `clotting.tab` file. Remember, we used the following to store the tab-separated values in the array `@columns` in the previous chapter:

```
my @columns = split (/\t/);
```

Instead of the tab-separator, we now face a variable number of blank spaces. Therefore, we use:

```
my @columns = split (/ +/);
```

The regular expression `/ +/` will match one or more spaces. Keep in mind that regular expression matching in Perl is *greedy*, so in this case the regular expression will match all the blank spaces up to the next column value.

Code 12.1 parse_hmmscan.pl

```perl
#!/usr/bin/perl -w

use strict;

print "protname\tlen\tdomname\tbegin\tend\n";
open(IN, 'clotting.tab') or die "could not open file\n";
while (<IN>) {
    unless (/^\#/) {        # avoid all lines beginning
                            # with the '#' character
        my @columns = split(/ +/);
        my $domname = $columns[0];
        my $protname = $columns[3];
        my $len = $columns[5];
```

```perl
        my $evalue = $columns[12];
        $protname =~ s/.*\|//;
        my $begin = $columns[17];
        my $end = $columns[18];
        if ( $evalue < 1e-5 ) {
            print "$protname\t$len\t$domname\t$begin\t$end\n";
        }
    }
}
close IN;
```

The column represented by `$columns[3]` contains values like `gi|119766|sp|P08709.1|FA7_HUMAN`. In this case we are interested only in the portion after the right-most pipe symbol (`|`). Therefore, we remove all characters in that column preceding and including that pipe symbol. Because the pipe has a special meaning in a regular expression, we need to put a backslash (\) in front of it to indicate that it is the actual pipe character we are matching. So, the replacement operation is:

```perl
s/.*\|//;
```

where `.*` refers to any character matched zero or more times. Again, remember that matching is greedy, so all characters preceding the right-most pipe symbol will be matched.

The results of our script in Code 12.1 to analyse `hmmscan` data may be saved in a file `parse_hmmscan.out`:

```
% perl parse_hmmscan.pl > parse_hmmscan.out
```

The first lines of the output file should look like this:

protname	len	domname	begin	end
FA7_HUMAN	466	Trypsin	213	447
FA7_HUMAN	466	Gla	65	106
FA7_HUMAN	466	DUF1986	223	326
FA8_HUMAN	2351	F5_F8_type_C	2055	2185
FA8_HUMAN	2351	F5_F8_type_C	2208	2342

These data are shown graphically in Fig. 12.5. We now have a view of the architecture of the different coagulation proteins. In general, this type of domain analysis is very useful when it comes to understanding the function of a protein. For instance, if we had analysed an anonymous protein sequence and had no prior information about its function, the presence of one or more Pfam domains in that protein may give us important clues as to its function. As an example,

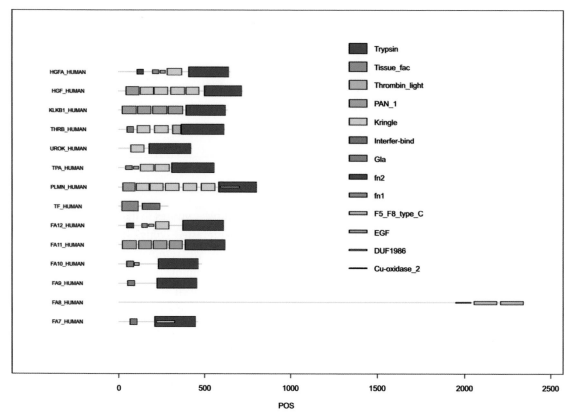

Fig. 12.5 *Domain architecture of clotting proteins*. The Pfam-based domain structure of proteins as described in Table 12.2 is shown. The plot was created with an R script (see the web resources for this book) using the results of an `hmmscan` analysis of the proteins. We may conclude from this plot that many of the proteins are related in structure. For instance, most of them contain a protease domain.

consider the proteins under study here. The presence of a protease domain as predicted by Pfam indicates to us that this protein has protease activity.

Evolution of blood clotting

Analysis based on Pfam also allows effective comparison between the members of a protein family. For instance, we may address questions such as: What proteins have at least one protease domain? What other proteins have exactly the domain structure (architecture) characteristic of coagulation factor VIII? Answers to questions like these will in general guide us as to the structure, function and evolution of protein molecules.

In general, proteins change during evolution in small steps by gain or loss of specific domains. Organisms of higher complexity often have more complex architectures compared to organisms of lower complexity, a phenomenon that

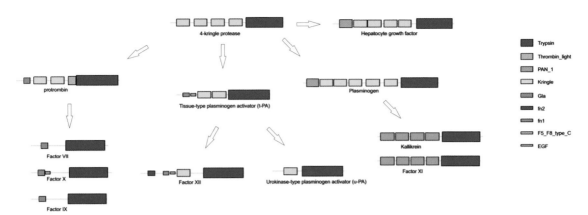

Fig. 12.6 *Possible evolution of blood-clotting proteins.* According to the model of evolution shown here (Doolittle, 2009), the ancestral sequence was a four-kringle protease. The domains shown here are Pfam domains as identified using `hmmscan`. See also Fig. 12.5.

has been referred to as *domain accretion* (Koonin *et al.*, 2000). A close examination of the results in Fig. 12.5, including the results of corresponding analyses of other vertebrate coagulation proteins, leads to models for the evolution of this group of proteins. From an examination of their architecture it would seem that many of the proteins are related by evolution and came about through gene duplication.

A quite remarkable evolution took place in the case of blood coagulation. Thus, in humans there are more than 20 different proteins involved in blood coagulation. A blood coagulation system is present in all vertebrates. However, it is lacking completely in all invertebrates, including species that are closely related to the vertebrates, such as amphioxus (*Branchiostoma floridae*) and the sea squirt (*Ciona intestinalis*). How was the blood coagulation system invented in vertebrates? The answer to this question is the same as when discussing almost any invention in nature: *the system developed in a stepwise manner.* In the case of the blood coagulation proteins, individual protein domains were already present in invertebrate proteins. New combinations of the domains, as well as changes to individual domains, were required in order to produce new functions. Once a certain blood coagulation protein had been invented it could be further modified through the gain or loss of domains, and paralogous proteins could be formed as a result of gene duplication.

Models have been put forward to show the evolution of the blood-clotting proteins (Doolittle, 2009), and these may be understood on the basis of our information that we obtained through the use of Pfam, as shown in Fig. 12.5. As one example, we may think of an ancestral protein with four kringle domains and one protease domain as shown in Fig. 12.6. This protein developed into different proteins by domain gain or loss. One example is prothrombin, which in turn developed into factors VII, IX and X. For more examples, see Fig. 12.6.

EXERCISES

12.1 In Chapter 7 we studied the human Swiss-Prot proteins BCR_HUMAN, ABL1_HUMAN and the fusion protein BCR–ABL. The sequences BCR_HUMAN and ABL1_HUMAN may be extracted from Swiss-Prot using `blastdbcmd`, and one BCR–ABL protein is in the file `bcrabl.fa`. Apply the procedures described in Chapter 12, including `hmmscan`, to find out what Pfam domains are present in these three proteins. Make a schematic drawing of the domain architectures.

12.2 Extract the domain referred to as SH2 from ABL1_HUMAN (see Exercise 12.1). Use this sequence as a query sequence in a PSI-BLAST search (Fig. 12.4) (see Appendix II for more on command-line PSI-BLAST). Specify 12 iterations. When no new sequences are found above the threshold, PSI-BLAST will report 'Search has CONVERGED!'. After how many iterations was this state reached in your search? How many hits above the threshold were found in the first BLAST search? In the final iteration?

12.3 Consider the proteins of Exercise 11.2 in the previous chapter. Examine the Pfam domain architecture of these proteins with `hmmscan`. What Pfam domain is present in all of these sequences? What domain(s) is/are specific for the GUNA_CLOLO sequence?

REFERENCES

Altschul, S. F., W. Gish, W. Miller, E. W. Myers and D. J. Lipman (1990). Basic local alignment search tool. *J Mol Biol* **215**(3), 403–410.

Altschul, S. F., T. L. Madden, A. A. Schaffer, *et al.* (1997). Gapped BLAST and PSI-BLAST: a new generation of protein database search programs. *Nucleic Acids Res* **25**(17), 3389–3402.

Doolittle, R. F. (2009). Step-by-step evolution of vertebrate blood coagulation. *Cold Spring Harb Symp Quant Biol* **74**, 35–40.

Durbin, R., S. R. Eddy, A. Krogh and G. Mitchison (2007). *Biological Sequence Analysis: Probabilistic Models of Proteins and Nucleic Acids*. Cambridge, Cambridge University Press.

Eddy, S. R. (1998). Profile hidden Markov models. *Bioinformatics* **14**(9), 755–763.

Finn, R. D., J. Mistry, J. Tate, *et al.* (2010). The Pfam protein families database. *Nucleic Acids Res* **38** (Database issue), D211–222.

Koonin, E. V., L. Aravind and A. S. Kondrashov (2000). The impact of comparative genomics on our understanding of evolution. *Cell* **101**(6), 573–576.

Papagrigoriou, E., P. A. McEwan, P. N. Walsh and J. Emsley (2006). Crystal structure of the factor XI zymogen reveals a pathway for transactivation. *Nat Struct Mol Biol* **13**(6), 557–558.

Pettersen, E. F., T. D. Goddard, C. C. Huang, *et al.* (2004). UCSF Chimera: a visualization system for exploratory research and analysis. *J Comput Chem* **25**(13), 1605–1612.

Rogaev, E. I., A. P. Grigorenko, G. Faskhutdinova, E. L. Kittler and Y. K. Moliaka (2009). Genotype analysis identifies the cause of the 'royal disease'. *Science* **326**(5954), 817.

Rogaev, E. I., A. P. Grigorenko, Y. K. Moliaka, *et al.* (2009). Genomic identification in the historical case of the Nicholas II royal family. *Proc Natl Acad Sci U S A* **106**(13), 5258–5263.

Steinberg, M. D. and V. M. Khrustalëv (1995). *The Fall of the Romanovs: Political Dreams and Personal Struggles in a Time of Revolution*. New Haven, CT, Yale University Press.

Tordai, H., L. Banyai and L. Patthy (1999). The PAN module: the N-terminal domains of plasminogen and hepatocyte growth factor are homologous with the apple domains of the prekallikrein family and with a novel domain found in numerous nematode proteins. *FEBS Lett* **461**(1–2), 63–67.

13

A function to every gene
A slimy molecule

Extensive sugar decoration

In the two previous chapters we dealt with problems of assigning function to proteins or protein domains. We concerned ourselves with methods relying on sequence similarity, i.e. methods based on local alignment or profiles. We will now see an example of a functional domain which is not easily identified on the basis of sequence similarity or with profiles. It has specific properties as it has a characteristic amino acid composition and functional properties based on that composition. However, the properties of the domain may not be captured on the basis of sequence alignment or by position-specific information. The domain in question is characteristic of a group of proteins referred to as *mucins* (Perez-Vilar and Hill, 1999; Hollingsworth and Swanson, 2004).

A characteristic property of all mucins is the ability to form gels. Mucins are a major component of the mucous layer that is present on the surface of epithelial cells of the lung and intestine. The proteins act as a diffusion barrier to prevent harmful microorganisms and substances having more intimate contact with the cell. Mucins also function as lubricants to protect epithelial cells from dehydration and physical and chemical injury. Because mucins protect against pathogens, they play an important role in immune defence. In addition, certain mucins are associated with colon cancer (Hollingsworth and Swanson, 2004) and there is a strong association between the mucin Muc2 and the inflammatory bowel disease *ulcerative colitis*. For example, inflammation of the large intestine similar to ulcerative colitis is observed in mice that are deficient in the mucin Muc2 (Heazlewood *et al.*, 2008).

Mucins are very large proteins. Not only are their polypeptide chains typically long, but they are also heavily glycosylated. The glycosylated domain is one rich in the amino acids proline, threonine and serine (Fig. 13.1) and here will be referred to as the PTS domain. The threonines and serines of this domain are the sites of sugar attachment. Through decoration by sugars a PTS domain adopts a 'bottle brush' type of structure, as shown in Fig. 13.2.

Mucins have been further classified as membrane-bound or secreted. In humans there are eight membrane-bound mucins (MUC1, MUC3, MUC4, MUC12, MUC13, MUC16, MUC17 and MUC20) and five secreted gel-forming mucins

STHTAPPITPTTSGTSQAHSSFSTNKTPTSLHSHTSSTHHPEVTPTSTTTITPNPTSTRTRTPVAHTNSATSSRPPPPFTT

PPTGSSPFSSTGPMTATSFKTTTTYPTPSHPQTTLPTHVPPFSTSLVTPSTHTVITPTHAQMATSASIHSMPTGTIPPPTT

ATGSTHTAPTMLTTSGTSQALSSLNTAKTSTSLHSHTSSTHHAEATSTSTTNITPNPTSTGTPPMTVTTSGTSQSRSSFS

Fig. 13.1 *Amino acid composition characteristic of mucins*. Mucins contain domains (PTS) that are rich in the amino acids serine, threonine and proline. A portion of one such domain in the human mucin MUC6 is shown here. An observant reader may also be able to identify non-identical repeats present in this sequence. In order to identify such repeats one may apply the manual method of 'sequence gazing', or even better make use of some bioinformatic method (see also Fig. 13.4).

(MUC2, MUC5B, MUC5AC, MUC6 and MUC19). The human MUC1, MUC3, MUC12, MUC13, MUC16 and MUC17 mucins contain SEA domains,[1] whereas MUC4 and the gel-forming mucins contain one or more VWD domains[2] (Lang *et al.*, 2007) (Fig. 13.3). When looking at this protein family we are reminded of the mechanism of protein evolution discussed in the previous chapter, in which proteins evolve by mutational events that present structural and functional domains in new combinations.

Fig. 13.2 *Structure of mucins*. Serines and threonines of the PTS domain in mucins are modified by glycosylation, giving rise to a 'bottle brush' type of structure for these PTS domains.

If we compare PTS domains from different human mucins, or from mucins of other animals, it is clear that these domains are not well conserved in sequence. For instance, when comparing the PTS domains of the human and mouse mucin homologues, there is no apparent sequence conservation at all. It would seem as if the actual sequence of amino acids in a PTS domain is not so important for its function. Rather, what matters is the overall amino acid composition, where serine, threonine and proline are particularly abundant.

Mucins and repeats

In addition to a characteristic amino acid composition, many mucin PTS domains have identical or near-identical repeats. The repetitive nature of mucin domains is illustrated in the dotplot analysis shown in Fig. 13.4. We looked at dotplots in Chapter 7 in the context of sequence alignments. Dotplots are useful to compare two sequences to each other, but they are also used to identify repeats in a sequence. For such a purpose the sequence to be analysed is compared to itself. In case it has repeated elements, these will be revealed by extra diagonal patterns. In mucins there are identical and near-identical repeats.

---

[1] The SEA domain got its name as it was a module first identified in three different proteins; *Sea* urchin sperm protein, *Enterokinase* and *Agrin* (Kitamoto *et al.*, 1994; Bork and Patthy, 1995).

[2] The VWD (von Willebrand factor type D) domain occurs in the von Willebrand factor, as well as in many other proteins, including mucins (Bork, 1993).

Fig. 13.3 *Domain structure of mucins.* In addition to PTS domains, most mucins contain either von Willebrand D domains (VWD) or SEA domains. Some of them, like MUC4, MUC1, MUC13 and MUC3 (shown here), are anchored to the plasma membrane through a transmembrane (TM) domain.

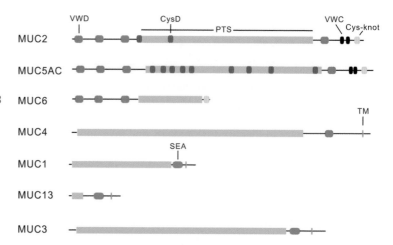

Mechanisms likely to generate the repeats are DNA recombination and slippage during DNA replication. We previously encountered replication slippage in the context of the CAG triplet repeats related to Huntington's disease, as discussed in Chapter 5.

Mucins are notoriously difficult to work with from a gene technology point of view because of the PTS domain repeats. There are at least two reasons why repeats are technically difficult. First, they are difficult to clone as recombination events tend to change the content of such a region. Second, the presence of repeats is awkward when it comes to the computational *assembly* process, i.e. the process of merging short sequencing reads into a longer sequence. For example, if a genomic DNA sequence has a very long region of tandem repeats and the sequencing reads cover only part of that region, it will be impossible to figure out how the genomic sequence should be reconstructed from the sequencing reads.

From what you sometimes hear about the human genome, you may get the impression that the complete nucleotide sequence of the human genome has been determined. This is not strictly true. Problematic regions are mainly duplicated segments or any kind of repeats. Therefore, some portions of the human genome have been left out in human genome sequencing projects. In the case of mucins, their repetitive nature is a significant problem as it has the consequence that some mucin genes have not been fully sequenced. For example, as this book is being written, the current human genome assembly is still lacking a complete version of the MUC5AC gene.[3]

---

[3] Then there are even more difficult repeat regions of the human genome. Regions referred to as *constitutive heterochromatin* are notably repetitive. For instance, the chromosomes 1, 9 and 16 and the Y chromosome contain large regions of this nature that are resistant to sequencing. However, the fact that we are lacking sequence information for these regions is not considered a big issue, as most of the constitutive heterochromatin is transcriptionally inactive and does not give rise to any products.

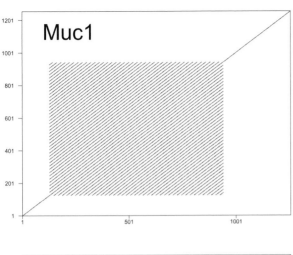

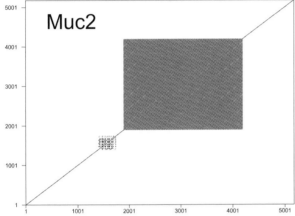

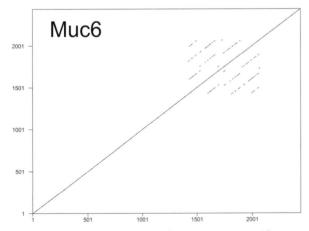

Fig. 13.4 *Repetitive nature of mucin amino acid sequences*. The figure was created using the EMBOSS program dottup. When a sequence is compared to itself, any repeat is revealed as a diagonal pattern in addition to the main diagonal. MUC1 and MUC2 contain many identical repeats, whereas MUC6 contains non-identical repeats. Compare also the prediction of PTS domains in Fig. 13.5.

BIOINFORMATICS

Tools introduced in this chapter	
Perl	Command-line arguments and @ARGV
	NCBI Entrez programming utilities

Computational identification of mucin domains

We now approach the problem of how to identify mucin proteins computationally in a large collection of protein sequences, such as those resulting from a genome sequencing project. The majority of mucins contain protein domains such as the VWD domain and the SEA domain. These could be identified using methods based on Pfam and HMMER, as described in the previous chapter. But non-mucin proteins also have VWD and SEA domains and we expect mucins to have one or more PTS domains. As PTS domains are not really conserved in sequence, a local alignment method like BLAST is not useful. Given a certain amino acid sequence, how can we predict whether it contains a PTS domain or not? A naïve solution to this problem is that in order for a region to qualify as a PTS domain we require specific amino acid composition characteristics. For instance, based on knowledge of previously known mucins, a typical mucin domain has a content of serine and threonine which is more than 40% and a content of proline which is at least 5%. We also require some minimum length for the PTS domain – say, 100 amino acids. This simple strategy is implemented in the Perl script shown in Code 13.1.

To examine an amino acid sequence with respect to potential PTS domains, consider the human mucin MUC6, contained within the file muc6.fa. This is a protein encoded by a gene on chromosome 11. The MUC6 gene is part of a cluster of mucin genes in the same chromosomal region, including MUC2, MUC5AC and MUC5B.

Code 13.1 pts.pl

```perl
#!/usr/bin/perl -w

use strict;

# Basic parameters used
my $wid = 100;      # size of sliding window

# check if argument to the script is there.
if ( $ARGV[0] eq '' ) {
     die "File in FASTA sequence format is to be
     used as argument to the script\n";
}

# read the sequence from the input file
my $seq = '';
```

```
open(IN, "$ARGV[0]") or die "Could not open file $ARGV[0]\n";
while (<IN>) {
    chomp;

        # in the identifier line all is captured
        # in the variable $id except for
        # the > character
        if (/>(.*)/) { my $id = $1; }
        else         { $seq .= $_; }
}
close IN;

# Now analyse the sequence in $seq
print "Position\tProline\tThreonine\tSerine\n";
for ( my $i = 0 ; $i < length($seq) - $wid + 1; $i++ ) {
    my $test = substr( $seq, $i, $wid );

        # Count proline, threonine and serine
        my $count_p = ( $test =~ tr/P// ) / $wid;
        my $count_t = ( $test =~ tr/T// ) / $wid;
        my $count_s = ( $test =~ tr/S// ) / $wid;
        my $pos      = $i + 1 + $wid/2;
        print "$pos\t$count_p\t$count_t\t$count_s\n";
}
```

When we make use of a sequence file in a Perl script, we can specify the name of the file within the script. We have seen examples of this earlier – for example, in the previous chapter. However, you may want to make your Perl script a bit more flexible than this. Thus, you may want to decide once you run the script what sequence is to be analysed. In such a case, you can specify on the Perl command line what file is to be analysed. We see an example of this procedure in the script in Code 13.1. We specify on the command line that muc6.fa is to be used:

```
% perl pts.pl muc6.fa
```

When writing the Perl code, how do we take care of stuff that follows after perl pts.pl on the command line? There is a specific array that stores any command-line arguments, @ARGV. If you have more than one argument to a script, like

```
% perl somecode.pl file1 file2 file3
```

the arguments file1, file2 and file3 will be stored as elements in the @ARGV array and are available as $ARGV[0], $ARGV[1] and $ARGV[2], respectively. In our example with the MUC6 sequence there is only one argument and this argument will consequently be available as $ARGV[0]. It is wise to have code in the script that checks whether an argument actually was supplied at the command line. For example:

```
if ($ARGV[0] eq '') {
        die "File in FASTA sequence format is to be
        used as argument to the script\n";
}
```

Looking more closely at Code 13.1, we want to analyse the input sequence (MUC6) with respect to its contents of serine, threonine and proline. To do this we use a sliding window as we have done previously (such as in Chapter 6 when identifying the iron responsive element). For the mucin example here we examine a window of size 100, but this could of course be set to some other value if required. The window is moved in steps of a single position.

As we discussed earlier in the book, characters in Perl may be counted with different methods. Here, the tr operator is used, as in this expression to count prolines:

```
my $count_p = ($test =~ tr/P//);
```

It should be noted that even though Code 13.1 achieves what we want and is simple to write, the procedure used for examining the amino acid composition is far from optimal from a computational point of view. The problem is that we are examining the same amino acid positions over and over again. Thus, we first analyse a window corresponding to positions 1–100 in the protein. Then we move to the next window, positions 2–101, and examine all these positions. Instead we could obtain new counts for the three amino acids simply by subtracting the count for position 1 and adding the count for position 101 (see also Exercise 13.2).

The first lines of the output of Code 13.1 are:

Position	Proline	Threonine	Serine
51	0.05	0.08	0.11
52	0.05	0.08	0.11
53	0.05	0.08	0.11
54	0.05	0.08	0.11
55	0.05	0.08	0.12
56	0.05	0.08	0.12

The output of the script is plotted in Fig. 13.5. In this plot we have also introduced a grey line to indicate where the content of serine and threonine is greater than 40% and the content of proline greater than 5%. Thus, we see evidence of a PTS domain in the C-terminal half of the protein. It would seem that it was correctly identified by this very simple Perl script. We also note that the PTS domain identified through its amino acid composition corresponds well to a region rich in repeated elements, as shown in the dotplot analysis in Fig. 13.4. Therefore, we could also think of combining the method based on amino acid composition with a method identifying repeats to further improve on the prediction of PTS domains.

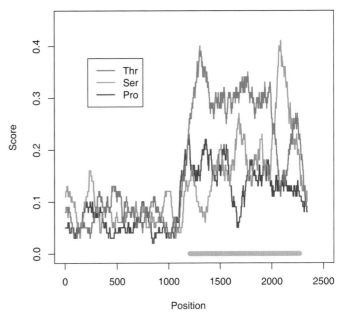

Fig. 13.5 *Prediction of mucin PTS domains.* The plot shows the output from Code 13.1. In addition, the grey line indicates a region where the content of serine and threonine is greater than 40% and the content of proline greater than 5%. The plot was obtained using an R script (see the web resources for this book).

EXERCISES

13.1 In Code 13.1 we count amino acids using the `tr///` operator. Modify the script to show that the counting could also be carried out using either of the constructs below:

```
$count_p = 0; while ($test =~ /P/g)  {$count_p++;}
$count_p = 0; while ($test =~ s/P//) {$count_p++;}
$count_p = ($seq =~ s/P//g);
```

13.2 The counting of amino acids in Code 13.1 is not optimal as we are analysing overlapping windows of the mucin sequence and are therefore examining the same amino acid positions several times. Modify Code 13.1 to avoid this situation.

13.3 The mucin sequence analysed in Code 13.1 (`muc6.fa`) contains repetitive sequences. Construct a Perl script to examine every possible word of size four (i.e. every sequence of four consecutive amino acids) and count the number of times that word occurs in the mucin sequence. What is the most common four-letter word and how many times does it occur in the mucin MUC6?

REFERENCES

Bork, P. (1993). The modular architecture of a new family of growth regulators related to connective tissue growth factor. *FEBS Lett* **327**(2), 125–130.

Bork, P. and L. Patthy (1995). The SEA module: a new extracellular domain associated with O-glycosylation. *Protein Sci* **4**(7), 1421–1425.

Heazlewood, C. K., M. C. Cook, R. Eri, *et al.* (2008). Aberrant mucin assembly in mice causes endoplasmic reticulum stress and spontaneous inflammation resembling ulcerative colitis. *PLoS Med* **5**(3), e54.

Hollingsworth, M. A. and B. J. Swanson (2004). Mucins cancer: protection and control of the cell surface. *Nat Rev Cancer* **4**(1), 45–60.

Kitamoto, Y., X. Yuan, Q. Wu, D. W. McCourt and J. E. Sadler (1994). Enterokinase, the initiator of intestinal digestion, is a mosaic protease composed of a distinctive assortment of domains. *Proc Natl Acad Sci U S A* **91**(16), 7588–7592.

Lang, T., G. C. Hansson and T. Samuelsson (2007). Gel-forming mucins appeared early in metazoan evolution. *Proc Natl Acad Sci U S A* **104**(41), 16209–16214.

Perez-Vilar, J. and R. L. Hill (1999). The structure and assembly of secreted mucins. *J Biol Chem* **274**(45), 31751–31754.

Information resources
Learning about flu viruses

14

All science is either physics or stamp collecting.

(Ernest Rutherford[1])

··

The 1918 flu pandemic, also referred to as the Spanish flu, was a devastating infectious disease. It is estimated that 50 million people, about 3% of the world's population at the time, died of the disease. About 500 million people were infected. The causative agent was an influenza virus. In this chapter we will learn more about these viruses. We will make use of highly significant molecular biology databases and bioinformatics tools. These are useful not only for learning about influenza viruses, but are widely used to explore just about any topic in biology.

Short history of sequence databases

A vast amount of information is collected by projects around the world designed to characterize genomes, genes and proteins. The development with respect to DNA sequencing is particularly remarkable. One important task in bioinformatics is to store all of this information in databases and, importantly, to make it available to the scientific community for downloading and analysis. Numerous dedicated individuals working on database projects are the unsung heroes of bioinformatics and molecular biology (see also the quotation on stamp collecting above).

Gene technology started out in the late 1970s. Methods were developed that allowed amplification and sequencing of DNA. Before that, we did not have access to any DNA sequences at all. All we knew about DNA was that it is

---

[1] Ernest Rutherford, famous for pioneering studies in nuclear physics and for his model of the atom, did not think highly of sciences that were not physics. Life scientists and bioinformaticians tend to see it differently. For example, Carl Linneaus and Charles Darwin are examples of 'stamp collectors' that also were great scientists.

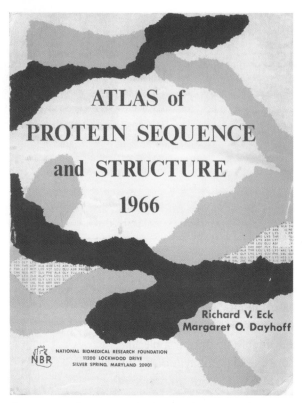

Fig. 14.1 *Early protein database*. The 1966 edition of *Atlas of Protein Sequence and Structure* by Margaret Dayhoff and Richard Eck (National Biomedical Research Foundation *et al.*, 1966). The amino acid sequences of some 70 different proteins were reported in this volume.

composed of the nucleotides A, T, C and G. But molecular sequences were generated before the days of cloning and gene technology,[2] although the methods used were very much more laborious than current DNA sequencing methods. RNA molecules were the first nucleic acids to be subject to sequence analysis. The first RNA to be sequenced was a yeast alanine tRNA, 75 nucleotides in length (Holley *et al.*, 1965). A heroic effort was the sequencing in 1976 of the bacteriophage MS2 RNA genome, with a total of 3569 nucleotides (Fiers *et al.*, 1976). The first protein sequence, that of insulin, was determined by Frederick Sanger in 1953 (described by Stretton, 2002). The first 'database' of protein sequences was a book, *Atlas of Protein Sequence and Structure*, assembled by Margaret Dayhoff and Richard Eck and published in 1965 (Dayhoff and National Biomedical Research Foundation, 1965; Strasser, 2009; Fig. 14.1). Margaret Dayhoff was an early pioneer in bioinformatics, also famous for other contributions to the subject, such as the first substitution matrices. The protein atlas of 1965 had 93 pages and listed around 70 proteins. Each page showed one protein and its sequence. Dayhoff's protein atlas was to be published in several editions and eventually gave rise to the *Protein Information Resource* (PIR). Whereas the original version of Dayhoff's database contained 70 proteins, we now have access to millions of protein sequences, as we shall see further below. The large majority of those are computationally derived from DNA sequences, and were not obtained by direct sequencing of protein molecules.

As DNA sequences were accumulating in the early 1980s it was realized that there was a need for a database collecting nucleotide sequences as well. Thus, the United States' GenBank was established in 1982 and the EMBL data library in Europe about the same time. In the first years of these databases all sequences were printed in physical books (Andersen, 1984; Armstrong *et al.*, 1985). For

---

[2] A period of biochemical science sometimes referred to as BC, 'before cloning'.

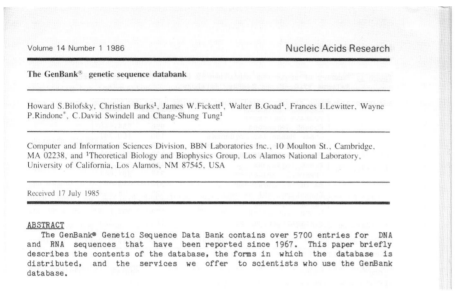

Volume 14 Number 1 1986   Nucleic Acids Research

The GenBank® genetic sequence databank

Howard S.Bilofsky, Christian Burks[1], James W.Fickett[1], Walter B.Goad[1], Frances I.Lewitter, Wayne P.Rindone*, C.David Swindell and Chang-Shung Tung[1]

Computer and Information Sciences Division, BBN Laboratories Inc., 10 Moulton St., Cambridge, MA 02238, and [1]Theoretical Biology and Biophysics Group, Los Alamos National Laboratory, University of California, Los Alamos, NM 87545, USA

Received 17 July 1985

ABSTRACT

The GenBank® Genetic Sequence Data Bank contains over 5700 entries for DNA and RNA sequences that have been reported since 1967. This paper briefly describes the contents of the database, the forms in which the database is distributed, and the services we offer to scientists who use the GenBank database.

Fig. 14.2 *Early days of nucleotide sequence databases.* A publication of 1986 describing the nucleotide sequence database GenBank (Bilofsky *et al.*, 1986).

an example from a publication from 1986 describing GenBank, see Fig. 14.2 (Bilofsky *et al.*, 1986).

The databases soon went electronic, with data first being distributed on magnetic tape and floppy discs. Data was also available on a computer at NCBI that could be 'accessed by direct long-distance dialup over the Telenet telecommunications network' (Bilofsky *et al.*, 1986).[3] GenBank has, since the early 1990s, been administered by the NCBI, Bethesda. Early on, collaboration was initiated with the EMBL database and with the Japanese DDBJ. The EMBL database is a European counterpart of GenBank, now administered by EBI, the European Bioinformatics Institute, Hinxton, UK. Release 3 of GenBank in 1982 had 606 sequences and 680 338 nucleotides, and the 1984 volume contained 3424 sequences with 2.8 million bases. This seemed a lot at the time, but the database has continued to grow exponentially ever since its conception; the February 2011 release of GenBank (Benson *et al.*, 2010) contained more than 130 million (130×10^6) sequences containing more than 124 billion (124×10^9) bases!

Features of nucleotide sequence databases

Each entry in the GenBank or EMBL DNA nucleotide sequence database is a distinct sequence, i.e. a string composed of the letters A, G, C and T. Each

---

[3] The young reader should note that there was a time before the days of the World Wide Web.

```
LOCUS       CY073775              1420 bp    cRNA    linear   VRL 07-SEP-2010
DEFINITION  Influenza A virus (A/Vienna/INS369/2010(H1N1)) segment 6, complete
            sequence.
ACCESSION   CY073775
VERSION     CY073775.1  GI:306412420   gi number
DBLINK      Project: 37813
KEYWORDS    .
SOURCE      Influenza A virus (A/Vienna/INS369/2010(H1N1))
  ORGANISM  Influenza A virus (A/Vienna/INS369/2010(H1N1))
            Viruses; ssRNA negative-strand viruses; Orthomyxoviridae;
            Influenzavirus A.
REFERENCE   1  (bases 1 to 1420)
  AUTHORS   Spiro,D., Halpin,R., Bera,J., Ghedin,E., Hostetler,J., Fedorova,N.,
            Hine,E., Overton,L., Kim,M., Szczypinski,B., Stockwell,T., Sitz,J.,
            Katzel,D., Li,K., Axelrod,N., Safford,T., Amedeo,P., Schobel,S.,
            Shrivastava,S., Wang,S., Resnick,A., Thovarai,V., Lynfield,R.,
            Losso,M., Davey,R., Dwyer,D., Neaton,J., Lane,X., Lin,X.,
            Wentworth,D.E., Bao,Y., Sanders,R., Dernovoy,D., Kiryutin,B.,
            Lipman,D.J. and Tatusova,T.
    TITLE   The NIAID Influenza Genome Sequencing Project
  JOURNAL   Unpublished
REFERENCE   2  (bases 1 to 1420)
  CONSRTM   The NIAID Influenza Genome Sequencing Consortium
    TITLE   Direct Submission
  JOURNAL   Submitted (07-SEP-2010) on behalf of JCVI/University of
            Minnesota/NCBI, National Center for Biotechnology Information, NIH,
            Bethesda, MD 20894, USA
FEATURES             Location/Qualifiers
     source          1..1420
                     /organism="Influenza A virus (A/Vienna/INS369/2010(H1N1))"
                     /mol_type="viral cRNA"
                     /strain="A/Vienna/INS369/2010"
                     /serotype="H1N1"
                     /host="human; gender M; age 31Y"
                     /db_xref="taxon:856498"
                     /segment="6"
                     /lab_host="P0 passage(s)"
                     /country="Austria: Vienna"
                     /collection_date="05-Jan-2010"
     gene            3..1412
                     /gene="NA"
     CDS             3..1412
                     /gene="NA"
                     /codon_start=1
                     /product="neuraminidase"
                     /protein_id="ADM86453.1"
                     /db_xref="GI:306412421"
                     /translation="MNPNQKIITIGSVCMTIGMANLILQIGNIISIWISHSIQLGNQN
                     QIETCNQSVITYENNTWVNQTYVNISNTNFAAGQSVVSVKLAGNSSLCPVSGWAIYSK
                     DNSVRIGSKGDVFVIREPFISCSPLECRTFFLTQGALLNDKHSNGTIKDRSPYRTLMS
                     CPIGEVPSPYNSRFESVAWSASACHDGINWLTIGISGPDNGAVAVLKYNGIITDTIKS
                     WRNNILRTQESECACVNGSCFTVMTDGPSDGQASYKIFRIEKGKIVKSVEMNAPNYHY
                     EECSCYPDSSEITCVCRDNWHGSNRPWVSFNQNLEYQIGYICSGIFGDNPRPNDKTGS
                     CGPVSSNGANGVKGFSFKYGNGVWIGRTKSISSRNGFEMIWDPNGWTGTDNNFSIKQD
                     IVGINEWSGYSGSFVQHPELTGLDCIRPCFWVELIRGRPKENTIWTSGSSISFCGVNS
                     DTVGWSWPDGAELPFTIDK"
ORIGIN
        1 aaatgaatcc aaaccaaaag ataataacca ttggttcggt ctgtatgaca attggaatgg
       61 ccaacttaat attacaaatt ggaaacataa tctcaatatg gattagccac tcaattcaac
      121 ttgggaatca aaatcagatt gaaacatgca atcaaagcgt cattacttat gaaaacaaca
      181 cttgggtaaa tcagacatat gttaacatca gcaacaccaa ctttgctgct ggacagtcag
      241 tggtttccgt gaaattagcg ggcaattctt ctctctgccc tgttagtgga tgggctatat
      301 acagtaaaga caacagtgta agaatcggtt ccaaggggga tgtgtttgtc ataagggaac
      361 cattcatatc atgctccccc ttggaatgca gaaccttctt cttgactcaa ggggccttgc
      421 taaatgacaa acattccaat ggaaccatta aagacaggag cccatatcga accctaatga
      481 gctgtcctat tggtgaagtt ccctctccat acaactcaag atttgagtca gtcgcttggt
      541 cagcaagtgc ttgtcatgat ggcatcaatt ggctaacaat tggaatttct ggcccagaca
      601 atggggcagt ggctgtgtta aagtacaacg gcataataac agacactatc aagagttgga
      661 gaaacaatat attgagaaca caagagtctg aatgtgcatg tgtaaatggt tcttgcttta
      721 ctgtaatgac cgatggacca agtgatggac aggcctcata caagatcttc agaatagaaa
      781 agggaaagat agtcaaatca gtcgaaatga atgcccctaa ttatcactat gaggaatgct
      841 cctgttatcc tgattctagt gaaatcacat gtgtgtgcag ggataactgg catggctcga
      901 atcgaccgtg ggtttctttc aaccagaatc tggaatatca gataggatac atatgcagtg
      961 ggattttcgg agacaatcca cgccctaatg ataagacagg cagttgtggt ccatgatcgt
     1021 ctaatggagc aaatggagta aaaggatttt cattcaaata cggcaatggt gtttggatag
     1081 ggagaactaa aagcattagt tcaagaaacg gttttgagat gatttgggat ccgaacggat
     1141 ggactgggac agacaataac ttctcaataa agcaagatat cgtaggaata aatgagtggt
     1201 caggatatag cgggagtttt gttcagcatc cagaactaac agggctggat tgtataagac
     1261 cttgcttctg ggttgaacta atcagagggc gacccaaaga gaacacaatc tggactagcg
     1321 ggagcagcat atccttttgt ggtgtaaaca gtgacactgt gggttggtct tggccagacg
     1381 gtgctgagtt gccatttacc attgacaagt aatttgttca
//
```

Fig. 14.3 *A GenBank nucleotide sequence record.* The entry is an H1N1 influenza virus sequence. Selected fields and values are highlighted by shading: definition, accession, version GI number, organism and feature table.

entry has a unique identifier as well as an accession code. Also belonging to the sequence is 'annotation' information, such as: (1) information about what species the sequence is derived from; (2) what scientific papers are associated with the sequence; and (3) cross-references to other database sources. Another important part of the annotation is the 'feature table', containing a description of different sequence elements, i.e. the different functional components found in the sequence. For instance, an expression like 'CDS 113..247' in a feature table means that a protein-coding sequence is between the nucleotide positions 113 and 247. An example Genbank entry is shown in Fig. 14.3, an influenza virus A sequence.

Two quantitatively important categories of sequences are deposited in Gen-Bank/EMBL. *Genomic* sequences are obtained by sequencing of chromosomal DNA, isolated from some source. *EST* (expressed sequence tag) sequences represent mRNA sequences. ESTs are produced in projects designed to characterize the mRNA population in a group of cells and thereby determine what genes are expressed. EST sequences are typically from one single sequencing experiment. As a result they are characterized by a high frequency of errors and they are relatively short and cover only part of the mRNA.

In the early days of DNA sequencing, relatively small pieces of DNA, like individual genes, were sequenced. Now that DNA sequencing technology is more powerful, extensive sequence information is presented in databases, such as complete genomes or chromosomes. The year 2001 offered the first report of the complete human genome (Lander *et al.*, 2001; Venter *et al.*, 2001, International Human Genome Sequencing Consortium, 2004). We now see a development in which many different human individuals are being sequenced. We are therefore able to analyse differences between individuals in a manner that was not possible only a few years ago (see also Chapters 18–20). Sequences are still being deposited to the DNA sequence databases like GenBank and EMBL, but there are also many other sites on the internet where one is able to identify, retrieve and analyse sequences. Examples are Ensembl (http://www.ensembl.org) and the UCSC Genome Bioinformatics (http://genome.ucsc.edu). Not only can a user retrieve information from such sites through a web browser interface; many of these sites have developed interfaces where more bioinformatically oriented users may use programming tools to access the databases. We will see in this chapter one such example for the NCBI query system.

Comparative genomics

Not only humans are analysed with respect to their genomes; many other species have been completely sequenced, including animals, fungi, plants, protozoa, bacteria and archaea. In addition to the classic databases GenBank and EMBL, there are many databases that house sequence information and that are

Table 14.1 *Database statistics*. The number of entries in February 2012 for different molecular biology databases. Counts for different NCBI databases were obtained with the query 'all[filter]'. Pfam and Rfam counts were obtained from the respective home pages.

Database	Number of entries
Nucleotide	164 739 613 (sequences) ($>10^{11}$ nucleotides)
Protein	46 853 302
Structure	78 304
Conserved domains	45 568
SNP	161 770 629
OMIM	22 123
Taxonomy	861 824
PubMed	21 515 764
Swiss-Prot	534 168
Pfam	13 672 families in Pfam-A
Rfam	1973 families

dedicated to specific species or phylogenetic groups, such as the Saccharomyces Genome Database (http://www.yeastgenome.org). Why study the genomes of so many species? One obvious reason is that from a basic scientific perspective we need to understand not only humans. Also, many species are of interest from an applied perspective. For instance, recent sequencing of chicken genomes has important applications when it comes to breeding these animals. Furthermore, many different animals such as mice, rats, zebrafish, the worm *C. elegans* and yeasts are important model systems for studies of human genes and disease. For this reason it is essential to clarify the relationship between humans and the different model organisms. In *comparative genomics* different species are compared with respect to their genomes in order to better understand how they are related. Comparative genomics aids in studies of the evolution of species and allows us to examine the function of different genomic elements. A common procedure in comparative genomics is to produce multiple genome alignments. From such alignments we may, for instance, identify conserved regions that are likely to be biologically significant, such as coding regions (exons) or regulatory regions.

Protein sequence and structure data

While nucleotide sequences are stored in the databases GenBank and EMBL, protein sequences – i.e. sequences of amino acids – are stored in separate databases. The NCBI protein sequence database now has a very large number of protein sequences (Table 14.1). It is certainly highly useful, but it is important

Table 14.2 *Information resources.* A selection of significant database resources in bioinformatics, most of them referred to in this book. See also web resources for this book.

NCBI	Nucleotide	http://www.ncbi.nlm.nih.gov/nucleotide
	Protein	http://www.ncbi.nlm.nih.gov/protein
	PubMed	http://www.ncbi.nlm.nih.gov/pubmed
	Taxonomy	http://www.ncbi.nlm.nih.gov/guide/taxonomy
	OMIM	http://www.ncbi.nlm.nih.gov/omim
	SNP	http://www.ncbi.nlm.nih.gov/snp
	CDD	http://www.ncbi.nlm.nih.gov/cdd
	BLAST	http://blast.ncbi.nlm.nih.gov/Blast.cgi
UCSC Genome Bioinformatics	UCSC genome browser/BLAT/ Table Browser	http://genome.ucsc.edu
EBI	UniProt/Swiss-Prot	http://www.uniprot.org
	InterPro	http://www.ebi.ac.uk/interpro
	Ensembl	http://www.ensembl.org/index.html
Wellcome Trust Sanger Institute	Pfam Rfam	http://pfam.sanger.ac.uk http://rfam.sanger.ac.uk
EMBL	SMART	http://smart.embl-heidelberg.de
GO Consortium	Gene ontology	http://www.geneontology.org
RCSB	Protein Data Bank (PDB)	http://www.pdb.org

to point out that some of the entries in such a large database may by incorrect or not biologically relevant at all. This is because many protein sequences are derived computationally from DNA sequences and rely on prediction of gene structure and translation start and stop sites. These are predictions that are not always correct. Another thing to note about the larger protein sequence collections is that they tend to be redundant, meaning the same protein may be present in different versions. To cope with this problem the NCBI non-redundant RefSeq collections have been created.

If you want to consider a non-redundant protein sequence database of even higher quality, the typical choice is Swiss-Prot, a component of UniProt. A staff of biologists and bioinformaticians working with Swiss-Prot closely inspect different protein sequences to make sure they are correct. They also provide extensive and useful annotation information to each sequence entry. The total number of proteins in Swiss-Prot is small compared to the NCBI protein sequence database (Table 14.2) (for more information on Swiss-Prot, see Chapter 7).

Databases like Swiss-Prot store the linear sequence of amino acids in proteins. (We commonly refer to such an amino acid sequence as the *primary structure* of the protein.) In the cell, the sequence of amino acids will govern the three-dimensional folding of the protein and this structure in turn will have distinct biological functions, as outlined in Chapter 1. The three-dimensional structures of proteins are determined with experimental methods known as X-ray crystallography and NMR. Such structures are deposited in the Protein Data Bank (PDB). Each structure has information about the three-dimensional position of each of the different atoms present in the protein. Not only proteins are found in PDB, but also nucleic acids and nucleic acid–protein complexes. It is still relatively time-consuming to determine the structure of a protein or nucleic acid. Therefore, the growth rate of PDB is moderate compared to the nucleotide and protein sequence databases of GenBank/EMBL.

A limited selection of significant bioinformatics databases and resources are shown in Table 14.2.[4] To give a sense of the level of complexity of different datasets, Table 14.1 shows the number of database entries present in a selection of databases, such as those housed by the NCBI.[5]

BIOINFORMATICS

Tools introduced in this chapter	
Perl	Internet access and LWP module one-liners
Unix	uniq grep

Exploring databases at the NCBI

What tools are available to download and examine data in molecular biology databases? Here we will focus on NCBI search tools, but there are numerous others. 'Genome browsers' such as the UCSC genome browser (http://genome.ucsc.edu) and the Ensembl genome browser (http://www.ensembl.org) (Table 14.2) enable a user to view a genomic region of choice and to explore the contents of that region. These browsers also allow retrieval of data that is the result of simple or advanced queries. Genome browsers, the NCBI query system and many others are freely available tools and are not only of interest in basic science. Thus, such tools are becoming increasingly important in the clinic, and for students of different disciplines such as medicine, biology and bioinformatics, as they are valuable complements to textbooks and other traditional learning material.

---

[4] There are numerous other databases that are also very helpful for the molecular biology and bioinformatics community and the authors of these should not be offended by the fact that they were left out here.

[5] As a bioinformatic comment, the numbers for the NCBI databases were obtained using the query 'all[filter]'. See below in the text for the syntax used in NCBI Entrez queries.

The query system at NCBI (also known as Entrez) allows effective exploration of molecular biology and biomedical topics with the help of a large number of different databases housed at NCBI (Baxevanis, 2008). Databases include those listed in Table 14.2. Not only sequence databases are available. The widely used PubMed is a database of biomedical literature abstracts. The Structure database has information about three-dimensional structure derived from PDB. The OMIM (Online Mendelian Inheritance in Man) database has information about genetic traits and disorders. The SNP database has a collection of single nucleotide polymorphisms from different species (further discussed in Chapters 18–20). Finally, the Taxonomy database, encountered in Chapter 10, contains organisms and their taxonomic classification.

All databases at NCBI have annotation information in which the data is structured into different fields. This information is exploited in queries using Entrez. The default is to search in the complete annotation section, but the search may also be restricted to specific fields of the database. For instance, when you are searching the nucleotide sequence database you may want to limit your search to a specific organism or to a specific publication date. When using the Entrez web interface (http://www.ncbi.nlm.nih.gov/nucleotide in the case of nucleotide records) you may access specific fields using the 'Limits' and 'Advanced Search' alternatives.

An important feature of Entrez is the efficient use of cross-links between databases, so when you have identified a specific record in one database, such as Nucleotide, you can easily find out what entries linked to that nucleotide sequence are available in the other databases.

Most users of NCBI databases make use of the web interface (http://www. ncbi.nlm.nih.gov), but for this chapter we will be somewhat more advanced and command-line oriented as we examine the Entrez Programming Utilities, *eUtils*, command-line and programming tools for making NCBI queries. Before we come to the actual use of these we need to introduce some basic concepts of the Entrez query system. We need to understand: (1) what databases are available for queries; (2) the NCBI query command syntax; and (3) the construction of server script URLs. Once we have covered these elements we are ready to proceed to the actual operation of eUtils. Should you for some reason not be interested in eUtils, but want to learn about NCBI Entrez in general, it is sufficient that you study the section 'NCBI query syntax' below.

NCBI databases in eUtils

The most important databases available for querying with eUtils are listed in Table 14.3. Each entry in the NCBI databases has a specific unique identifier – for example, a GI number for Nucleotide and Protein, and a PMID number for

Table 14.3 *Examples of databases searchable with eUtils.*

Entrez database	Primary ID	eUtils database name
Nucleotide	GI number	nucleotide
		– nuccore
		– nucest
		– nucgss
Protein	GI number	protein
Structure	MMDB ID	structure
Domains	PSSM-ID	cdd
OMIM	MIM number	omim
PubMed	PMID	pubmed
SNP	SNP ID	snp
Taxonomy	TAXID	taxonomy

PubMed entries (Table 14.3). A database entry can always be directly retrieved using such an identifier, as shown further below in the context of the eUtils program EFetch.

NCBI query syntax

Advanced queries at NCBI may be carried out by typing them in the query window of the web interface to the NCBI databases (for instance, http://www.ncbi.nlm.nih.gov/nucleotide), but this usage assumes you know the proper syntax to use. Searches are in the form

```
term1[field1] Op term2[field2] Op term3[field3] Op ...
```

where `Op` is any of the logical operators AND, OR and NOT. The `term1[field1]`, `term2[field2]`, etc. are search terms applied to specific fields. The most important fields available are shown in Table 14.4. If a field is not specified, the search is carried out in all fields. The set of fields available are specific to each database. For the fields listed in Table 14.4, all are available for the Nucleotide and protein sequence databases, except for 'Feature key', which does not apply for the Protein database, and the UID, which is specific for PubMed. Here are now some example queries using this syntax:

(1) Find records in the protein sequence database where any of the fields contain the words 'brca1', 'human' and 'cancer'. Querying the Protein database we can use:

```
brca1 AND human AND cancer
```

Table 14.4 *A selection of search fields using NCBI Entrez.* Information is as provided by NCBI.

Search field	Definition	Qualifier
Accession	Contains the unique accession number of the sequence or record, assigned to the nucleotide, protein, structure, genome record or PopSet by a sequence database builder. The Structure database accession index contains the PDB IDs but not the MMDB IDs.	[ACCN]
All fields	Contains all terms from all searchable database fields in the database.	[ALL]
Author name	Contains all authors from all references in the database records. The format is last name space first initial(s), without punctuation (e.g. marley jf).	[AUTH]
Feature key	Contains the biological features assigned or annotated to the nucleotide sequences and defined in the DDBJ/EMBL/GenBank feature table (http://www.ncbi.nlm.nih.gov/projects/collab/FT/index.html). Not available for the Protein or Structure databases.	[FKEY]
Journal name	Contains the name of the journal in which the data were published. Journal names are indexed in the database in abbreviated form (e.g. J Biol Chem). Journals are also indexed by their by ISSNs. Browse the index if you do not know the ISSN or are not sure how a particular journal name is abbreviated.	[JOUR]
Modification date	Contains the date that the most recent modification to that record is indexed in Entrez, in the format YYYY/MM/DD (e.g. 1999/08/05). A year alone (e.g. 1999) will retrieve all records modified for that year; a year and month (e.g. 1999/03) retrieves all records modified for that month that are indexed in Entrez.	[MDAT]
Organism	Contains the scientific and common names for the organisms associated with protein and nucleotide sequences.	[ORGN]
Properties	Contains properties of the nucleotide or protein sequence. For example, the Nucleotide database's Properties index includes molecule types, publication status, molecule locations and GenBank divisions. A Properties index is not available in the Structure database.	[PROP]

(cont.)

Table 14.4 *(cont.)*		
Search field	Definition	Qualifier
Publication date	Contains the date that records are released into Entrez, in the format YYYY/MM/DD (e.g. 1999/08/05). It is the date the entry first appeared in GenBank explicitly indexed in Entrez. A year alone (e.g. 1999) will retrieve all records for that year; a year and month (e.g. 1999/03) will retrieve all records released into GenBank for that month.	[PDAT]
Sequence length	Contains the total length of the sequence. Sequence length indexes are not available in the Structure or PopSet databases.	[SLEN]
Title word	Includes only those words found in the definition line of a record. The definition line summarizes the biology of the sequence and is carefully constructed by database staff. A standard definition line will include the organism, product name, gene symbol, molecule type and whether it is a partial or complete CDS (a sequence of nucleotides that code for amino acids of the protein product (coding sequence)). Title word indexes are not available in the Structure or PopSet databases.	[TITL]
Uid	Contains the Medline unique identifier for records that contain published references that are linked to PubMed. The Uid index is not browsable.	[UID]

Note that if we leave out the field it is assumed that we search all fields. If the operator is left out the 'AND' operator will be assumed. As an alternative to the expression above we can simply type:

```
brca1 human cancer
```

(2) Find all human nucleotide sequences that have the word 'brca1' in their title. We use the following expression using the Nucleotide database:

```
brca1[title] AND human[orgn]
```

(3) Find all mouse protein sequences that have the expression 'factor IX' in their titles and where the sequence length is within the range 300–500. We do this query using the Protein database:

```
'factor ix'[titl] AND 'mus musculus'[orgn] AND 300[slen]:500[slen]
```

Note the use of quotes in this query. When searching for a phrase we need to put that phrase within quotes, because otherwise an AND operator is assumed between the words. Thus, in this case 'factor AND xi' would search for all entries that contain these words, independent of their relative position.

Table 14.5 *The seven eUtils programs.* Descriptions are from the NCBI.

Program	Function
EInfo	Provides the number of records indexed in each field of a given database, the date of the last update of the database and the available links from the database to other Entrez databases.
EGQuery	Responds to a text query with the number of records matching the query in each Entrez database.
ESearch	Responds to a text query with the list of UIDs matching the query in a given database, along with the term translations of the query.
ESummary	Responds to a list of UIDs with the corresponding document summaries.
EPost	Accepts a list of UIDs, stores the set on the History Server, and responds with the corresponding query key and web environment.
EFetch	Responds to a list of UIDs with the corresponding data records.
ELink	Responds to a list of UIDs in a given database with either a list of related IDs in the same database or a list of linked IDs in another Entrez database.

The Entrez Programming Utilities

The Entrez Programming Utilities (eUtils) are a set of seven programs that provide an interface to the Entrez query and database system at NCBI. The programs are at the server side, so while you supply search terms and parameters, the actual search is carried out at the NCBI server. The eUtils are practical tools for retrieving data from databases such as the GenBank nucleotide database. You could use the standard web interface (http://www.ncbi.nih.gov) for such purposes, but there are situations when you want to be able to do things in a command-line fashion. Thus, you may want to retrieve a specific sequence from within a script, or you may want to automate complex queries such as one in which you use certain text terms and want to retrieve the corresponding sequences in a FASTA format. Such applications are illustrated further below.

The seven eUtils are shown in Table 14.5. In our Perl examples below we will make use of EFetch, ESearch and Elink. Only these eUtils will be described here.

Parameters supplied to eUtils: scripts and construction of URLs

All eUtils scripts have the following basic URL:

http://eutils.ncbi.nlm.nih.gov/entrez/eutils

This is followed by the name of the eUtils program. For instance, as applied to EFetch:

http://eutils.ncbi.nlm.nih.gov/entrez/eutils/efetch.fcgi?

Table 14.6 *URL parameters used with the eUtils.* Information is based on NCBI documentation.

Parameter	Description/comment	Example
Database	Examples of valid eUtils database names are in Table 14.3	db=nucleotide
Web environment (WebEnv)	History link value previously returned in results from ESearch	WebEnv=WgHmIcDG]
Query key	The value used for a history search number or previously returned in results from Esearch	query_key=6
Email address		email=someone@somewhere
Record identifier	IDs required if WebEnv is not used	id=1234,U12345
Display numbers (retstart retmax)	* retstart – sequential number of the first id retrieved – default is 0, which will retrieve the first record * retmax – number of items retrieved	retstart=100&retmax=50
Parameters specific to sequence databases	strandseq_startseq_stop	strand=2&seq_start=50&seq_stop=2000
Retrieval mode or output format	Output format Current values: * xml * html * text * asn.1	retmode=text
Retrieval type	Output types based on database. Note that not all retrieval modes are possible with all retrieval types (see full documentation at NCBI)	rettype=gbwithparts (full record) rettype=fasta (sequence in FASTA format) rettype=ft (feature table report) rettype=acc (to convert list of gis to list of accession codes)

We also need to supply a number of parameters and values. The most important parameters available are listed in Table 14.6. The final query URL then has the form:

http://eutils.ncbi.nlm.nih.gov/entrez/eutils/efetch.fcgi?key1=value1&key2=value2& . . .

where the key–value pairs are separated by an & character. Spaces should be avoided in such a URL. In case they are needed, a plus sign (+) should be used in place of the blank space. Here are two examples of URLs using EFetch:

(1) Returning the sequence with accession code 'U01305.1' in a FASTA format:
    http://eutils.ncbi.nlm.nih.gov/entrez/eutils/efetch.fcgi?db=
        nucleotide&id[PDAT]=U01305.1&rettype=fasta&retmode=text
(2) Getting the PubMed IDs (PMIDs) for articles about breast cancer published in *Science* in 2008:
    http://eutils.ncbi.nlm.nih.gov/entrez/eutils/esearch.fcgi?db=
        pubmed&term=science[journal]+AND+breast+cancer+AND+
        2008[pdat]

You do not actually need a script for these particular queries. You can try the URLs out by pasting them into the address (location) bar of your web browser. In addition, you can use `wget` at the Unix command line (see Appendix I), but now things get somewhat awkward as you need to introduce a backslash in front of the ? and & characters:

```
% wget URL
```

where `URL` is any of the URLs above, but where all ? and & are replaced by \? and \&.

EFetch

We are now ready to see how EFetch is used in a Perl script to retrieve a set of entries for which we know the accession code or the unique identifier (accession in the example below). The base URL is:

http://eutils.ncbi.nlm.nih.gov/entrez/eutils/efetch.fcgi?

Our first short example script is designed to retrieve the feature tables of three influenza virus nucleotide entries. The retrieval is based on their accession codes (Code 14.1).

Code 14.1 efetch.pl

```perl
#!/usr/bin/perl -w

use strict;

use LWP::Simple;

my $ac = 'CY073775.1, CY022055.1, U47817.1';

my $efetch = "?".
"http://www.ncbi.nlm.nih.gov/entrez/eutils/efetch.fcgi?".
```

```
"db=nucleotide&id=$ac&rettype=ft&retmode=text" ;

my $efetch_result = get($efetch);
print "$efetch_result\n";
```

What is LWP::Simple in Code 14.1? It is an example of a Perl module, just like strict, which we encountered before. The LWP::Simple module is one that introduces web client functions.[6] The only LWP function being used in the eUtils scripts is the function named get. This function takes as an argument a document identified by a URL and returns it. For instance, we may in a Perl program use the following:

```
$page = get (http://www.perl.org/index.html);
```

The result of this operation is that the variable $page will receive the contents of the web page http://www.perl.org/index.html.

In our script the variable $efetch_result will, with the help of get, contain the result returned from the server efetch.cgi script. What did we learn from the output of this script? Not so much, but according to the feature table all three sequences encode a neuroaminidase gene.

ESearch

We have seen in the previous example that retrieving documents with EFetch is really simple, assuming we know the unique identifier or the accession code. However, in most cases we do not know these. In such a case we first need to carry out a search based on other fields and we use ESearch to do that. ESearch searches and retrieves unique identifiers. In addition, it is able to 'remember' these and use them in another operation such as with EFetch. Here we will explore how a script is constructed with ESearch.

This chapter started by referring to influenza viruses. For more background information on influenza viruses, see Box 14.1. We will, for the following example, think of a specific application for which we want to explore influenza virus entries present in the Nucleotide database. More specifically, we want to identify sequences of H1N1, the virus subtype that was responsible for the 1918 Spanish flu, as well as the 2009 swine flu. Furthermore, there are eight different segments of the influenza virus genome (see Box 14.1); for this example we assume we want to retrieve the fragments described as 'segment 6'. Finally, we would like to have the sequences in a FASTA format.

---

[6] LWP is short for 'Library for WWW in Perl'.

Code 14.2 esearch.pl

```perl
#!/usr/bin/perl -w

use LWP::Simple;

my $db     = "nucleotide";
my $query  = "h1n1 \"segment 6\" influenza a virus";
my $report = "fasta";
my $retmax = "100";   # this number should be made larger to
                      # retrieve a complete set of entries

my $Base_URL = "http://www.ncbi.nlm.nih.gov/entrez/eutils";

my $esearch =
  "$Base_URL/esearch.fcgi?" .
  "db=$db&retmax=1&usehistory=y&term=$query";

my $esearch_result = get($esearch);
$esearch_result =~

m|<Count >(\d+)</Count>.*<QueryKey>(\d+)</QueryKey>.*<WebEnv>(\S+)
</WebEnv>|s;

my $Count    = $1;
my $QueryKey = $2;
my $WebEnv   = $3;

# uncomment these print statements in case you
# want see the values in the output
# print "Count $Count\n";
# print "QueryKey $QueryKey\n";
# print "WebEnv $WebEnv\n";

my $efetch = "$Base_URL/efetch.fcgi?" .
             "db=$db&rettype=$report&retmax=$retmax".
             "&query_key=$QueryKey&WebEnv=$WebEnv";

my $efetch_result = get($efetch);
print "$efetch_result\n";
```

Our script to use ESearch is shown in Code 14.2. First, let us examine the code for the ESearch part of this script. The URL parameters we are using for ESearch are db, retmax, history and term (see Table 14.6). The retmax parameter specifies the maximum number of entries to retrieve. You typically do not know how many entries will match a certain query. Therefore, when you run a script like this for the first time it may be a good idea to restrict the number of entries to retrieve. For this reason, retmax is here set to 100. If you uncomment the line # print "Count $Count\n"; you will see in the output from the script that the number of hits in the search is much greater (10 899 in February 2012, but this may have changed as you read this). In order to retrieve all of these hits, you need to change the `retmax` value to something equal to or greater than this number.

Box 14.1 Influenza viruses

Influenza viruses are the cause of influenza,[7] or the flu, not to be confused with the common cold. Mammals and birds are affected. The most common symptoms are fever, sore throat and coughing. A human influenza epidemic occurs when an influenza virus appears in the human population and effectively is passed on between human individuals. Typically the virus originates from a non-human source, like birds or pigs. A critical step therefore in the development of an epidemic is the transfer of the virus from a non-human animal to humans. Between 250 000 and 500 000 human individuals die each year as a result of influenza virus infection. There have been three major pandemics in the twentieth century; the most famous and devastating was the Spanish flu, which lasted from 1918 to 1919. It killed about 50 million people.

There are three major types of influenza virus: A, B and C. A is the most important in terms of human disease. There are many variants of the influenza A virus. One example is the H1N1 virus, which was responsible for the Spanish flu of 1918 and the 2009 swine flu pandemic. Other variants are H2N2 (Asian flu of 1957), H3N2 (Hong Kong flu of 1968) and H5N1 (Bird flu of 2004).

Influenza virus isolates are described with a nomenclature showing, from left to right: (1) the major type: (2) the origin (animal or geographical location in case of a human virus); (3) the strain number; (4) year of isolation; and (5) virus subtype (within parentheses). Examples are:

```
A/Jiangsu/2/2009(H1N1)
A/swine/Italy/290271/2009(H1N1)
```

The genomes of influenza viruses are RNA molecules, like the HIV retrovirus previously encountered in Chapter 9. All influenza A virus genomes are divided up into eight different segments, or molecules of RNA (Fig. 14.4). Segment size ranges from 890 to 2341 nt, and the total size of the genome is 13 500 nt.

The genome encodes 11 different proteins, haemagglutinin (HA), neuraminidase (NA), nucleoprotein (NP), M1, M2, NS1, NS2 (NEP), PA, PB1, PB1-F2 and PB2 (Fig. 14.4). The HA and NA are two glycoproteins located on the outside of the virus particle. HA is important in the process by which the virus binds and enters target cells. NA is an enzyme that is important for the release of viruses from infected cells. There are inhibitors of NA, such as *zanamivir*, which block the release of the virus and are used to combat influenza virus infection. The HA and NA are important antigenic determinants. The virus subtype, e.g. 'H1N1', refers to the categories of HA and NA present in that virus.

---

[7] The word 'influenza' is an Italian word, meaning 'influence' and initially referring to the disease as being the result of unfavourable astrological influences.

Influenza viruses have a complex mode of evolution. If a host cell is infected with two different viruses, a new virus particle may obtain some of its segments from one of the viral strains and other segments from the other strain. This mixing of genetic information is known as *reassortment*. There are human pandemic influenza A viruses that were created by reassortment of avian and human viruses. For instance, in 1957 a novel human–avian H2N2 reassortant virus appeared in the human population ('Asian flu'). This virus had the HA, NA and PB1 segments of an avian virus, but the other segments originated from human H1N1.

The 'swine flu' of 2009 was a result of an H1N1 virus with a complex evolutionary history. The older history of the virus is in birds, but it evolved further in pigs and humans. The 2009 virus was derived from a North America swine virus that had received its N and M segments from European swine lineages (Brockwell-Staats *et al.*, 2009; Garten *et al.*, 2009; Smith *et al.*, 2009).

The result of the query is returned in the variable named `$esearch_result`. By default this result is in XML format. XML has the basic syntax `<name>content</name>`. For instance, the number of entries identified in the search is represented in the following manner:

```
<Count>10 899</Count>
```

We want to capture the count value, as well as the QueryKey and WebEnv variables (see Table 14.6). We do this with a pattern-matching operation:

```
$esearch_result =~
m|<Count>(\d+)</Count>.*<QueryKey>(\d+)</QueryKey>.*
<WebEnv>(\S+)</WebEnv>|s;
```

We are used to having a slash (/) as the delimiter in pattern matching. Other symbols may actually be used, but then we must have an 'm' in front, as above, indicating that we want to use the matching operator. In this example a pipe (|) is used instead of a slash because we want to avoid confusion with the slash symbols being part of the pattern. (Had we instead used slashes as pattern delimiters, all other slash symbols would need to be preceded by a backslash, so each one would look like this: \/.)

In the pattern above, the \d is a Perl meta-symbol that refers to a digit and \S refers to any non-whitespace character (see also Appendix III).

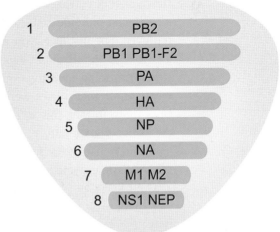

Fig. 14.4 *Segments of an influenza virus genome.* An influenza virus genome is divided up into eight different RNA molecules. For more information on the genome, see Box. 14.1.

We now change the `$retmax` variable to 10 899 (or a larger number) in Code 14.2 and run the script:

```
% perl esearch.pl > influenza.fa
```

A larger number of sequences will be retrieved. These are segment 6 of the H1N1 influenza viruses. With a script like that shown in Code 14.2 it is easy to modify the search. For instance, we may want instead to download the sequences in a GenBank format, with all the annotation information. This may be accomplished simply by changing the `$report` variable to `"gbwithparts"`.

Extensive data may be downloaded with eUtils. Before going too wild with eUtils, the reader is encouraged to read the user restrictions available on the NCBI web pages. The major take-home message is: 'Do not overload NCBI's systems'!

So far we have reached the conclusion that eUtils are useful when downloading lots of data in a Unix environment. It is important to note that the downloading is independent of a web browser, so we do all this from the command line in Unix. There is also the advantage that retrieval of entries from the NCBI database may be done from within a more complex script that may contain a whole pipeline of events. Thus, the script may be generating the search terms to be used, initiating the actual query and processing further the results of the query.

Further analysis of influenza viruses: extracting and filtering information with Perl and Unix tools

Downloading entries from the NCBI databases alone does not normally answer a particular biological question. Rather, you need to analyse the data you have retrieved. For instance, you may want to subject sequences to a phylogenetic analysis, as described in Chapters 9 and 10. The coming examples show possible steps prior to an investigation of phylogeny. We first want to understand from what source the viruses were isolated. We can easily find out using Perl and Unix tools. In fact, this information would have been difficult to extract using the Entrez system only.

We make use of the FASTA definition line, which is of the format

```
>gi|306412420|gb|CY073775.1| Influenza A virus (A/Vienna/INS369/
2010(H1N1)) segment 6, complete sequence
```

Here, the location (Vienna) is found as the second element in the expression within parentheses. We may want to print all locations that are present in our collection of sequences. We can do that with the simple script in Code 14.3.

Code 14.3 short.pl

```
while (<>) { # while reading standard input
  m|\(A/(.*?)/|; # Match the string which is between 'A/' and the
                 # next '/'. Because there are multiple '/' we need
                 # a '?' to specify minimal matching instead of
                 # greedy.
  print "$1\n";
}
```

This script illustrates how data produced on the command line may be fed into a Perl script. We use the Unix `cat` command to produce the contents of the file `influenza.fa`. But instead of these contents being printed to the screen, they are 'piped' to the Perl code using the following command line:

```
% cat influenza.fa | perl short.pl
```

In fact, this being a Perl script of very few lines, we can also make use of the following command line:

```
% cat influenza.fa | perl -ne 'm|\(A/(.*?)/|;print "$1\n" '
```

This line looks a bit messy, but the `"perl -ne ...."` expression is an example of a Perl *one-liner* (see also Appendix III). It will achieve exactly the same thing as the previous Unix command line, without having to use Perl code in a separate file.

The result of our analysis of the file `influenza.fa` is a very long list of locations, so we may want to count the number of occurrences of each. We could do that by modifying the Perl code above. However, another possibility is to direct the output from the previous Perl code to the Unix commands `sort` and `uniq` (enter in one line):

```
% cat influenza.fa | perl -ne 'm|\(A/(.*?)/|;print "$1\n" ' | sort |
uniq -c | sort -n
```

The first `sort` command will sort the lines coming from Perl. Having such a sorted list, `uniq -c` will produce the unique lines together with information about how many instances there are of each. The final `sort -n` will sort numerically the output of `uniq`. If we examine the output we observe a large number of different locations, such as New York, Wisconsin, Texas, etc. Whenever the sequences are of non-human origin the species name, such as 'swine', will be listed instead of a location. We also note that all of the non-human sequences begin with a lower-case letter. This means that if we want to identify these non-human animals specifically we can direct output of the command above to yet another Unix command, `grep ^[a-z]*`. `grep` is a utility to search for strings

Fig. 14.5 *Swine flu virus phylogeny*. A selection of H1N1 segment-6 sequences were subjected to phylogenetic analysis by alignment with ClustalW, followed by a neighbour-joining procedure of the same program. Human isolates of 2009 (cyan background) are seen to be closely related to swine isolates of the same year (grey background). In contrast, bird isolates (green background) are more distantly related. These results are consistent with the idea that the human 'swine flu' outbreak in 2009 originated from swine.

or regular expressions in files. It uses a syntax for regular expressions similar to that in Perl. The expression `^[a-z]*` means we are searching all words that start with a lower-case letter. For more information on `grep`, see Appendix I.

The reader is encouraged to experiment further with Perl code and the different Unix commands to extract information from the `influenza.fa` file. Of course, instead of using all the different Unix functions `cat`, `sort`, `uniq` and `grep`, we could have used one single Perl script to accomplish these tasks. However, the examples above illustrate that, rather than writing a separate extensive Perl script, it is sometimes more convenient to work with a command line that is modified as you work and discover new ways you want to process the data produced by the original command.

Once we are done with analysis of the definition lines, we may be ready to extract a number of sequences for phylogenetic analysis. The result of one such analysis in Fig. 14.5 shows that the human viral strains of the 2009 'swine flu' pandemic isolates are closely related to the swine isolates, whereas the avian strains are more distant. By this type of analysis for the different segments of the H1N1 virus we may further explore the complex mode of evolution of the 2009 'swine flu' virus, as alluded to in Box 14.1.

ELink

Finally, as an example of the ELink utility, the script in Code 14.4 shows how protein records associated with nucleotide sequences may be retrieved. We input GI numbers for three different nucleotide sequences. The Protein database links to those sequences are identified by the ELink script. These are, in turn, used by EFetch to retrieve the protein sequences.

Code 14.4 elink.pl

```perl
#!/usr/bin/perl

use strict;
use LWP::Simple;

# Download protein records linked to a set
# of nucleotide records corresponding to a list
# of GI numbers.

my $dbfrom = 'nuccore';          # linking from
my $db = 'protein';              # linking to
my $linkname = 'nuccore_protein'; # desired link name

# input UIDs in $dbfrom (protein GI numbers)
my $id_list = '306412420,306412402,306412365';

# elink URL
my $base = 'http://eutils.ncbi.nlm.nih.gov/entrez/eutils/';
```

```
my $url = $base . "elink.fcgi?dbfrom=$dbfrom&db=$db&id=$id_list";
$url .= "&linkname=$linkname&cmd=neighbor_history";

my $elink_result = get($url);

# parse WebEnv and QueryKey
my $web = $1 if ($elink_result =~ /<WebEnv>(\S+)<\/WebEnv>/);
my $key = $1 if ($elink_result =~ /<QueryKey>(\d+)<\/QueryKey>/);

# assemble the efetch URL
$url = $base . "efetch.fcgi?db=$db&query_key=$key&WebEnv=$web";
$url .= "&rettype=fasta&retmode=text";

my $data = get($url);
print "$data";
```

EXERCISES

14.1 Design NCBI Entrez queries with the proper syntax to retrieve documents as specified below. For the syntax used, see text, Table 14.4 and NCBI documentation as listed at the end of this chapter. You may build the queries and try them out at the NCBI Entrez web interface (http://www.ncbi.nlm.nih.gov/pubmed) with the 'advanced search' option.

    (a) Retrieve from the Nucleotide database sequences from *Thylacinus cynocephalus* (thylacine) with a maximum length of 1000 nucleotides.

    (b) Retrieve from the Protein database all mammalian proteins with 'foxp2' in the title.

    (c) Retrieve from PubMed all articles of Eugene V. Koonin published during the years 2005–2010.

    (d) Retrieve from PubMed articles with the word 'cancer' in the title and that have been published in *Nature* or *Science* during the years 2008–2011.

14.2 Create a Perl script based on ESearch and EFetch to carry out the nucleotide database query below (compare Code 14.2). Design the script such that the sequences in FASTA format are retrieved.

```
"factor ix"[titl] "mus musculus"[orgn] 1000[slen]:3000[slen]
```

14.3 The output of Code 14.2 based on the ESearch utility contains a number of sequences in FASTA format (`influenza.fa`). Write a Perl program to extract from the output file `influenza.fa` all H1N1 sequences of 2009 isolates with a minimum length of 1000 nucleotides.

NCBI DOCUMENTATION RESOURCES

Entrez: http://www.ncbi.nlm.nih.gov/books/NBK3837
eUtils: http://www.ncbi.nlm.nih.gov/books/NBK25501

REFERENCES

Andersen, J. S. (1984). *Nucleotide Sequences 1984: A Compilation from the Genbank and EMBL Data Libraries – A Special Supplement to Nucleic Acids Research*. Oxford, IRL Press.

Armstrong, J., GenBank and Laboratory of European Molecular Biology (1985). *Nucleotide Sequences 1985: A Compilation from the GenBank and EMBL Data Libraries – A Special Supplement to Nucleic Acids Research*. Oxford and Washington, DC, IRL Press.

Baxevanis, A. D. (2008). Searching NCBI databases using Entrez. *Curr Protoc Bioinformatics*, Chapter 1, Unit 1.3.

Benson, D. A., I. Karsch-Mizrachi, D. J. Lipman, J. Ostell and E. W. Sayers (2010). GenBank. *Nucleic Acids Res* **36**(1), D25–D30.

Bilofsky, H. S., C. Burks, J. W. Fickett, *et al.* (1986). The GenBank genetic sequence databank. *Nucleic Acids Res* **14**(1), 1–4.

Brockwell-Staats, C., R. G. Webster and R. J. Webby (2009). Diversity of influenza viruses in swine and the emergence of a novel human pandemic influenza A (H1N1). *Influenza Other Respi Viruses* **3**(5), 207–213.

Dayhoff, M. O. and National Biomedical Research Foundation (1965). *Atlas of Protein Sequence and Structure*. Silver Spring, MD, National Biomedical Research Foundation.

Fiers, W., R. Contreras, F. Duerinck, *et al.* (1976). Complete nucleotide sequence of bacteriophage MS2 RNA: primary and secondary structure of the replicase gene. *Nature* **260**(5551), 500–507.

Garten, R. J., C. T. Davis, C. A. Russell, *et al.* (2009). Antigenic and genetic characteristics of swine-origin 2009 A(H1N1) influenza viruses circulating in humans. *Science* **325**(5937), 197–201.

Holley, R. W., J. Apgar, G. A. Everett, *et al.* (1965). Structure of a ribonucleic acid. *Science* **147**, 1462–1465.

International Human Genome Sequencing Consortium (2004). Finishing the euchromatic sequence of the human genome. *Nature* **431**(7011), 931–945.

Lander, E. S., L. M. Linton, B. Birren, *et al.* (2001). Initial sequencing and analysis of the human genome. *Nature* **409**(6822), 860–921.

National Biomedical Research Foundation, R. V. Eck and M. O. Dayhoff (1966). *Atlas of Protein Sequence and Structure, 1966*. Silver Spring, MD, National Biomedical Research Foundation.

Smith, G. J., D. Vijaykrishna, J. Bahl, *et al.* (2009). Origins and evolutionary genomics of the 2009 swine-origin H1N1 influenza A epidemic. *Nature* **459**(7250), 1122–1125.

Strasser, B. J. (2009). Collecting, comparing, and computing sequences: the making of Margaret O. Dayhoff's *Atlas of Protein Sequence and Structure*, 1954–1965. *J Hist Biol* **43**(4), 623–660.

Stretton, A. O. (2002). The first sequence: Fred Sanger and insulin. *Genetics* **162**(2), 527–532.

Venter, J. C., M. D. Adams, E. W. Myers, *et al.* (2001). The sequence of the human genome. *Science* **291**(5507), 1304–1351.

15 Finding genes
Going ashore at CpG islands

Elucidation of the human genome sequence was a significant milestone in the life sciences (Lander *et al.*, 2001; Venter *et al.*, 2001; International Human Genome Sequencing Consortium, 2004). However, with access to this information an obvious but entirely non-trivial problem was encountered. What does all the genetic information in the form of some three billion bases represent in biological terms? One important category of information is the sequences that specify genes, i.e. regions that give rise to mRNAs that in turn encode specific protein molecules. Not only proteins are specified by the genome; also a large number of RNAs are transcribed from DNA that do not give rise to mRNA, but have other functions. (These are *non-coding RNAs* and will be discussed in Chapter 17.) In the next three chapters we will deal with the computational problems of finding proteins and non-coding RNA genes, starting out with a genomic sequence.

When it comes to the protein-coding genes of a mammalian genome, only a very small fraction, about 1.5–2%, of the genome codes for protein. For these genomes we are faced with the problem of identifying relatively small and scattered coding regions in a vast sea of non-coding material. There is a striking difference in this respect between mammals and a bacterium like *Escherichia coli*, whose genome contains as much as 83% of coding sequence. In the next chapter the focus will be on prediction of exon regions of protein-coding genes. Here we will address another sub-problem of finding protein-coding genes. We will see how a simple prediction of regions known as CpG islands will help us to locate sites in the genome that are close to the transcription start sites of genes.

Transcription and its regulation

Transcription of genes takes place with the help of the enzyme RNA polymerase. In eukaryotes there are three different classes of RNA polymerases, named I, II and III. RNA polymerase II is responsible for the transcription of protein-coding genes. In addition to this enzyme, other factors, referred to as *transcription factors*, are required for initiation of RNA polymerase II transcription. We may distinguish two different categories of transcription factors. First, the 'general' transcription factors that are part of a basic machinery of RNA polymerase II transcription, responsible for initiation of transcription of *all* protein genes. Second, there are more specific transcription factors that are involved in the regulation of a subset of genes, or perhaps even a single gene. Such factors may either enhance or repress gene expression.

Important for the efficiency of transcription is also the availability of DNA sequences for the transcription apparatus. Eukaryotic DNA molecules are very long. Imagine that you would be able to take a single DNA molecule making up a human chromosome and then extend it fully (but while maintaining its double-helical structure).[1] In such a case it would be as much as several centimetres in length. Clearly, in order to fit within a nucleus where it normally belongs, the DNA needs to be extensively compacted. There are different levels of compaction, but one important mechanism is the packing of DNA into *nucleosomes*, which is possible with the aid of *histone* proteins. In each nucleosome unit the DNA is wound around an octamer of histone proteins. In addition to histones, other proteins are attached to DNA. We refer to the complex of proteins and DNA in the nucleus of a eukaryotic cell as *chromatin*. The structure of chromatin is important for transcriptional efficiency. When DNA is in a highly condensed form, referred to as *heterochromatin*, gene expression is typically turned off.

The packing of DNA and the structure of chromatin are far from static. Chromatin structure is subject to control, and important players are a vast number of enzymatic activities that may introduce chemical modifications to histones. Thus, histones may, for example, be modified by acetylation, methylation, phosphorylation and ubiquitylation. As one example, the acetylation of lysines in the N-terminal tails of histones tends to have the effect of opening up chromatin structure, thereby stimulating gene expression. In addition, the different modifications of histones attract a number of protein factors that can influence the expression of a gene.

DNA sequences that influence transcription

What about the role of DNA sequences in transcription? For the problem of computationally identifying genes in genomes, and in order to improve our understanding of the process of gene transcription, we want to investigate characteristics of sequences associated with transcription initiation. Such sequences could either be sequences that attract the basic machinery of transcription initiation, or more specific sequences that regulate the expression of a subset of genes.

---

[1] In practical terms this is a very difficult operation.

In the case of the more gene-specific transcription factors, there are many examples in which these proteins recognize specific sequence elements in a region upstream of the transcription start site. One example is the oestrogen receptor, which recognizes a sequence AGGTCANNNTGACCT (where N is any nucleotide), referred to as the oestrogen response element. The binding of the oestrogen receptor will in this case activate gene expression.

For the problem of identifying the transcription start site of all genes, however, we would be more interested in knowing about DNA sequences that are characteristic of the basal transcription machinery. One example of a sequence motif commonly associated with genes is the 'TATA box'. This motif is recognized by the TATA binding protein, a subunit of one of the general transcription factors, TFIID. However, only 10–25% of all human genes have a recognizable TATA box, and for this reason it is of limited value for the problem of identifying transcription start sites.

CpG islands

A more common feature of transcription initiation regions are the *CpG islands* (Illingworth and Bird, 2009). These are regions close to the transcription start site of approximately 60–70% of all human genes. 'CpG' refers to the dinucleotide sequence CG, where the 'p' represents the phosphodiester linkage between C and G. The 'p' is there to avoid confusion with a base pair formed between C and G, which is something entirely different. As yet more nomenclature, a single dinucleotide CpG is often referred to as a *CpG site*.

In regions described as CpG islands, the frequency of CpG sites is much higher than the background frequency of this dinucleotide. By and large, CpG sites are very rare in vertebrate genomes. As an example, consider the human chromosome 4. The frequencies of C and G are both 0.19. If these bases were randomly distributed you would expect the frequency of CpG sites to be $0.19 \times 0.19 = 0.036$. However, the observed frequency is 0.0078, i.e. only 20% of the expected. It is commonly believed that the low frequency of CpG sites is because the C in a CpG sequence is frequently methylated and the methylated C will occasionally be converted through spontaneous deamination to a T (thymine).[2] Over evolutionary time the CpG sequence will tend to be converted to TpG. The reason CpG islands near the transcription start site in a large fraction of genes have been retained during evolution could be that in regions close to transcription start sites, CpG sites tend not to be methylated. Hence, they are less likely to be changed by deamination from C to T.

---

[2] This is unlike unmethylated C, where deamination will result in a U (uracil), an error typically corrected with the help of a specific DNA repair mechanism.

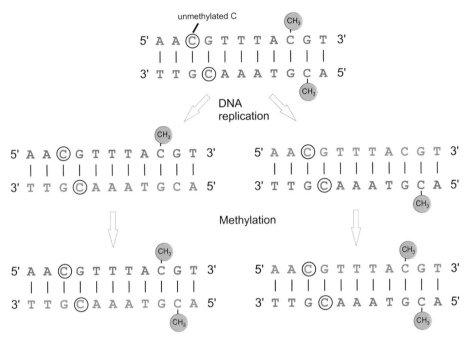

Fig. 15.1 *DNA methylation patterns are epigenetically inherited.* In a chromosome, some CpG sites are methylated and others are not. In a CpG site, both strands are methylated (top). A newly synthesized strand is not methylated but the enzyme DNA methyltransferase can methylate a CpG sequence which is paired with a methylated CpG. However, a non-methylated CpG which is paired to another unmethylated CpG cannot be methylated by this enzyme. In this manner the methylation pattern of a parental DNA is copied to the DNA of a daughter cell.

Methylation of C in CpG sites is carried out by enzymes – DNA methyltransferases. The majority of CpG sites (about 70%) in a mammalian genome are methylated, but the CpG islands associated with transcription start sites are typically not methylated. What is the role of DNA methylation? It seems that in general such modification will effectively turn gene expression off. Therefore, DNA methylation is important in the control of gene expression. There are also examples where disregulation involving CpG islands gives rise to disease. For instance, CpG islands play a role in cancer development in which tumour suppressor genes may be inactivated by DNA methylation.

CpG sites are also interesting from the perspective that the methylation pattern in a specific cell may be transmitted to the progeny cells, an example of *epigenetic inheritance.* A likely mechanism for the inheritance of DNA methylation is shown in Fig. 15.1. In fact, it would seem that one of the most important functions of DNA methylation is that information for stable repression of a set of genes may be passed on to the next generation. There is actually a close relationship between such modifications and DNA methylation. They work in concert to regulate gene expression and, like DNA methylation patterns, the patterns of histone modifications may be epigenetically inherited.

DNA methylation is also an important basis for *genomic imprinting*, i.e. the phenomenon that a gene may be differently expressed depending on whether it is of maternal or paternal origin. As an example consider the human gene H19, which encodes a non-coding RNA. In male germ cells this gene is methylated, but it is not in female germ cells. After fertilization the resulting embryo contains a paternal allele which is methylated (and inactive) together with a maternal allele which is unmethylated. Thus, the maternal allele is active whereas the paternal allele is not.

BIOINFORMATICS

Finding CpG islands

CpG islands are located close to the transcription start site of many genes and they are of great importance in regulating transcription in vertebrates. Indeed, different methods have been employed to predict them computationally (Zhao and Han, 2009). What is characteristic of the CpG islands? Originally they were defined as regions at least 200 base pairs in length with a GC content of at least 50% and an observed CpG/expected CpG ratio greater than 0.6 (Gardiner-Garden and Frommer, 1987). This definition is rather arbitrary, as is almost any definition of a CpG island you may encounter. A drawback with this original definition is that many *Alu elements* of the human genome will be incorrectly predicted as CpG islands. Alu elements are mobile elements with a GC and CpG site composition similar to CpG islands; there is in the order of one million copies of them in the human genome. To account for this problem as well as to make the prediction more specific in general, modifications to the original method of CpG island prediction were suggested by Takai and Jones (2002). It is their protocol that we will be using here. There are also alternative methods that have been used for predicting CpG islands, such as methods based on HMMs[3] (Durbin *et al.*, 2007). It is important to realize that any method for CpG island prediction is unreliable in the sense that we do not for any larger region of a genome have information about the location of all functional CpG islands. The most important criterion for optimizing specificity in methods of CpG island prediction is that CpG islands should be located close to known transcription

---

[3] HMMs applied to CpG island prediction is a common textbook example to illustrate the use of HMMs in sequence analysis. Here, we will brutally avoid such probabilistic models and use a more commonplace method.

start sites. A problem with this criterion is that it is most likely that not all transcription start sites have been experimentally identified.

We will now apply the Takai and Jones method to design a simple method for CpG island prediction. Although you will find that this method has been already implemented and is available as a web resource (Takai and Jones, 2003), our discussion here will illustrate that the very simple Perl script in Code 15.1 will do the trick. The criteria to be implemented are:

(1) The total length of the CpG island should be at least 500 nt;
(2) The content of C + G should be at least 55%;
(3) The ratio between the frequency of observed CpG sites and the frequency of expected CpG sites should be at least 0.65.

The ratio in (3) is calculated as:

$$\frac{fc_{pG}}{fc\, f_G} = \frac{SN_{CpG}}{N_C\, N_G}$$

where S is the length of the sequence, f represents frequency and N counts.

Code 15.1 cpg.pl

```perl
#!/usr/bin/perl -w

use strict;

my $win = 500;
my $step = 10;
my $seq = '';

my $infile = 'short.fa';          # a shorter sequence
# my $infile = 'chr4_region.fa'; # alternative sequence
open(IN, $infile) or die "Could not open $infile\n";
while (<IN>) {
        unless (/>/) {
        chomp;
        $seq.=$_;
        }
}
close IN;

print "pos\tcpg\tcg_ratio\tcg_obs_exp\n";

for (my $i = 0; $i < length($seq) - $win + 1; $i += $step) {
        my $testseq = substr($seq, $i, $win);
        my $c = ($testseq =~ s/C/C/g);
        my $g = ($testseq =~ s/G/G/g);
        my $cg = ($testseq =~ s/CG/CG/g);
        my $cg_ratio = ($c + $g) * 100 / length($testseq);
```

```
        my $cg_obs_exp = ($cg * length($testseq)) / ($c * $g);
        my $pos = $i + $win/2;
        if (($cg_ratio >= 55) && ($cg_obs_exp >= 0.65)){
            print "$pos\t1\t$cg_ratio\t$cg_obs_exp\n";
        }
        else {print "$pos\t0\t$cg_ratio\t$cg_obs_exp\n";}
    }
```

Code 15.1 has few technical novelties compared to what we have encountered already in this book. The sequence in FASTA format is being read from a file. We are using a sliding window to analyse the sequence, as in previous chapters. In this case the window size is 500 (this is required as a minimum size for the CpG island) and we move the window in steps of 10. We use a `for` loop:

```
for (my $i = 0; $i < length($seq) - $win + 1; $i += $step) {}
```

We increase `$i` in steps of `$step` (which is 10). The expression `$i += $step` is shorthand for `$i = $i + $step`. (Remember, Perl programmers are lazy typists!) The loop is terminated when we reach `$i` at equal to or greater than `length($seq) - $win`.

We count the Cs, Gs and CpG sites using the substitution `s///` operator, as in this example:

```
my $cg = ($testseq =~ s/CG/CG/g);
```

Previously in this book (Chapter 3) we used an expression with the transliteration operator `tr///` for counting. However, in this case `tr///` will not be useful at all, because the transliteration only works with individual characters, not with a regular expression or set of characters like 'CG'. Thus, if you were to use `tr/CG/CG/`; you would simply count the Cs and Gs individually (see also Appendix III).

We finally print in a tab-separated manner, for each window tested, (1) the nucleotide position, (2) '1' if the criteria for a CpG island are met, '0' if they are not, (3) the GC content and (4) the observed/expected ratio of CpG sites. We save the output of the script in a file:

```
% perl cpg.pl > cpg_short.out
```

The output from the Perl script is illustrated graphically in Figure 15.2, in which a predicted CpG island is shown as the shaded region.

Please note that Code 15.1 is not optimal from a computational point of view as we use overlapping windows and count all Cs, Gs and CGs multiple times. See Exercise 15.3 for more on this topic.

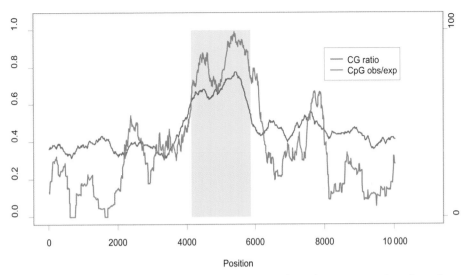

Fig. 15.2 *Prediction of CpG islands in a shorter nucleotide sequence*. A region of
10 000 nucleotides was analysed with respect to CpG islands as described in the text. For a region to
qualify as a CpG island we require that the content of C + G ('CG ratio' in the plot) should be at least
55%. Furthermore, the ratio between the frequency of observed CpG sites and the frequency of
expected CpG sites should be at least 0.65. Such a region is shown here with an orange background. The
plot was obtained using an R script (for details, see web resources for this book).

As the next step, we use the same method to analyse a longer sequence, in
this case a region of the human chromosome 4, containing the known genes
SLC4A4, GC, NPFFR2 and ADAMTS3. With this example we will see how the
predicted CpG islands compare to the transcription start sites of these genes.

To achieve reading of the file `chr4_region.fa` we simply modify the script
by appropriately commenting and commenting out:

```
# my $infile = 'short.fa';        # a shorter sequence
  my $infile = 'chr4_region.fa';   # alternative sequence
```

We save the output in another file:

```
% perl cpg.pl > cpg_chr4.out
```

You will notice that the script now takes a while to complete, as we are considering about 1.5 million nucleotides, which is to say that we are analysing the
base composition of about 150 000 windows. The output is shown graphically in
Fig. 15.3. The exons of the individual genes are shown, as well as their relative
orientation. We note that for three of the four genes, a CpG island is predicted
close to the transcription start site. It would seem that our Perl script to predict
CpG islands performs well.

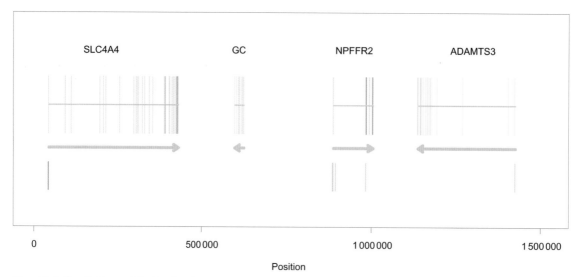

Fig. 15.3 *Prediction of CpG islands in a region of a human chromosome with four protein genes.* A region of the human chromosome 4, containing the known genes SLC4A4, GC, NPFFR2 and ADAMTS3 was analysed with respect to potential CpG islands. At the top is shown the exon–intron structure of the four genes; the arrows indicate the polarity of the genes. In the lower part of the figure (below the arrows) are the results of the CpG island prediction. All vertical lines represent regions predicted to be a CpG island (compare Fig. 15.2). Three of the four genes clearly have a predicted CpG island which is close to the transcription start site. The plot was obtained using an R script (for details, see the web resources for this book).

EXERCISES

15.1 It was stated in this chapter that in the human chromosome 4 the observed frequency of CpG sites is only about 20% of the expected frequency based on GC content. Simplify Code 15.1 to calculate the ratio between the observed and expected frequency of CpG sites in the entire nucleotide sequence in `chr4_region.fa`.

15.2 In the output from Code 15.1 there is a '1' in the second column in the case that the position is part of a CpG island. Design a Perl script that uses the output file `cpg_chr4.out` and prints the begin and end positions of the different CpG islands. For instance, the first CpG island could be printed as 43070–44360.

15.3 As in Chapter 13, there is an efficiency problem with Code 15.1, which means we are counting the same Cs, Gs and CpGs multiple times. Modify the script to avoid this problem. When analysing the `chr4_region.fa` sequence you can use the Unix `time` command to determine the speed of your new code as compared to Code 15.1:

```
% time cpg.pl > cpg.out
```

REFERENCES

Durbin, R., S. R. Eddy, A. Krogh and G. Mitchison (2007). *Biological Sequence Analysis: Probabilistic Models of Proteins and Nucleic Acids*. Cambridge, Cambridge University Press.

Gardiner-Garden, M. and M. Frommer (1987). CpG islands in vertebrate genomes. *J Mol Biol* **196**(2), 261–282.

Illingworth, R. S. and A. P. Bird (2009). CpG islands: 'a rough guide'. *FEBS Lett* **583**(11), 1713–1720.

International Human Genome Sequencing Consortium (2004). Finishing the euchromatic sequence of the human genome. *Nature* **431**(7011), 931–945.

Lander, E. S., L. M. Linton, B. Birren, *et al*. (2001). Initial sequencing and analysis of the human genome. *Nature* **409**(6822), 860–921.

Takai, D. and P. A. Jones (2002). Comprehensive analysis of CpG islands in human chromosomes 21 and 22. *Proc Natl Acad Sci U S A* **99**(6), 3740–3745.

Takai, D. and P. A. Jones (2003). The CpG island searcher: a new WWW resource. *In Silico Biol* **3**(3), 235–240.

Venter, J. C., M. D. Adams, E. W. Myers, *et al*. (2001). The sequence of the human genome. *Science* **291**(5507), 1304–1351.

Zhao, Z. and L. Han (2009). CpG islands: algorithms and applications in methylation studies. *Biochem Biophys Res Commun* **382**(4), 643–645.

16 Finding genes
In the world of snurps

I fell in love with RNA in one of my first jobs as an undergraduate.

(Joan Steitz, quoted by Sedwick, 2011[1])

Methods of gene prediction

We saw in the previous chapter how prediction of CpG islands may be used to identify transcription start sites of protein-coding genes. However, there are many other elements and statistical properties of such genes that we may exploit for gene finding.

What are the methods available for computational gene finding? In general, one may distinguish between two major categories: *de novo* or *ab initio* methods and *homology*-based methods. The *de novo*[2] methods make use of statistical signals in DNA sequences that are characteristic of protein-coding genes; the homology-based methods rely on the identification of exons by matching known mRNA or protein sequences or even profile HMMs to a genomic sequence. The homology-based methods are powerful but they require that mRNA or protein sequence information is available. Here we will focus on the *de novo* type of gene finding. What are the signals characteristic of protein-coding genes? In addition to the transcriptional signals that we dealt with in the previous chapter, the most useful features are:

(1) Coding sequences often have nucleotide or *word frequencies* that are different from non-coding sequences. This is mainly because the amino acid composition of proteins is not random and this is reflected in the nucleotide composition of protein-coding sequences. Methods have been exploited that

---

[1] Joan Steitz is famous for a number of discoveries involving RNA, such as snRNPs and their role in splicing (Lerner *et al.*, 1980).

[2] *Ab initio* and *de novo* are Latin terms that both mean 'from the beginning'. The *de novo* gene-finding methods are based on analysis of DNA sequence only and do not make use of any external information such as mRNA and protein sequences.

examine the distribution of six-letter words in genomic sequences (Claverie *et al.*, 1990; Bishop, 1994).

(2) In humans, as well as all other eukaryotes, genes are mosaics of *exons* and *introns*, as discussed in Chapter 1. The exons are regions that are combined in the process of splicing to form a mature mRNA. Important for the problem of identifying exons and introns computationally is the fact that the exon–intron border regions are conserved in sequence. We will plunge into the details of this below.

(3) The *length of exons and introns* are not randomly distributed. Thus, exons are typically 50–300 nt and introns are normally longer than 70 nt.[3]

(4) Termination of transcription of a vast majority of eukaryotic mRNA molecules occurs close to a *polyadenylation site* with a distinct consensus sequence, AAUAAA (or AATAAA if you want to think of the DNA sequence). As this signal is encountered during transcription, the RNA product will be cleaved downstream of the AAUAAA sequence. To the 3′ end of the RNA will then be added adenosine monophosphate units to form a *polyA tail*. Newly formed mRNAs have polyA tails of length 200–250 nt.

We will cover in more detail below the problem of how to computationally identify the exon–intron border regions. The conserved sequences characteristic of those regions are very much the result of basic properties of the splicing machinery. We therefore need to know a few things about the biochemistry of splicing.

The splicing machinery

Splicing occurs by two sequential *trans-esterification* reactions and is catalysed by a multi-component complex, the *spliceosome* (Nilsen, 2003; Fig. 16.1). In the first step a branch site adenosine attacks the *5′ splice site* (also referred to as the *donor site*). In the second step the free 3′ end of the first exon attacks the *3′ splice site* (*acceptor site*) and the two exons become covalently joined. The intron is released in the form of a *lariat* structure.[4]

The spliceosome contains a number of RNA–protein complexes referred as snRNPs (pronounced 'snurps'). There are five different snRNPs and the RNA components of these are named U1, U2, U4, U5 and U6, respectively.[5] Spliceosome assembly is initiated by the interaction of U1 snRNP with the 5′ splice

---

[3] But many introns are very much longer. In a typical mammalian gene, the total length of the introns is much greater than the total length of the exons.

[4] The intron resembles a lariat; a rope in the form of a lasso.

[5] There is actually also another spliceosome responsible for catalysing a minor form of introns (Patel and Steitz, 2003), but we will not discuss that sort of splicing here.

Fig. 16.1 *Splicing of mRNA precursors is a two-step reaction*. Splicing occurs by two different trans-esterification reactions. In the first reaction the 3' hydroxyl of the branch site adenosine attacks the phosphodiester bond at the junction between the 3' end of the 5' exon (Exon 1) and the 5' end of the intron. As a result, the branch site adenosine is bonded to three different nucleotides. The 3' OH of the 5' exon (Exon 1) is thus free, and in the next reaction it attacks the phosphodiester bond of the 3' splice site. The end products are the covalently joined exons and the intron is in a 'lariat' form.

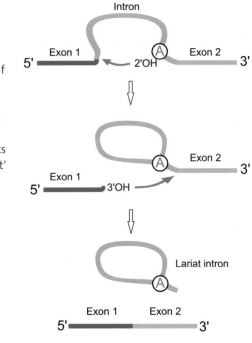

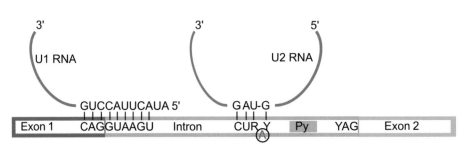

Fig. 16.2 *Pairing of spliceosomal RNAs to precursor mRNA*. The spliceosomal U1 RNA is part of the U1 snRNP and pairs with the 5' splice site region. The U2 RNA, being part of the U2 snRNP, pairs with a sequence at the adenosine branch site. The 3' splice site sequence AG does not pair with a spliceosomal RNA. Instead, it is recognized by a protein component of the U2 snRNP.

site and U2 snRNP with the branch site (see Fig. 16.2). Here, the U1 and U2 RNAs play important roles as they pair with the 5' splice site and branch site sequences, respectively. The other snRNPs then associate with U1, U2 and the pre-mRNA to form a spliceosome. Structural rearrangements finally take place and a U6/U2 complex plays an important role in the catalytic reaction. Hence, during splicing the U RNAs play important roles, both in the recognition of the splice sites and in the actual catalytic reactions.

A comparison of exon and intron regions in eukaryotic genes reveals that there are certain well-conserved positions in the exon–intron border regions, i.e. the 5′ and 3′ splice site regions (Fig. 16.2). The sequence conservation of the 5′ splice site is related to the fact that it pairs with a region of the U1 RNA. There is also pairing between the branch site and the U2 RNA, but the canonical sequence AG of the 3′ splice site is, in contrast, recognized by a protein component of the U2 snRNP (Kent *et al.*, 2003).

The consensus features of protein or DNA sequences are often illustrated by *sequence logos* (Schneider and Stephens, 1990; Crooks *et al.*, 2004). Sequence logos are based on multiple alignments and

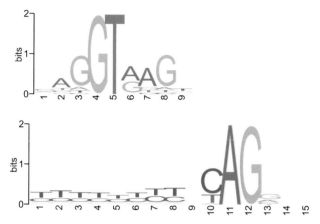

Fig. 16.3 *Sequence conservation at splice sites.* Sequence logos of the 5′ or donor site (top) and 3′ or acceptor splice site (bottom) are shown. Note the strongly conserved GT (or GU at the level of RNA) and AG of the 5′ and 3′ splice sites, respectively.

show for each position in the alignment the extent of conservation (for details, see Schneider and Stephens, 1990). For the splice sites they show that the 5′ end of the intron has a strongly conserved sequence GU (or GT at the level of DNA sequence) and an equally conserved AG at the 3′ end (Fig. 16.3).

BIOINFORMATICS

We now turn to a method for identifying the 5′ and 3′ splice sites in a gene or genomic sequence. We could make use of the strongly conserved positions and decide to assign all GT dinucleotides to be potential 5′ splice sites and AG dinucleotides to be potential 3′ splice sites. However, this would be a poor prediction as it results in

Tools introduced in this chapter		
Perl	two-dimensional arrays
	functions uc and lc
	function log
	rounding numbers

far too many false positives. A better method is to use a *profile*, or *position-specific scoring matrix* (PSSM) to describe the splice site. We touched upon these in the context of blood coagulation proteins in Chapter 12, but now we will have a more detailed look at how the PSSM is constructed. The basic principles of constructing the PSSM, as well as scoring with the PSSM, is shown with a specific example in Box 16.1. Two different Perl scripts will be presented; one of them is designed to construct a PSSM (Code 16.1) and the other (Code 16.2) is for scoring a sequence with that PSSM.

Box 16.1 Identification of splice sites with a PSSM

The construction of a 5′ splice site PSSM is shown here with a simple example. The scoring with this PSSM is also illustrated. We start out with an alignment of 5′ splice site sequences. In the sequences below the three left-most positions belong to the 5′ exon and the remaining positions form the 5′ end of the intron.

```
 1  C A G G T A G G G
 2  C A G G T T A C A
 3  A A G G T A T G T
 4  G A G G T G A G C
 5  G A G G T A A A C
 6  A G A G T A A G G
 7  C G G G T G G G T
 8  G T G G T G A T T
 9  A C A G T A A C T
10  C T T G T A A G T
```

Based on the alignment we construct a two-dimensional count matrix (C_{ij}) by simply counting the occurrences of each base. Columns 1–9 in the table below correspond to the columns in the original alignment.

	1	2	3	4	5	6	7	8	9
A	3	5	2	0	0	6	7	1	1
T	0	2	1	0	10	1	1	1	5
C	4	1	0	0	0	0	0	2	2
G	3	2	7	10	0	3	2	6	2

We then modify the count matrix by adding a pseudocount of one.

	1	2	3	4	5	6	7	8	9
A	4	6	3	1	1	7	8	2	2
T	1	3	2	1	11	2	2	2	6
C	5	2	1	1	1	1	1	3	3
G	4	3	8	11	1	4	3	7	3

The next step is to transform the counts to frequencies (F_{ij}), in this case by dividing the counts in the previous matrix with the total number of bases in each column, which is $10 + 4 = 14$ ($F_{ij} = C_{ij} / N$).

	1	2	3	4	5	6	7	8	9
A	0.29	0.43	0.21	0.07	0.07	0.50	0.57	0.14	0.14
T	0.07	0.21	0.14	0.07	0.79	0.14	0.14	0.14	0.43
C	0.36	0.14	0.07	0.07	0.07	0.07	0.07	0.21	0.21
G	0.29	0.21	0.57	0.79	0.07	0.29	0.21	0.50	0.21

A log-odds matrix M_{ij} is constructed by taking the log of observed frequency divided by the expected frequency, $M_{ij} = \log (F_{ij} / F_{exp})$; in this case we set F_{exp} to be 0.25.

	1	2	3	4	5	6	7	8	9
A	0.13	0.54	−0.15	−1.25	−1.25	0.69	0.83	−0.56	−0.56
T	−1.25	−0.15	−0.56	−1.25	1.15	−0.56	−0.56	−0.56	0.54
C	0.36	−0.56	−1.25	−1.25	−1.25	−1.25	−1.25	−0.15	−0.15
G	0.13	−0.15	0.83	1.15	−1.25	0.13	−0.15	0.69	−0.15

The resulting matrix may now be used to score sequences. Consider, for instance, the sequence TCCTGTCCCAGGTAGGGAA. We score each window the size of the PSSM, which is 9. The score is obtained simply by adding up the relevant values from the PSSM. As an example, consider the subsequence CAGGTAGGG:

	1	2	3	4	5	6	7	8	9
A	0.13	0.54	−0.15	−1.25	−1.25	0.69	0.83	−0.56	−0.56
T	−1.25	−0.15	−0.56	−1.25	1.15	−0.56	−0.56	−0.56	0.54
C	0.36	−0.56	−1.25	−1.25	−1.25	−1.25	−1.25	−0.15	−0.15
G	0.13	−0.15	0.83	1.15	−1.25	0.13	−0.15	0.69	−0.15

This gives the total score for this sub-sequence, which is $0.36 + 0.54 + 0.83 + 1.15 + 1.15 + 0.69 − 0.15 + 0.69 − 0.15 = 5.1$.

Constructing a PSSM

For the construction of a PSSM we will make use of a larger collection of splice sites, assembled from information on human genes. We will consider both 5′ and 3′ splice sites, but explicit examples will be shown here only for the 5′ splice site. Sequences of 5′ splice sites are collected in a file `splice5.txt`. These sequences are derived from data downloaded with the Table Browser at UCSC Genome Bioinformatics (http://genome.ucsc.edu). Our script to produce a PSSM from this collection of 5′ splice sites is shown in Code 16.1.

Code 16.1 make_matrix5.pl

```perl
#!/usr/bin/perl -w

use strict;

my @msa_matrix;    # two-dimensional array
                   # to store multiple alignment

my $number_of_sequences = 0;    # we want to count the number of
                                # sequences in the file splice5.txt

my $infile = 'splice5.txt';
open(IN, $infile) or die "Oops, could not open $infile\n";
while (<IN>) {
    chomp;
    for ( my $j = 0 ; $j < 9 ; $j++ ) { # nine positions in alignment
        $msa_matrix[$number_of_sequences][$j] = substr( $_, $j, 1 );
    }
    $number_of_sequences++;
}
close IN;

# produce count matrix
my @bases = ( 'A', 'T', 'C', 'G' );
my @pssm;
for ( my $i = 0 ; $i < 4 ; $i++ ) {
    for ( my $j = 0 ; $j < 9 ; $j++ ) {
        # add pseudocount = 1 to each of the values in the matrix
        $pssm[$i][$j] = 1;

        # add counts to the pssm matrix
        for ( my $k = 0 ; $k < $number_of_sequences ; $k++ ) {
            if ( $msa_matrix[$k][$j] eq $bases[$i] ) {
                $pssm[$i][$j]++;
            }
        }
    }
}

# from count matrix produce PSSM by
# calculating the log odds values
for ( my $i = 0 ; $i < 4 ; $i++ ) {
    for ( my $j = 0 ; $j < 9 ; $j++ ) {
        $pssm[$i][$j] =
            log( ( $pssm[$i][$j] / ( $number_of_sequences + 4 ) * 4 ) )
            / log(2);
        print "$pssm[$i][$j] "; # print PSSM
        # print with two decimal places:
```

```
        # printf( "%.2f\t", $pssm[$i][$j] );
    }
    print "\n";
}
```

In the file `splice5.txt` the sequences are in a 'raw' format (i.e. in a format with sequence and nothing else), with one sequence in each line:

```
CAGGTAGGG
CAGGTAACA
AAGGTAAGT
GAGGTGAGC
```

and so on.

In the first part of Code 16.1 we read this file, one line at a time, and put the individual characters in the two-dimensional array `@msa_matrix`. The matrix ends up as shown in Table 16.1. In a two-dimensional array the elements may be referred to by an expression of type `$msa_matrix[$i][$j]`. The way we have constructed this matrix the `$i` variable is the index reflecting the splice site sequence considered. The first sequence has index 0. The `$j` variable represents the base position index, where 0 is the first position. As an example, the G highlighted in Table 16.1 is referred to as `$msa_matrix[1][2]`.

In the next part of the script we create a PSSM from the data available in the array `@msa_matrix`. We first create the count matrix (see Box 16.1) and store it in another two-dimensional array, `@pssm`. The `@pssm` matrix has four rows (one for each of the bases A, T, C and G) and nine columns for each of the nucleotide positions. For example, in the count matrix element `$pssm[2][8]` the 2 corresponds to one of the four bases (which is 'C' from the way we defined things in the script) and the 8 refers to the nucleotide position. With three different `for` loops we calculate all elements of the `@pssm` array. Each element is first assigned a pseudocount value 1. Then the counts are updated according

Table 16.1. *First rows and columns of the matrix* `@msa_matrix`. *The highlighted element is referred to as* `$msa_matrix[1][2]`.

| Index of sequence | Index of nucleotide position | | | | |
	0	1	2	3	...
0	C	A	G	G	...
1	C	A	G	G	...
2	A	A	G	G	...
...	...	...	...	...	...

to the information in the matrix `@msa_matrix`. We also make use of an array `@bases`, which contains the four bases in the same order as in the `@pssm` array.

We finally transform the count matrix to a PSSM. We loop through the indices of `@pssm` and calculate the log-odds scores:

```
$pssm[$i][$j] =
log(($pssm[$i][$j] / ($number_of_sequences + 4 ) * 4))/log(2);
```

The number 4 is added to the total number of sequences because we added a pseudocount of 1 to each of the four bases (see also Box 16.1). The factor 4 comes from the fact that we assume a background probability of 0.25 for each of the bases. We use the Perl function `log`, which returns the natural logarithm of a number and the division with `log(2)` is because we want to have log2 values. Finally, the output of Code 16.1 is stored in a file:

```
% perl make_matrix5.pl > matrix5.txt
```

The numbers of the PSSM have a lot of decimal places, but in the representation below they have been rounded to two decimal places for clarity (see an example of how to round numbers in Perl in a commented line in Code 16.1):

```
 0.42    1.28   -1.41  -11.89  -11.89   1.08    1.50   -1.80   -0.67
-1.11   -0.81   -1.78  -11.89    2.00   -3.30   -1.39   -2.07    0.88
 0.56   -0.93   -2.93  -11.89  -11.89   -3.18   -1.70   -2.16   -0.59
-0.47   -1.05    1.68    2.00  -11.89    0.74   -1.05    1.70   -0.21
```

In this output, the rows correspond to the four bases A, T, C and G, respectively, and the columns are the different nucleotide positions (compare Box 16.1).

Scoring with a PSSM

As an important final step we will now use the PSSM from above to *score* a sequence. We will make use of a human gene that encodes a protein named 'serum amyloid P component'. Considering that this is a human gene, it is unusually small, with only one short intron. Its sequence, in FASTA format, is in the file `amyloid.fa`. The scoring is carried out with the Perl script Code 16.2.

Code 16.2 score5.pl

```perl
#!/usr/bin/perl -w

use strict;

# read matrix
my $i      = 0;
my @pssm   = ();
```

```perl
my $infile = 'matrix5.txt';
open(MATRIX, $infile) or die "Could not open matrix file $infile\n";
while (<MATRIX>) {
    chomp;
    my @array = split;
    for ( my $j = 0 ; $j < 9 ; $j++ ) { $pssm[$i][$j] = $array[$j]; }
    $i++;
}
close MATRIX;

# read sequence to be analysed
$infile = 'amyloid.fa';
open(IN, $infile) or die "Could not open $infile\n";
my $seq = '';
while (<IN>) {
    unless (/>/) {
        chomp;
        $seq .= $_;
    }
}
close IN;

print "pos\tscore\n";     # print header

$seq = uc($seq);
my @bases = ( 'A', 'T', 'C', 'G' );

# score with the matrix
for ( my $k = 0 ; $k < length($seq) - 8 ; $k++ ) {
    my $test = substr( $seq, $k, 9 );
    my $score = 0;
    for ( my $j = 0 ; $j < 9 ; $j++ ) {
        my $base = substr( $test, $j, 1 );
        for ( my $b = 0 ; $b < 4 ; $b++ ) {
            if ( $bases[$b] eq $base ) {
                $score += $pssm[$b][$j];
            }
        }
    }

    $score = 2 ** $score;    # convert log2 to real values
                             # ** is the exponentiation operator

    my $pos = $k + 3;        # We want to print a position
                             # next to the exon-intron
                             # junction
    print "$pos\t$score\n";
}
```

In Code 16.2 we start by reading the file containing the PSSM matrix information and store the matrix as @pssm. We then read the file containing the sequence to be analysed. This sequence is in lower-case letters. What if we want to convert it to upper-case letters? We here use the function uc. The corresponding function to convert to lower case is lc. You sometimes also see the tr/// operator being used for such operations. Consider this operation:

```
$seq =~ tr/a-z/A-Z/;
```

The expression a-z refers to all lower-case letters in the alphabet and A-Z to all upper-case letters. The result of the tr operation is that all 'a' characters will be replaced by 'A' characters', all 'b' characters with 'B', etc. This is equivalent to the uc function.

In the last part of the script we do the actual scoring. A sliding window is used at the size of the PSSM (nine). Finally, we run the Perl code:

```
% perl score5.pl > score5.txt
```

The first lines of the resulting file are:

```
pos       score
3         3.50757912033014e-11
4         6.77423367528269e-06
5         5.77683587574513e-09
6         6.27879781292501e-10
```

The output is shown graphically in Fig. 16.4. Also shown in this graph are the results from scoring with a similar procedure developed for the 3′ splice site (corresponding files are make_matrix3.pl and score3.pl). One difference is that the 3′ PSSM has 15 instead of 9 positions.

We may conclude from these graphs that the splice site predictions make sense. For instance, the strongest signal of the 5′ splice site PSSM matches exactly the expected 5′ splice site. There is also a strong 3′ signal at the end of the intron. However, at the same time there are additional strong signals with the 3′ splice site PSSM that seem to be false positives. Had we analysed a much larger gene we would have encountered many more false positive signals. In general, splice site prediction in the human genome as implemented here is far from perfect and we need a lot more information to correctly predict splice sites. But you should not be dejected; a method like this is useful when integrated with other statistical signals that are characteristic of protein-coding genes. Furthermore, the simple splice site finder implemented here is also useful in special cases, such as when trying to spot a potential splice site that has been missed in the standard annotation of a gene.

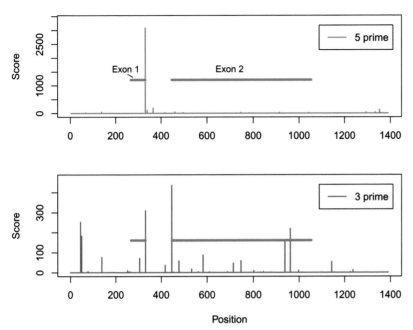

Fig. 16.4 *Prediction of splice sites*. Plot shows the prediction of splice sites using PSSMs as described in the text. The genomic sequence analysed is the human gene APCS, which is located on chromosome 1 and encodes the protein serum amyloid P component. This gene contains only one intron. The plot was created with an R script (see the web resources for this book) using the results of the prediction with a Perl script as described in the text, as well as information regarding the location of the two exons of this gene (green). Both 5′ and 3′ splice sites of this gene are correctly predicted, although there are also false positives, in particular for the 3′ splice site.

EXERCISES

16.1 Use NCBI Entrez to download the entry with accession L78833 (human BRCA1 gene). From the feature table we may glean information about exons and introns. Thus, the following type of information may be observed:

```
intron          3999..4619
exon            4620..4718
exon            complement(96693..98031)
```

In the last line the word 'complement' is to indicate that the exon is on the complementary strand. Write a Perl script that will read the information in the feature table and calculate the *mean length* of all exons, as well as the mean length of all introns.

16.2 In order to discriminate between coding and non-coding sequences, we may take advantage of the fact that codons are not randomly distributed in a protein-coding sequence. For instance, the glycine codons GGU and GGC are more frequently used than GGA and GGG in the bacterium *Escherichia coli*. This exercise illustrates how codon usage may be studied with the help of a Perl script. Consider a portion of the mRNA sequence of *E. coli* elongation factor Tu in file `eftu.fa`. Make a script to analyse the three different reading frames of this sequence with respect to the frequency of glycine codons. Which reading frame is most likely the correct one, considering how glycine codons are being used?

16.3 Sequence logos may be used to represent the extent of conservation in a DNA or protein sequence (Schneider and Stephens, 1990; Fig. 16.3). The following mathematics may be used to construct such logos. We assume with this example that we are dealing with DNA sequences and we start out with a multiple alignment. The *amount of uncertainty* may be stated as:

$$H_u = -\sum_a f_{u,a} \log_2 f_{u,a}$$

where H_u is the uncertainty at position u and a is one of the four bases in a DNA sequence. $f_{u,a}$ is the frequency of base a in column u. *Total information* in the position u is represented by the decrease in uncertainty, which for DNA is:

$$I_u = \log_2 4 - H_u$$

Using these equations, make a Perl script to derive the total information for the five positions of the simple multiple alignment below.

```
1    GCATT
2    GCATA
3    GTGTG
4    GATTC
```

REFERENCES

Bishop, M. (1994). *Guide to Human Genome Computing*. London and San Diego, CA, Academic Press.

Claverie, J. M., I. Sauvaget and L. Bougueleret (1990). K-tuple frequency analysis: from intron/exon discrimination to T-cell epitope mapping. *Methods Enzymol* **183**, 237–252.

Crooks, G. E., G. Hon, J. M. Chandonia and S. E. Brenner (2004). WebLogo: a sequence logo generator. *Genome Res* **14**(6), 1188–1190.

Kent, O. A., A. Reayi, L. Foong, K. A. Chilibeck and A. M. MacMillan (2003). Structuring of the 3′ splice site by U2AF65. *J Biol Chem* **278**(50), 50572–50577.

Lerner, M. R., J. A. Boyle, S. M. Mount, S. L. Wolin and J. A. Steitz (1980). Are snRNPs involved in splicing? *Nature* **283**(5743), 220–224.

Nilsen, T. W. (2003). The spliceosome: the most complex macromolecular machine in the cell? *Bioessays* **25**(12), 1147–1149.

Patel, A. A. and J. A. Steitz (2003). Splicing double: insights from the second spliceosome. *Nat Rev Mol Cell Biol* **4**(12), 960–970.

Schneider, T. D. and R. M. Stephens (1990). Sequence logos: a new way to display consensus sequences. *Nucleic Acids Res* **18**(20), 6097–6100.

Sedwick, C. (2011). Joan Steitz: RNA is a many-splendored thing. *J Cell Biol* **192**(5), 708–709.

17 Finding genes
Hunting for the distant RNA relatives

At some point a particularly remarkable molecule was formed by accident. We will call it the Replicator. It may not have been the biggest or the most complex molecule around, but it had the extraordinary property of being able to create copies of itself.

(Richard Dawkins, 1989)

The RNA world

So far this book has focused on proteins and the genes that encode them. The human genome encodes some 21 000 different proteins and the vast majority of them are important. On the other hand, there is a whole range of RNAs transcribed from the human genome that do not code for proteins, but have other functions. We refer to these RNAs as *non-coding RNAs* (ncRNAs). In fact, a major portion of the human genome is transcribed, although only about 1.5% of it corresponds to coding regions. We still do not know the function of many of these RNAs, but there are a large number of ncRNA families that have been characterized. Classic examples are tRNAs and ribosomal RNAs, which are part of the translation machinery. A set of U RNAs are involved in splicing (Chapter 16) and there are catalytically important RNA molecules of the RNA-processing enzymes RNases P and MRP. A vital and highly populated class of ncRNA is the RNAs involved in gene silencing as described in Chapter 3.

It is remarkable that in many central cellular processes RNA molecules have critical functions. For instance, RNA molecules are most likely involved in the actual catalytic mechanism during splicing. Furthermore, the critical step in protein synthesis, where amino acids are covalently joined, is a catalytic event mediated by RNA. Why do many RNA molecules have critical roles in cellular activities?

The *RNA world* hypothesis was put forward to discuss the role of RNA during early evolution of life (Woese, 1967; Gilbert, 1986; Gesteland *et al.*, 1999; Yarus, 2010). According to this hypothesis RNA had an essential role at an early stage in the development of living organisms, and predated proteins and DNA. RNA was able to have such a prominent role because of two critical properties. First, because it is a nucleic acid it is able to carry genetic information. Indeed, today we observe viruses that have their genetic information in the form of RNA, such as HIV and other retroviruses. We may therefore think of RNA as being the carrier of genetic information in the RNA world, and this was at a time when DNA was not yet around. The other important property of RNA is that it is structurally and functionally versatile. It is able to catalyse a variety of chemical reactions. We are able to observe in nature today many examples of RNAs having catalytic properties. Catalytic RNAs were first discovered by Sidney Altman and Thomas Cech; in 1989 they received the Nobel Prize in Chemistry for their important pioneering work (Kruger *et al.*, 1982; Guerrier-Takada *et al.*, 1983; Guerrier-Takada and Altman, 1984). Furthermore, there are many examples of synthetic RNAs that have been identified by a procedure of in vitro replication and selection to have desired catalytic properties. For example, RNA molecules with the capacity to replicate other RNA molecules have been produced. An important conclusion from this is that RNA molecules that were present in the RNA world were not only able to store genetic information, but may also have been able to carry this information onto the next generation with the help of an RNA-copying mechanism.

In summary, RNA molecules may have had an essential role in early molecular evolution, without too much help from proteins. As evolution proceeded, RNA as a carrier of genetic information was replaced by DNA because it has the advantage of being chemically more stable than RNA. Proteins are more versatile than RNA in terms of catalysis, and for this reason proteins enhanced or replaced many of the catalytic functions of RNA. However, we see remnants of the RNA world in the form of several RNAs with critical functions, such as the ribosomal RNAs that catalyse peptide bond formation. There are also many chemical compounds in the living cell that are either precursors of RNA or compounds related to these precursors, and that have prominent roles in cellular metabolism. One striking example is the ATP (adenosine triphosphate) molecule, which is not only required for the synthesis of RNA, but is also a universal carrier of energy and an important regulator of enzymes and other proteins. ATP and other molecules with ribonucleotide components may also be regarded as 'molecular fossils' of the RNA world.

Properties of RNA and computational RNA finding

We now turn to the problem of identifying ncRNAs in genomic sequences. We may distinguish between two categories of ncRNA gene finding. In one category we want to find *any* ncRNA, including classes of ncRNA that were not previously known (this may be referred to as a *de novo* method – see the previous chapter). In the other category we want to find a member of a specific class of RNA. A general difficulty with the first category is that ncRNA genes have very few elements in common, if any. Many protein genes are characterized by a variety of statistical signals, such as promoters, translation start and stop sites, properties of coding sequences, exon–intron borders and polyadenylation sites, as we have seen in previous chapters. In comparison, most ncRNA genes do not have such useful information. Therefore, *de novo* ncRNA gene finding is quite difficult.

The second category of ncRNA gene finding is less painful, and this is the one that we deal with in this chapter. Remember, we have been discussing evolutionary relationships between protein molecules as well as methods to reveal such relationships. Important concepts are *homology, orthology* and *paralogy* (see Chapter 7). These concepts also apply to ncRNAs. We are faced with problems such as: if we know the sequence of a certain RNA, like the RNA component of human RNase P, how do we find related RNAs – orthologues and paralogues – in other species? What are the properties of ncRNAs that may be exploited to solve this problem?

For proteins we have seen that a local alignment method like BLAST may be used to find orthologues. Can we do this for ncRNAs? To some extent we can, but a problem with ncRNAs as compared to proteins is that the primary structure of ncRNAs (the linear sequence of nucleotides) tends to be much less conserved than that of proteins. As a consequence, standard tools such as BLAST may be used only in the identification of orthologues or paralogues in fairly closely related species.

However, while primary sequence is poorly conserved, each class of ncRNA has a characteristic *secondary structure*. 'Secondary structure' refers to the fact that the RNA may be represented in two dimensions, showing a structure based on internal base-pairing. Examples of such secondary structure are shown for four selected ncRNAs in Fig. 17.1. The concepts of RNA primary, secondary and *tertiary* structure are illustrated in Fig. 17.2, which shows a part of a phenylalanine tRNA containing the anticodon loop and stem. The secondary or tertiary structure of this part of the tRNA represents a common theme in RNA structure that we may refer to as a *stem-loop* or a *hairpin* structure, meaning that two paired RNA strands are connected through an unpaired loop sequence. A tRNA anticodon stem and loop is also shown in Fig. 17.3, in this case from four different species and illustrating that the sequence in a helical region in RNA is often allowed to change during evolution without affecting its structure and function significantly. This is in fact one important reason why RNAs, as compared to proteins, are poorly conserved in terms of primary structure.

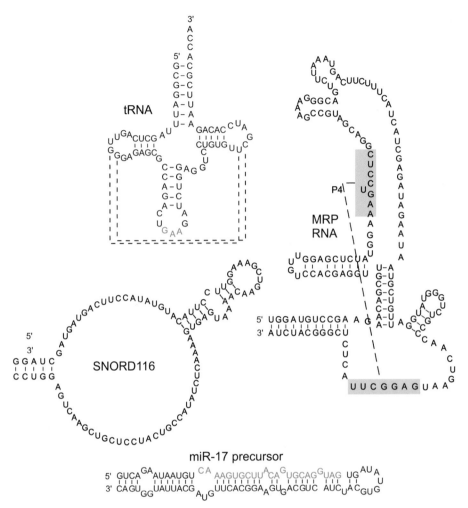

Fig. 17.1 *Examples of non-coding RNAs.* Four different examples of non-coding RNA are shown: (1) phenylalanine tRNA from yeast (*Saccharomyces cerevisiae*); (2) MRP RNA from the microsporidian *Encephalitozoon cuniculi*; (3) human snoRNA SNORD116; and (4) human mir-17 precursor RNA (the mature miRNA sequence is shown in red). The tRNA also contains modified nucleosides, but these are not shown in this figure. The anticodon sequence GAA is highlighted in colour. Examples of tertiary interactions are indicated with dashed lines. For instance, a 'pseudoknot' type of structure, named the P4 helix, is formed in MRP RNA. Many non-coding RNAs are associated with human disease. A chromosomal deletion of the human SNORD116 is associated with Prader–Willi disease and mutations in the human MRP RNA gene give rise to a disease known as cartilage-hair hypoplasia. The miRNA miR-17 is related to cancer and there are many genetic defects that are caused by mutations in mitochondrial tRNAs.

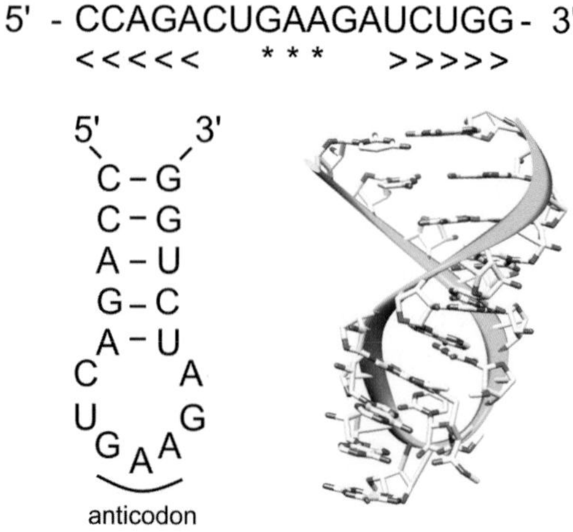

Fig. 17.2 *Three representations of RNA structure.* Yeast phenylalanine tRNA anticodon and stem are shown as *primary structure* together with bracket notation of base-pairing (top), *secondary structure* (bottom-left) and *tertiary* or three-dimensional structure as determined by X-ray crystallography (bottom-right, PDB entry 1EHZ (Shi and Moore, 2000)). The three-dimensional structure figure was produced with the software UCSF Chimera (Pettersen *et al.*, 2004).

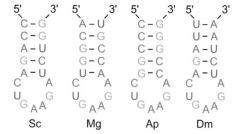

5' - CCAGACUGAAGAUCUGG - 3'
<<<<<   * * *   >>>>>

5'   3'
C – G
C – G
A – U
G – C
A – U
C   A
U   G
 G A A

anticodon

Fig. 17.3 *Variation in an RNA helix to maintain base complementarity.* Yeast tRNA anticodon loop and stem are shown from species *Saccharomyces cerevisiae* (baker's yeast), the eubacterium *Mycoplasma genitalium*, the archaeon *Aeropyrum pernix* and the fruit fly *Drosophila melanogaster*. Note the strong conservation in the anticodon loop, whereas the primary sequence in the stem is variable.

5'   3'   5'   3'   5'   3'   5'   3'
C – G   A – U   C – G   U – A
C – G   C – G   C – G   U – A
A – U   G – C   C – G   A – U
G – C   G – C   G – C   G – C
A – U   U – A   G – C   A – U
C   A   C   A   C   A   C   A
U   G   U   G   U   G   U   G
 G A A   G A A   G A A   G A A

Sc   Mg   Ap   Dm

Not only bona fide ncRNAs have the property of adopting a specific secondary structure. There are also many examples of local RNA structure in mRNAs, particularly in the 5′ and 3′ UTR regions, that are important in the regulation of translation or mRNA stability. Examples include riboswitches, the selenocysteine insertion sequence (SECIS) and the iron responsive element discussed in Chapter 6. From a bioinformatics perspective such cis-acting regulatory elements are analogous to the bona fide ncRNAs, and the same computational methods are used to identify and analyse them.

As the examples in Fig. 17.1 show, most classes of ncRNAs have a characteristic secondary structure, which is a property that may be exploited in gene finding. It is important to note that the most effective methods of ncRNA gene finding use a *combination of primary and secondary structure*. Methods that have been used so far include a form of pattern matching where base-pairing schemes are also taken into account. We have seen in Chapter 6 one example of this type of method for the identification of the iron responsive element.

There are also probabilistic methods that have been developed for ncRNA gene finding. The *covariance models* developed by Sean Eddy (Eddy and Durbin, 1994) are stochastic context-free grammar models (SCFGs) that describe RNA primary sequence and secondary structure in a probabilistic framework. These have been extremely useful for identification of RNA orthologues. The RNA covariance models may be thought of as the RNA equivalent of the protein profile HMMs referred to in Chapter 12.

The Infernal software (Nawrocki *et al.*, 2009) uses covariance models to search sequence databases. Infernal and covariance models are used in the *Rfam* database (Gardner *et al.*, 2010), which is a database that aims to catalogue ncRNAs. The database has a collection of RNA families, each represented by multiple sequence alignments, consensus secondary structures and covariance models. It is one of the most useful resources for ncRNA, and in Rfam version 10.1 (June 2011) a total of 1973 ncRNA families were represented there.

In the bioinformatics section below we will be looking into a problem of ncRNA homologue identification and we will see how sequence (primary structure) is poorly conserved, and how for a successful identification we need to make use of information about secondary structure. A Perl script and the Infernal software will be used. As our test case we will examine one of the 1973 families in Rfam, a non-coding RNA that plays an important role in protein transport.

An RNA involved in protein transport

The *signal recognition particle* (SRP) is an RNA–protein complex that is responsible for targeting proteins to the endoplasmic reticulum (ER) membrane in eukaryotic cells (Egea *et al.*, 2005; Shan and Walter, 2005; Grudnik *et al.*, 2009). Proteins that are to be targeted to the membrane have an N-terminal signal peptide. As this peptide emerges from the ribosome during translation it is recognized by the SRP. The resulting complex of SRP with the ribosome then docks to a receptor (SR) in the membrane. Here, the translation resumes and the protein is eventually translocated across the membrane. There is evidence that one of the functions of SRP RNA is to accelerate the interaction of SRP and SR (Peluso *et al.*, 2000). The SRP was initially thought of as being specific to eukaryotes, but it was discovered in the late 1980s that there is an SRP also in bacteria (Poritz *et al.*, 1988, 1990). In fact, SRP is highly ubiquitous and present in all kingdoms of life. Although the structure of the SRP RNA is quite variable, all species have one common helical region, known as helix 8. The structure of a bacterial SRP RNA, that of *Escherichia coli*, is shown in Fig. 17.4 – consensus features at the apical end are highlighted. This RNA, as most other bacterial SRP RNAs, forms a long hairpin structure.

The organelles and their evolution

In all eukaryotes the SRP RNA is encoded by genes in the nuclear genome. But eukaryotes also contain *mitochondria*, organelles specialized in oxidation, which generate energy in the form of ATP molecules. The mitochondria have

Fig. 17.4 *Structure of a bacterial signal recognition particle RNA*. Upper panel: structure of *Escherichia coli* SRP RNA and conserved apical region (shaded background). Lower panel: elements of SRP RNA apical region that are conserved in bacteria. The region with a shaded background highlights the consensus features used in the Perl pattern-matching script of this chapter. N is any of the four bases, R is either A or G, Y is either U or C.

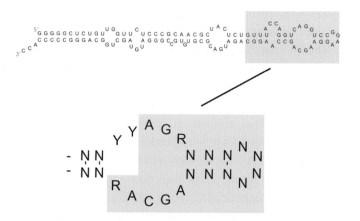

their own DNA, although this is a very small DNA encoding only 13 different proteins. In fact, most mitochondrial proteins are encoded by the nucleus and are imported into mitochondria. In addition to mitochondria, plant and algae cells have *chloroplasts*, organelles that are responsible for *photosynthesis*, i.e. the process by which the energy of sunlight is used to produce ATP and carbon dioxide and water are used to build carbohydrates. What about mitochondria and chloroplasts; do they also contain an SRP RNA? In such a case, is that RNA encoded by the organellar genome? In this chapter we will find out, at least about chloroplasts. But first we need to know a few things about the evolution of mitochondria and chloroplasts.

Both mitochondria and chloroplasts have a bacterial origin and are the result of *endosymbiotic* events. Thus, the mitochondria were originally Proteobacteria that came to exist within cells of another organism (Lane, 2005). This happened about two billion years ago. The chloroplasts also arose by endosymbiosis, but originated from the photosynthetic cyanobacteria, also known as blue-green algae. Chloroplast evolution is often quite complex and involves more than one endosymbiotic event (McFadden, 2001). In *secondary endosymbiosis* the result of a primary symbiosis event is taken up by another eukaryotic organism. For instance, a green alga (with chloroplasts) was engulfed by another eukaryote to produce a group of organisms known as *chlorarachniophytes*. These organisms have retained the chloroplasts derived from the green alga, while the original green algal nucleus exists as a remnant in the form of a *nucleomorph*. Even more complex evolutionary paths may be observed, and in *tertiary* endosymbiosis an alga with a secondary plastid is taken up by another eukaryote. An example is dinoflagellates such as *Durinskia baltica* and *Kryptoperidinium foliaceum*, which harbour a diatom endosymbiont (Imanian *et al.*, 2010).[1] Characteristic of both mitochondrial and chloroplast evolution is that many of the bacterial

---

[1] The diatoms (a common type of phytoplankton) are algae originally derived from a primary endosymbiosis involving red algae. Dinoflagellates are a kind of marine plankton.

genes were transferred to the nuclear genome of the host. Thus, the organellar genomes that we observe today are quite small. For the algae, the secondary and tertiary endosymbiosis was associated mainly with gene transfer from the nucleus of the endosymbiont alga to the nucleus of the new host.

For the coming exercises the most important conclusion from the discussion about chloroplast evolution is that the genes of the chloroplast genome were originally derived from cyanobacteria. For this reason we should expect the chloroplast genes to be bacterial-like. Any chloroplast SRP RNA is likely to resemble the bacterial members of this RNA family. When trying to find a chloroplast SRP RNA gene, it may be a good idea to first examine more closely all bacterial SRP RNAs and see what they have in common. The results of such an investigation reveal that most of these RNAs are in the order of 100 nucleotides in length and each of them form a long (though imperfect) hairpin structure. The consensus features of bacterial SRP RNAs are shown in Fig. 17.4, with strongly conserved primary and secondary structure features at the apical end of the RNA. We will exploit these features in our gene-finding strategy below.

BIOINFORMATICS

The quest for chloroplast RNAs

We will illustrate how an RNA gene may be identified with a procedure that uses a combination of pattern-based searches and covariance models. At the same time, this exercise illustrates how Perl may be used in a pipeline in sequence analysis. In such a case Perl does one or more of the following tasks:

Tools introduced in this chapter	
Perl	building a pipeline
	function system
Unix	Unix shell script
Software running in a Unix environment	Infernal

(1) Analysis on its own, as in the example for this chapter in which it performs useful pattern-matching operations.
(2) Processes the output of some external program to extract useful information, or to convert the output into a suitable format.
(3) Converts data into a format suitable for the input to some external program or for another Perl script.

We want to identify potential SRP RNA genes in chloroplast genomes. We will use a strategy in which we first identify candidates with the help of pattern matching in a Perl script. Then, the output of that Perl script will be directed to analysis by the Infernal software. The reason we do not subject all sequences

directly to Infernal is that this program is relatively demanding from a computational point of view. Therefore, we may save a lot of time with a filtering step
based on patterns.

As a first step we want to obtain a number of chloroplast genomes. For
practical reasons we restrict our analysis to the phylogenetic group Chlorophyta
(green algae). These genomes may be downloaded using NCBI Entrez using a
query to the database of nucleotide sequences. We specify the organism group
as 'chlorophyta', and in the title we require the words 'complete' and 'genome',
as well as either 'plastid' or 'chloroplast' (see Chapter 14 for more details on
how to perform Entrez queries):

```
chlorophyta[orgn] AND genome[titl] AND complete[titl] AND
(plastid[titl] OR chloroplast[titl])
```

A total of 41 genomes were identified in this search (at the time this book
was written). The dataset was filtered to remove duplicated species. We were
then left with 23 sequences. These are in the file chlorophyta_genomes.fa. We
are now ready to analyse these sequences. As a first step of this analysis
we identify SRP RNA candidates by pattern matching. The regular expression
we use is somewhat arbitrarily composed, but we assume that after analysing
a large number of available bacterial SRP RNA sequences we are left with the
consensus features as depicted in Fig. 17.4. These are also summarized here –
the angular brackets indicate the base-pairing:

```
5' ---AGRNNNNNNNNNNNAGCAR--- 3'
            <<      >>
```

The Perl code for analysing genomic sequences with respect to this pattern is
shown in Code 17.1. Six different elements of the code (also numbered in the
code) are brought to the attention of the reader as described below.

(1) We have constructed a function named pattern for carrying out the pattern
matching. The arguments passed to this function are (a) the identifier, (b) the
sequence to be analysed and (c) the strand being analysed (plus or minus,
see also under item (2)).

(2) An SRP RNA gene may be encoded on either strand of the double-stranded
DNA. Therefore, we need to analyse both strands and here we are producing
the reverse complement of the input DNA sequence using a procedure similar
to that used in Chapter 1. We keep track of the orientation of the strand
with the $strand variable, which is either '+' (original input strand) or '-'
(complementary strand).

(3) In the pattern function we encounter a pattern-matching operation:

```
$id =~ /gi\|(\d+)\|.*\|.*\| (.*)(chloroplast|plastid)/
```

This is used because we know the identifier line in this case is of this nature:

```
>gi|229915480|gb|FJ968740.1| Pedinomonas minor culture-collection
UTEX:LB 1350 chloroplast, complete genome
```

In the pattern above we need to put a backslash (\) in front of the pipe symbol because it has a special meaning in patterns (logical OR), but here we want to match that exact symbol. We enclose in parentheses (1) the gi number (following `gi|`) and (2) all text that is between the right-most pipe symbol and either of the words 'chloroplast' or 'plastid'. We can recall these matches with `$1` and `$2`, respectively.

(4) As discussed under the previous item, we have captured information about the organism in `$2`. We store that information in `$org`. We eventually want to pass this information to the output from the script as part of the identifier. As blank spaces may not be part of an identifier, we want to avoid them and instead replace them with underscores:

```
$org =~ s/ /_/g;
```

For instance, `Bryopsis hypnoides` will become `Bryopsis_hypnoides`.

(5) We use a sliding window of 100 nucleotide positions because we expect an SRP RNA to be in this size range. The sequence to be tested ends up in the variable `$subseq`.

(6) For each of the windows of size 100 we use a regular expression to describe the consensus properties in SRP RNA as shown above and in Fig. 17.4:

```
$subseq =~ /.{42}AG[AG].(.)(.).{4}(.)(.).AGCA[AG].{40}/
```

We are enclosing within parentheses bases in specific positions (`(.)`) because we want to recall them later using `$1`, `$2`, etc. and test whether they are able to form base pairs. The whole length of the sequence `$subseq` is covered by the pattern as we have also represented flanking sequences as `.{42}` and `.{40}`.

Code 17.1 pattern_search.pl

```perl
#!/usr/bin/perl -w

use strict;

my $id     = '';
my $seq    = '';
my $rev    = '';
my $strand = '';
my $x      = 0;

my $infile = 'chlorophyta_genomes.fa';
```

```perl
open(IN, $infile) or die "Oops, could not open $infile\n";
while (<IN>) {
    chomp;
    if (/^>/) {
        if ($id ne '') {

            # 1 #
            pattern( $id, $seq, '+' );

            # 2 #
            $rev = $seq;
            $rev = reverse($seq);
            $rev =~ tr/AGCT/TCGA/;
            pattern( $id, $rev, '-' );
        }
        $id = $_;
        $seq = '';
    }
    else {
        $seq .= $_;
    }
}
close IN;

pattern( $id, $seq, '+' );
$rev = $seq;
$rev = reverse($seq);
$rev =~ tr/AGCT/TCGA/;
pattern( $id, $rev, '-' );

sub pattern {
    my ( $id, $seq, $strand ) = @_;
    my $gi = '';

    # 3 #
    if ( $id =~ /gi\|(\d+)\|.*\|.*\| (.*)(chloroplast|plastid)/ ) {
        $gi = $1;
        my $org = $2;

      # 4 #
        $org =~ s/ /_/g;
        $gi = $gi . '_' . $org;
    }
    my $len = length($seq);

    # 5 #
    for ( my $i = 0 ; $i < $len - 100 ; $i++ ) {
```

```
    my $subseq = substr( $seq, $i, 100 );

    # 6 #
    if ( $subseq =~ /.{42}AG[AG].(.)(.).{4}(.)(.).AGCA[AG].{40}/ ) {
        if ( ( pair( $1, $4 ) ) && ( pair( $2, $3 ) ) ) {
            $x++;
            my $newid = $x . '_' . $gi . "$strand";
            print ">$newid\n$subseq\n";
        }
    }
  }
}

sub pair {
    my ( $base1, $base2 ) = @_;
    if (    ( ( $base1 eq 'G' )   && ( $base2 eq 'C' ) )
         || ( ( $base1 eq 'G' ) && ( $base2 eq 'T' ) )
         || ( ( $base1 eq 'A' ) && ( $base2 eq 'T' ) )
         || ( ( $base1 eq 'C' ) && ( $base2 eq 'G' ) )
         || ( ( $base1 eq 'T' ) && ( $base2 eq 'A' ) )
         || ( ( $base1 eq 'T' ) && ( $base2 eq 'G' ) ) )
    {
        return 1;
    }
}
```

The Perl script in Code 17.1 is executed and the output is saved in a file:

```
% perl pattern_search.pl > pattern_search.out
```

Examination of the output reveals that 97 RNA candidates were identified as a result of this search. The next step is to analyse them with cmsearch of the Infernal software. This program searches one or more sequences with an RNA covariance model. For this example we will use the Rfam model of the bacterial SRP RNA, which is in the file SRP_bact.cm.

The cmsearch program is executed in our example with (enter in one line):

```
% cmsearch --toponly --tabfile chlorophyta.tab SRP_bact.cm
pattern_search.out
```

The parameter --toponly means we are only analysing the input strand, not the reverse complement. We do not need to analyse the complementary strand because all the candidates should be on the input strand.

The parameter --tabfile chlorophyta.tab is an option to produce the output in a tabular format. We prefer this option as we are going to parse the output with another Perl script. In the file chlorophyta.tab there are 13 different sequences with a score greater than 20. This means that 84 sequences out of the 97 identified by the script in Code 17.1 were false positives, at least according to cmsearch.

The full-length RNA sequences are not shown in the output from cmsearch. In order to obtain these, we write yet another Perl script, shown in Code 17.2.

Code 17.2 extract_chloroplast_rnas.pl

```perl
#!/usr/bin/perl -w

use strict;

# first read the sequences that were used as input
# to cmsearch from the file pattern_search.out
my %origseq;
my $id;
my $strand;
my $infile = 'pattern_search.out';
open(IN, $infile) or die "Oops, could not open $infile\n";
while (<IN>) {
    chomp;
    if (/>(.*)/) { $id = $1; }
    else         { $origseq{$id} = $_; }
}
close IN;

# read the output from cmsearch
$infile = 'chlorophyta.tab';
open(IN, $infile) or die "Oops, could not open $infile\n";
while (<IN>) {
    if (/SRP_bact +(\S+) +(\S+) +(\S+) +(\S+) +(\S+) +(\S+) +(\S+)/) {
        $id = $1;
        my $beg = $2;
        my $end = $3;
        my $score = $6;
        my $len = $end - $beg + 1;
        my $ret = substr( $origseq{$id}, $beg - 1, $len );
        if ( $score > 20 ) {
            print ">$id BEG:$beg END:$end\n$ret\n";
        }
    }
}
close IN;
```

In Code 17.2 we first read the original sequences from the file `pattern_search.out`. These sequences are stored in the hash `%origseq`, where the keys are the identifiers and the values are the sequences.

In the second part of the script we read the information in the file `chlorophyta.tab`, and use that information to extract the relevant sequences contained within `%origseq`. Pattern matching is used to extract the information we need. The lines of interest start with the name of the covariance model, in this case `SRP_bact`. One example of such a line is:

```
SRP_bact 9_7524759_Chlorella_vulgaris_+ 4 100 1 101 33.81 5.07e-09 35
```

The data are in a tabular, space-delimited format in which the third and fourth columns have the begin and end positions, respectively, of the sequence analysed by cmsearch. We can capture the column values with the regular expression below:

```
/SRP_bact +(\S+) +(\S+) +(\S+) +(\S+) +(\S+) +(\S+) +(\S+)/
```

Here, `\S` is a metasymbol to denote any non-whitespace character, as introduced in Chapter 9.

The script in Code 17.2 is executed and the result is saved:

```
% perl extract_chloroplast_rnas.pl > extract_chloroplast_rnas.out
```

The output file contains the sequences of the 13 SRP RNA candidates identified by cmsearch using the score cut-off of 20. One important question is: are they really chloroplast SRP RNAs? We can actually gather some more evidence in case you are not convinced already. First, these sequences should adopt a hairpin structure. Therefore, we may subject them to an RNA secondary structure prediction using, for example, the software MFOLD (Zuker, 1989, 2003) or UNAFold (Markham and Zuker, 2008). Such structures are shown for a selection of the predicted RNAs in Fig. 17.5 and a cyanobacterial SRP RNA is included for reference. Indeed, all of the RNAs fold up nicely into a hairpin structure, with properties similar to those of the cyanobacterial RNA.

If you are still not convinced about these predictions, we may actually gather more evidence that the different chloroplast RNAs are homologous. This is because homologous genes in closely related species often have the same genomic localization. Where are the different SRP RNAs positioned in the chloroplast genome? The gene organization of a genomic region in which the predicted SRP RNA gene is located is shown for different chlorophyta species in Fig. 17.6 (see also Rosenblad and Samuelsson, 2004). There is, in fact, a striking similarity in gene organization. Thus, in all the chlorophyta species the SRP RNA gene is located a relatively short distance upstream of the *rpoB* gene, a gene encoding the beta subunit of a chloroplast RNA polymerase. This strictly

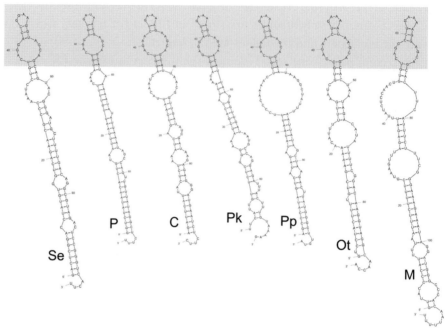

Fig. 17.5 *Predicted chloroplast SRP RNAs.* Chloroplast RNAs were predicted using a combination of pattern matching and searches based on covariance models. Secondary structures of these RNAs were then predicted with UNAFold (Markham and Zuker, 2008). Species are, from left to right, the cyanobacterium *Synechococcus elongatus* (Se) and the chlorophyta *Pedinomonas* (P), *Chlorella vulgaris* (C), *Parachlorella kessleri* (Pk), *Pyramimonas parkeae* (Pp), *Ostreococcus tauri* (Ot) and *Monomastix* (M).

conserved gene arrangement provides strong support that the SRP RNA genes we predicted are homologous.[2]

In conclusion it would seem that the Perl script, in combination with cmsearch, was able to identify a number of SRP RNA homologues in chloroplasts. Could these homologues have been identified using a simple local alignment? The answer is no, the primary structures are poorly conserved. This is seen in Fig. 17.7, in which the sequences shown in Fig. 17.6 have been used to produce a multiple alignment. In comparison, a profile-based search based on this alignment using, for example, HMMER (Chapter 12), would have done a

---

[2] We cannot, however, use gene order information to conclude on the homology between cyanobacterial and chloroplast SRP RNAs. The gene order of a cyanobacterial genome is not at all related to that of chloroplasts, because many genes were lost and many rearrangements took place in the chloroplast genome during its evolution from a cyanobacterial-like species. We should therefore not expect the chloroplasts and cyanobacteria to be related with respect to the position of the SRP RNA gene.

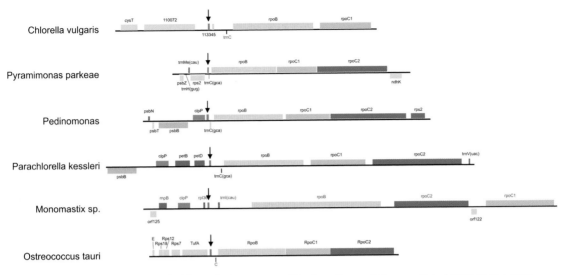

Fig. 17.6 *Gene organization of chloroplast genomes.* The locations of the predicted SRP RNA genes (green) in chloroplast genomes are indicated with arrows. For all chlorophyta species considered here this gene is adjacent to a tRNA gene and to the *rpoB* gene.

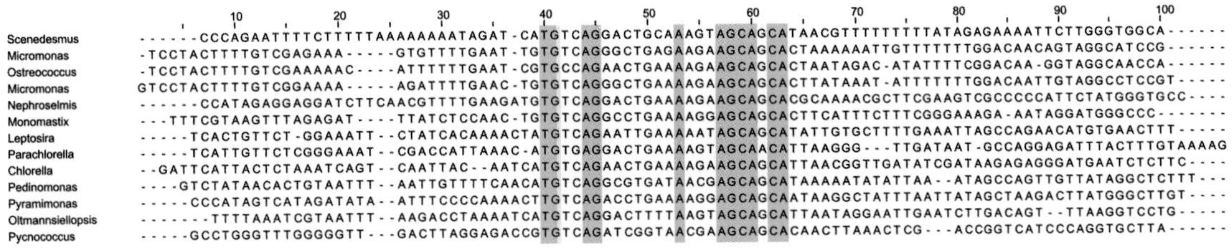

Fig. 17.7 *Multiple sequence alignment of chlorophyta chloroplast SRP RNA gene sequences.* Strictly conserved sequences are highlighted with a blue background. Compare the structures in Fig. 17.5.

much better job in identifying the SRP RNA genes. However, it would still not have been as effective as Infernal.

Automating tasks with Unix and Perl

The exercises of this chapter illustrate how Perl and bioinformatics software may be used together to form a pipeline. For example, we might be interested in automating the procedure of RNA finding if we wanted to repeatedly analyse many different genomes. In such a case we would like to use a single command typed at the Unix prompt to carry out the whole operation without any intervention. Two ways of achieving this are briefly described below.

(1) The whole procedure is specified within a Unix *shell script*. Create a file named `rnafind.sh` with the commands as we have used them above:

```
#!/bin/bash
perl pattern_search.pl > pattern_search.out
cmsearch --tabfile chlorophyta.tab SRP_bact.cm pattern_search.out
perl extract_chloroplast_rnas.pl > extract_chloroplast_rnas.out
```

This script may be executed with:

```
% bash rnafind.sh
```

Alternatively, the script file is made executable with `chmod +x rnafind.sh` and then it is executed with `./rnafind.sh`.

(2) The whole procedure is carried out with the help of a single Perl script. In such a case the two Perl scripts used above (Code 17.1 and Code 17.2) may, with some minor modifications, be merged into a single script. The cmsearch-based search may be initiated from within the Perl script with the `system` function:

```
system ("cmsearch --toponly --tabfile chlorophyta.tab SRP_bact.cm
pattern_search.out");
```

The `system` function executes any program that is available in the system (like a Unix system). The `system` function does not return the output of the program called; instead it returns the exit status. For more information on the `system` function, and for the use of *backticks* as a method of capturing output from an external program, see Appendix III.

EXERCISES

17.1 Chloroplast genomes contain a number of different tRNA genes. Download a covariance model for tRNA (RF00005) from the Rfam database (http://rfam.sanger.ac.uk) and use it with the Infernal software to search the same chlorophyta chloroplast genomes that we have used to identify SRP RNAs. How many tRNA genes are there typically in a chlorophyta chloroplast genome according to the Infernal software?

17.2 Combine the procedures contained within Code 17.1, Code 17.2 and the Infernal search into a *single* Perl script, as discussed above.

17.3 In Code 17.1 we read sequences of chloroplast genomes. When doing that we have neglected the fact that although the sequences are presented in a linear form, the chloroplast genomes are *circular* molecules. This means that in reality the 5′ and 3′ ends of the sequences are joined. Therefore, in

Code 17.1 we may be missing RNA candidates close to the 5' and 3' ends of the linear sequence. Modify the code to take this into account.

REFERENCES

Dawkins, R. (1989). *The Selfish Gene*. Oxford and New York, Oxford University Press.

Eddy, S. R. and R. Durbin (1994). RNA sequence analysis using covariance models. *Nucleic Acids Res* **22**(11), 2079–2088.

Egea, P. F., R. M. Stroud and P. Walter (2005). Targeting proteins to membranes: structure of the signal recognition particle. *Curr Opin Struct Biol* **15**(2), 213–220.

Gardner, P. P., J. Daub, J. Tate, *et al*. (2010). Rfam: Wikipedia, clans and the 'decimal' release. *Nucleic Acids Res* **39** (Database issue), D141–D145

Gesteland, R. F., T. Cech and J. F. Atkins (1999). *The RNA World: The Nature of Modern RNA Suggests a Prebiotic RNA*. Cold Spring Harbor, NY, Cold Spring Harbor Laboratory Press.

Gilbert, W. (1986). Origin of life: the RNA world. *Nature* **319**(6055), 618.

Grudnik, P., G. Bange and I. Sinning (2009). Protein targeting by the signal recognition particle. *Biol Chem* **390**(8), 775–782.

Guerrier-Takada, C. and S. Altman (1984). Catalytic activity of an RNA molecule prepared by transcription in vitro. *Science* **223**(4633), 285–286.

Guerrier-Takada, C., K. Gardiner, T. Marsh, N. Pace and S. Altman (1983). The RNA moiety of ribonuclease P is the catalytic subunit of the enzyme. *Cell* **35**(3 Part 2), 849–857.

Imanian, B., J. F. Pombert and P. J. Keeling (2010). The complete plastid genomes of the two 'dinotoms' *Durinskia baltica* and *Kryptoperidinium foliaceum*. *PLoS One* **5**(5), e10711.

Kruger, K., P. J. Grabowski, A. J. Zaug, *et al*. (1982). Self-splicing RNA: autoexcision and autocyclization of the ribosomal RNA intervening sequence of Tetrahymena. *Cell* **31**(1), 147–157.

Lane, N. (2005). *Power, Sex, Suicide: Mitochondria and the Meaning of Life*. Oxford and New York, Oxford University Press.

Markham, N. R. and M. Zuker (2008). UNAFold: software for nucleic acid folding and hybridization. *Methods Mol Biol* **453**, 3–31.

McFadden, G. I. (2001). Primary and secondary endosymbiosis and the origin of plastids. *J Phycol* **37**(6), 951–959.

Nawrocki, E. P., D. L. Kolbe and S. R. Eddy (2009). Infernal 1.0: inference of RNA alignments. *Bioinformatics* **25**(10), 1335–1337.

Peluso, P., D. Herschlag, S. Nock, *et al*. (2000). Role of 4.5S RNA in assembly of the bacterial signal recognition particle with its receptor. *Science* **288**(5471), 1640–1643.

Pettersen, E. F., T. D. Goddard, C. C. Huang, *et al*. (2004). UCSF Chimera: a visualization system for exploratory research and analysis. *J Comput Chem* **25**(13), 1605–1612.

Poritz, M. A., H. D. Bernstein, K. Strub, *et al*. (1990). An *E. coli* ribonucleoprotein containing 4.5S RNA resembles mammalian signal recognition particle. *Science* **250**(4984), 1111–1117.

Poritz, M. A., K. Strub and P. Walter (1988). Human SRP RNA and *E. coli* 4.5S RNA contain a highly homologous structural domain. *Cell* **55**(1), 4–6.

Rosenblad, M. A. and T. Samuelsson (2004). Identification of chloroplast signal recognition particle RNA genes. *Plant Cell Physiol* **45**(11), 1633–1639.

Shan, S. O. and P. Walter (2005). Co-translational protein targeting by the signal recognition particle. *FEBS Lett* **579**(4), 921–926.

Shi, H. and P. B. Moore (2000). The crystal structure of yeast phenylalanine tRNA at 1.93 A resolution: a classic structure revisited. *RNA* **6**(8), 1091–1105.

Woese, C. R. (1967). *The Genetic Code: The Molecular Basis for Genetic Expression*. New York, Harper & Row.

Yarus, M. (2010). *Life from an RNA World: The Ancestor Within*. Cambridge, MA, Harvard University Press.

Zuker, M. (1989). On finding all suboptimal foldings of an RNA molecule. *Science* **244**(4900), 48–52.

Zuker, M. (2003). MFOLD web server for nucleic acid folding and hybridization prediction. *Nucleic Acids Res* **31**(13), 3406–3415.

Personal genomes
The differences between you and me

The genome revolution is only just beginning.

(J. Craig Venter, 2010)

Personal genomes

The first versions of the human genome sequence were presented in 2001 (Lander *et al.*, 2001; Venter *et al.*, 2001). They resulted from projects highly demanding in terms of resources and financing. The publicly funded sequencing project was supported by a $3 billion grant allocated to the Human Genome Project, and Craig Venter's project is reported to have cost $300 million. Since 2001, however, DNA sequencing technology has been made much more effective (Metzker, 2010) and the cost of sequencing a human genome has dropped dramatically. Now, in 2012, it is less than $5000 and is expected to become even less expensive. Furthermore, a human genome is now being sequenced in less than one week. Current DNA sequencing machines are able to produce gigabytes of data every day, and large sequencing centres are able to produce data at an unprecedented rate. For instance, the Beijing Genomics Institute (BGI) in Hong Kong is reported to have, as of December 2010, a total sequencing capacity of an astounding five terabases (5×10^{12}, equivalent to 1000 human genomes) per day.

As a consequence of this technology development we see many different organisms being analysed with respect to their genomes. We also see numerous human genomes being sequenced. The first individuals whose complete genomes were sequenced include Craig Venter and James Watson, as well as anonymous individuals from Asia and Africa. The most important of these projects are listed in Table 18.1. In addition to these efforts there are many ongoing projects such as the 1000 Genomes Project (http://www.1000genomes.org; Durbin *et al.*, 2010), as well as commercial efforts (see, for instance, Drmanac *et al.*, 2010).

What can we learn from the genome of a human individual, a 'personal' genome? Most important, there is a large number of medical applications.

Table 18.1 *A selection of first personal genomes.*

Human individual	Year	Reference
First versions of human genome (mixture of anonymous individuals)	2001–2004	(Lander *et al.*, 2001; Venter *et al.*, 2001; International Human Genome Sequencing Consortium, 2004)
Craig Venter	2007	(Levy *et al.*, 2007)
James Watson	2008	(Wheeler *et al.*, 2008)
AML (leukaemia) patient, normal and cancer tissue	2008	(Ley *et al.*, 2008)
Yoruba, Ibadan, Nigeria (anonymous)	2008	(Bentley *et al.*, 2008)
YanHuang* (Han Chinese individual, anonymous)	2008	(Wang *et al.*, 2008)
Stephen Quake (researcher at Stanford University)	2009	(Pushkarev *et al.*, 2009)
Seong-Jin Kim (Korean researcher)	2009	(Ahn *et al.*, 2009)
James Lupski (Researcher at Baylor College of Medicine)	2010	(Lupski *et al.*, 2010)
Desmond Tutu and !Gubi,** Bantu and Khoi-San individuals, respectively	2010	(Schuster *et al.*, 2010)
Glenn Close (actress)	2010	

*YanHuang is an anonymous Han Chinese; refers to Yan and Huang, two emperors thought to be ancestors of Han Chinese. **The exclamation point represents one of the click sounds in Bushmen language, see also Chapter 19.

Some of the projects listed in Table 18.1 were directly medically motivated, with the objective of identifying disease-related genes. One example is the study of a leukaemia patient, which allowed the identification of a number of cancer-specific changes in genomic DNA (Ley *et al.*, 2008). Another example is the analysis of the genome of James Lupski, who is affected by the Charcot–Marie–Tooth disease, an inherited neurological disorder (Lupski *et al.*, 2010).

Individual variation and SNPs

Now we have access to several personal genomes, and we are able to compare them using different criteria. What are the differences we are able to observe between individual genomes? We may first distinguish different categories of individual variation. One significant and useful category is *single nucleotide polymorphisms*, or *SNPs*. SNPs are single nucleotide/base changes in the DNA sequence[1] (Fig. 18.1). Another category of differences are *small insertions* or *deletions*. An example would be a 'G' in one site in the genome of one individual that is replaced by 'GAC' in another individual. Such differences are much less common than SNPs. Furthermore, there is *copy-number variation*, i.e. variation in the number of tandem repeats in a specific chromosomal site. These are individual differences of medical importance and that are being exploited in many forensic applications. Finally, there are *larger structural changes* such as insertions or deletions of longer sequences.

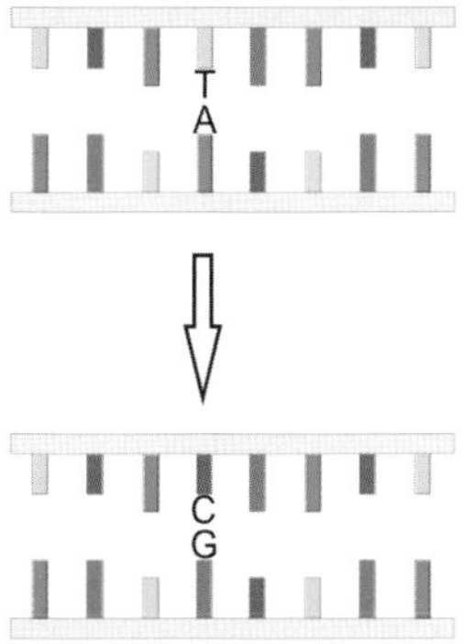

For this chapter we will focus on SNPs – they will also be the objects of the programming exercise for this chapter. SNPs are highly predominant, well-characterized and often used for comparative studies. They may occur at any location in the genome. Some of the SNPs in non-coding parts of a genome may affect the expression of a gene. However, the SNPs that are in coding regions of genes are particularly interesting. An SNP in a coding region can have the effect of changing the amino acid sequence of the encoded protein. In such a case the SNP is likely to alter the biological properties of the protein. An SNP may also leave the encoded amino acid sequence unchanged. We refer to *synonymous* and *non-synonymous* mutations to describe the two categories of changes.[2]

SNPs are sometimes associated with disease. For instance, the protein factor V is involved in blood coagulation (see Chapter 12) and there is an SNP in the gene of that protein which causes a change from arginine to glutamine. The resulting variant protein 'factor V Leiden' gives rise to a disorder in which clotting occurs more easily (Bertina *et al.*, 1994). SNPs are generally useful in diagnostics and in order to predict how individuals respond to drugs. Furthermore, they play an important role in identifying regions in a genome that are important in disease. In such cases a typical strategy is to compare cohorts with and without disease with respect to their SNPs.

In February 2012 there were about 60 million SNPs of the human genome according to the records of the NCBI SNP database.[3] This database stores SNPs

Fig. 18.1 *Single nucleotide polymorphism (SNP).* SNPs are a common type of variation between individuals and are the result of single nucleotide/base changes in the DNA sequence. In this example the T–A base pair in the double-stranded DNA has been replaced with the C–G base pair.

---

[1] To be more precise, an SNP is defined as a single nucleotide change which is present among human individuals and that has been well characterized in the sense that it is known to occur in many individuals. A less well-characterized single nucleotide change, such as one that occurs in one individual or family only, is referred to as *single nucleotide variant*, SNV.

[2] These concepts are related to the degeneracy of the genetic code, meaning that many amino acids are encoded by more than one codon. For instance, the amino acid glycine has four synonymous codons GGA, GGT, GGC and GGG. A mutation in the third position of those codons does not change the amino acid encoded (a synonymous mutation), whereas a mutation in any of the other positions will result in a different amino acid (a non-synonymous mutation; for the genetic code, see also Chapter 1).

[3] It should be noted, however, that a fraction of the records in dbSNP are not single nucleotide changes but other changes like small insertions and deletions.

of humans, as well as many other species. However, new SNPs are continually being discovered as new individuals are sequenced. When comparing two randomly selected individuals there are in the order of three million SNP differences.[4] In our programming example below we will actually examine these differences more closely.

When considering SNP differences between individuals, it has to be kept in mind that each human individual has two copies of each chromosome, one originating from the father and one from the mother. In terms of SNPs this has the consequence that at a specific location of the SNP, the individual may be *homozygous* – for instance, with the base pair G-C in both chromosomes. Or, the individual may be *heterozygous*, with, as an example, G-C in one of the chromosomes and A–T in the other. In the language of genetics the G–C and A–T sites are referred to as two different *alleles*. In the programming example below we are to make use of SNPs to measure the genetic distance between human individuals. In doing that we will count positions where at least one allele is different, but there are other ways of considering the distance, as illustrated in Exercise 18.1.

It should be noted that the SNP differences we discuss in this chapter are relative in the sense that in many cases we do not know what is the more common allele in human populations. We regard the original human genome sequence (the first version published in 2001) as a reference sequence, but it is important to note that this sequence is based on only a few individuals. Therefore, the different nucleotide positions in that genome in no way represent the most common allele among humans.

BIOINFORMATICS

Counting SNPs

Tools introduced in this chapter	
Perl	repeat operator

We are to make use of a number of human individual genomes and compare them in a global fashion by simply counting the differences in terms of SNPs. SNP data of the type used here may be downloaded using the table browser at the UCSC Genome Bioinformatics site (http://genome.ucsc.edu/cgi-bin/hgTables?command=start).[5] Genomes of eight different human individuals will be considered, but for simplicity our analysis

---

[4] Three million may sound a lot, but looking at it from the perspective of similarity, two humans would be about 99.9% identical.

[5] The SNP data for this chapter was, however, generously provided by Adam Siepel and Melissa Jane Hubisz, Department of Biological Statistics and Computational Biology, Cornell University.

will be restricted to chromosome 4 only. Six of the individuals are listed in Table 18.1 and two are from the 1000 Genomes Project. The individuals are:

(1) YanHuang (Han Chinese individual, anonymous)
(2) Seong-Jin Kim (Korean)
(3) James Watson
(4) Craig Venter
(5) YRI NA18507 (Yoruba, anonymous of the 1000 Genomes Project)
(6) NA12891 (Central European origin, anonymous of the 1000 Genomes Project)
(7) ABT, Desmond Tutu
(8) KB1, Khoisan/Bushmen individual.

The Southern African individuals ABT and KB1 are further discussed in Chapter 19. Data on the human individuals was processed to generate a tab-delimited file, snp.txt. Each line in this file has tab-separated values in which the first value is a nucleotide position in chromosome 4. Then follows characters that describe the nucleotide(s) in that position in the different individuals. The order of the individuals is the same as listed above. Sequence data of the chimpanzee genome was also included for reference, and the chimpanzee nucleotide is shown as the last value in each line.

Excluded in the file snp.txt are positions that are not useful for our analysis. Certain positions in a genome may not have been covered by the sequencing project. For such positions, comparison to the other genomes is obviously not meaningful. Thus, whenever at least one genome has an unknown composition at a certain position, this position has been left out. Furthermore, all positions containing the same nucleotide in all nine genomes have been removed. This is an example of a line in the snp.txt file:

```
38357 T T Y T T T Y Y T
```

Whenever we encounter any of the four symbols A, T, C or G in a human individual, this means that individual is homozygous at that position. In the event that a human individual is heterozygous at a position we use the same nucleotide ambiguity symbols as we have encountered in Chapters 2 and 4. In the example above, for instance, Y refers to C or T, meaning that one allele is C and the other is T.

We are now going to make use of this SNP information in order to compare the different genomes. The Perl script in Code 18.1 is designed to create a distance matrix to show how the genomes are related. Recall how we used a distance matrix in the context of the neighbour-joining method discussed in Chapters 9–10. Details of Code 18.1 are discussed in the following, in which the numbers refer to numbered sections in the code.

Code 18.1 snp.pl

```perl
#!/usr/bin/perl -w

# obtain pairwise distances from snp data
# counting sites where at least one allele is different

use strict;

my @humans = (
    # SNPs appear in the SNP data file in columns in this order
    'YH',           # Han Chinese
    'SJK',          # Seong-Jin Kim
    'JW',           # James Watson
    'CV',           # Craig Venter
    'NA18507',      # Yoruban of 1000 Genomes project
    'NA12891',      # Of Central European origin
    'ABT',          # Archbishop Desmond Tutu
    'KB1',          # Bushmen individual
    'chimp'         # Chimpanzee
);

my @diff;

# 1 #
# initialize the distance matrix with zero values
# for the diagonal cells
for ( my $i = 1 ; $i < 10 ; $i++ ) {
    $diff[$i][$i] = 0;
}

# read the snp data from file
my $infile = 'snp.txt';
open(IN, $infile) or die "Oops, could not open file";
while (<IN>) {
    chomp;
    my @columns = split;

    # 2 #
    for ( my $i = 1 ; $i < 9 ; $i++ ) {
        for ( my $j = $i + 1 ; $j < 10 ; $j++ ) {

            # 3 #
            if ( $columns[$i] ne $columns[$j] ) {
                $diff[$i][$j]++;

                # 4 #
                # to produce a symmetric matrix
                $diff[$j][$i]++;
            }
        }
```

```
        }
    }
    close IN;

    # 5 #
    # print a header for PHYLIP format
    # with the number of species

    print " 9\n";

    # print the matrix data

    for ( my $i = 1 ; $i < 10 ; $i++ ) {

        # 6 #
        my $txt = $humans[ $i - 1 ];
        $txt = substr( $txt, 0, 7 );
        print "$txt";
        my $len = 10 - length($txt);
        my $short = ' ' x $len;
        print "$short";
        for ( my $j = 1 ; $j < 10 ; $j++ ) {
            print "$diff[$i][$j] ";
        }
        print "\n";
    }
```

(1) The variable @diff is used to store the counts of pairwise distances. It is a two-dimensional array. We begin by initializing this variable with zero values for the diagonal values, i.e. $diff[$i][$i], where $i is in the range 1–9.

(2) We use for loops to go through all pairs of genomes. The total number of genomes is nine. We consider all pairs with indices i and j, where i and j are in the range 1–9. However, a pair like 1,4 is the same as 4,1 and therefore we need to consider only those pairs where $i < j$. To accomplish this we use a loop construction:

```
    for ( my $i = 1 ; $i < 9 ; $i++ ) {
        for ( my $j = $i + 1 ; $j < 10 ; $j++ ) {...}
    }
```

Note that we start at $i = 1 because $columns[1] corresponds to the first genome, whereas $columns[0] is the chromosomal position.

(3) We want to count instances where the two genomes are different with respect to at least one allele. For instance, 'T' is different from 'C' as well as from 'Y' (C/T). Therefore we simply test whether the two character strings are unequal:

```
    if ( $columns[$i] ne $columns[$j] ) {...}
```

As we have referred to above, there are other ways of estimating the distance between two genomes based on SNPs, and this is illustrated in Exercise 18.1.

(4) If we find that two genomes are different at a specific nucleotide position, the array @diff will be updated with $diff[$i][$j]++;. As we have a symmetric matrix $diff[$j][$i] should be the same, hence we also use $diff[$j][$i]++;.

(5) The distance matrix output of the script is later used as input to the program Neighbor of the PHYLIP package. Therefore, to fit with the PHYLIP distance matrix format, we need to add a top line with the number of species involved.

(6) In this part of the script we want to achieve a truncation of the names of the different individuals as they appear in our original file snp.txt. This is because we are to use the PHYLIP program Neighbor, and in this program a name is not allowed to be more than ten characters in length. If the name is shorter than ten characters it needs to be padded out with blanks. In this script we first truncate the name to seven characters:

```
$txt = substr( $txt, 0, 7 );
```

Then we determine the number of blank spaces required for the padding. We need to do this because the length of the original name may have been less than seven characters:

```
my $len = 10 - length($txt);
```

Finally, we create the padding with blanks. We use the string operator 'x', which is a kind of 'multiply' operator for strings referred to as the *repeat* operator. It is not to be confused with the multiplication operator '*' used for numbers. The variable $short will have the number of blanks specified by the numeric $len variable:

```
my $short = ' ' x $len;
```

For the execution of the script we use:

```
% perl snp.pl > snp.out
```

The distance matrix will be:

```
        9
YH       0 44597 53594 53913 67914 53710 68837 77272 593367
SJK      44597 0 54192 54537 68826 55281 69404 76929 593496
JW       53594 54192 0 50859 70284 51260 70256 77590 592751
CV       53913 54537 50859 0 70149 51009 69659 77369 592632
NA18507  67914 68826 70284 70149 0 69245 70057 79508 599102
NA12891  53710 55281 51260 51009 69245 0 69941 78130 594831
ABT      68837 69404 70256 69659 70057 69941 0 77707 599292
KB1      77272 76929 77590 77369 79508 78130 77707 0 600776
chimp    593367 593496 592751 592632 599102 594831 599292 600776 0
```

What are the conclusions from our analysis of the different genomes? For chromosome 4 we arrive at a number of SNP differences between human individuals which is in the range 45 000–80 000. If we allow extrapolation this would correspond to about 700 000–1 250 000 differences for the whole genome. We said earlier that we should expect approximately three million differences between any two individuals. The reason we end up with a smaller number here is that our original dataset does not have information for all possible SNPs. For instance, there are one or more individuals where we are lacking sequence information for portions of the genome. All such regions have been filtered out in our dataset.

We may also use the distance matrix to construct a tree, showing how the different human individuals are related. If we use the distance matrix above as input to a neighbour-joining tree construction method the tree in Fig. 18.2 is obtained. The root of this tree is the chimpanzee. The deepest branch among humans is KB1, a Khoisan or Bushmen individual. This is consistent with the idea referred to in the following chapter that the Bushmen represent an old lineage in the human tree. The other two African individuals are also deeply branching. We also note that the three Europeans group together as well as the two Asian individuals.

The tree in Fig. 18.2 suggests that the SNP data is able to correctly classify the different individuals into human populations or ethnic groups. We would have obtained a similar tree if we had made use of a multiple sequence alignment of mitochondrial DNA sequences as shown in Chapter 10. There are indeed SNPs that are specific to a certain population. At the same time, it is interesting to note that most of the human genetic variation is not between ethnic groups or between geographically distinct populations. Rather, most of the variation is found *within* populations (Lewontin, 1972; Barbujani *et al.*, 1997;

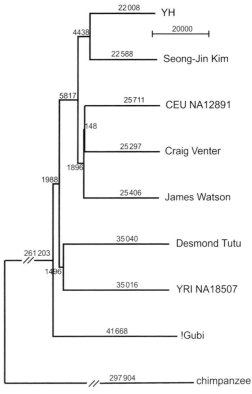

Fig. 18.2 *Relationship of human individuals based on SNP differences.* A distance matrix was obtained by counting the number of SNP differences between eight human individuals and a chimpanzee. This distance matrix was used to generate the rooted neighbour-joining tree shown. The human individuals are: Han Chinese anonymous (YH); Seong-Jin Kim (Korean); anonymous individual of Central Europe of the 1000 Genomes Project (CEU NA12891); James Watson; Craig Venter; Archbishop Desmond Tutu; anonymous Yoruba individual of the 1000 Genomes Project (YRI NA18507); and Khoisan individual !Gubi.

Rosenberg *et al.*, 2002). For instance, differences between continents represent only 10% or even less of human genetic diversity (Barbujani *et al.*, 1997; Rosenberg *et al.*, 2002). These observations of human genetic diversity are also reflected in our distance matrix. For instance, consider the two African genomes

ABT and KB1 with a differences count from our distance matrix of 77 707. This number is about the same or even higher as compared to just any pair in our collection of individuals. Our current understanding of human genetic diversity should be distressing for any human individual advocating racism, not to mention racial biology and eugenics (Black, 2003).

EXERCISES

18.1 In Code 18.1 we have used a distance measure where we say that two positions are different if at least one allele is different. An alternative approach is one in which we set a distance which is:

1 if both alleles are different

0.5 if one of the alleles is shared

0 if both alleles are the same

As an example, consider two individuals where one is R and the other is A. We then compare the heterozygous AG with the homozygous AA. In this case one of the alleles is shared and the distance is 0.5. Another example is when one individual is A and the other is G. In this case both alleles are different and the distance is counted as 1. Modify Code 18.1 to implement this method. Also use the results to construct a tree like that shown in Fig. 18.2.

18.2 The file `chr4snp.txt` (see the web resources for this book) is a list of SNPs in the human chromosome 4, according to dbSNP build 130. Write a Perl script that will list the SNPs (positions) that are present in this file but that are not found in the file `snp.txt` used in Code 18.1. The file `chr4snp.txt` uses 'zero-based' numbering as described in Chapter 19. From a practical point of view this means that the third column positions in that file are comparable to the position numbers in `snp.txt`.

18.3 Modify Code 18.1 to produce a *lower triangular matrix* instead of a symmetrical matrix. An example of such a matrix in a PHYLIP format and with five OTUs is shown here:

```
Alpha       5
Beta        10
Gamma       30 30
Delta       30 30 20
Epsilon     30 30 20 10
```

REFERENCES

Ahn, S. M., T. H. Kim, S. Lee, *et al.* (2009). The first Korean genome sequence and analysis: full genome sequencing for a socio-ethnic group. *Genome Res* **19**(9), 1622–1629.

Barbujani, G., A. Magagni, E. Minch and L. L. Cavalli-Sforza (1997). An apportionment of human DNA diversity. *Proc Natl Acad Sci U S A* **94**(9), 4516–4519.

Bentley, D. R., S. Balasubramanian, H. P. Swerdlow, *et al.* (2008). Accurate whole human genome sequencing using reversible terminator chemistry. *Nature* **456**(7218), 53–59.

Bertina, R. M., B. P. Koeleman, T. Koster, *et al.* (1994). Mutation in blood coagulation factor V associated with resistance to activated protein C. *Nature* **369**(6475), 64–67.

Black, E. (2003). *War Against the Weak: Eugenics and America's Campaign to Create a Master Race*. New York, Four Walls Eight Windows.

Drmanac, R., A. B. Sparks, M. J. Callow, *et al.* (2010). Human genome sequencing using unchained base reads on self-assembling DNA nanoarrays. *Science* **327**(5961), 78–81.

Durbin, R. M., G. R. Abecasis, D. L. Altshuler, *et al.* (2010). A map of human genome variation from population-scale sequencing. *Nature* **467**(7319), 1061–1073.

International Human Genome Sequencing Consortium (2004). Finishing the euchromatic sequence of the human genome. *Nature* **431**(7011), 931–945.

Lander, E. S., L. M. Linton, B. Birren, *et al.* (2001). Initial sequencing and analysis of the human genome. *Nature* **409**(6822), 860–921.

Levy, S., G. Sutton, P. C. Ng, *et al.* (2007). The diploid genome sequence of an individual human. *PLoS Biol* **5**(10), e254.

Lewontin, R. (1972). The apportionment of human diversity. *Evol Biol* **6**, 381–398.

Ley, T. J., E. R. Mardis, L. Ding, *et al.* (2008). DNA sequencing of a cytogenetically normal acute myeloid leukaemia genome. *Nature* **456**(7218), 66–72.

Lupski, J. R., J. G. Reid, C. Gonzaga-Jauregui, *et al.* (2010). Whole-genome sequencing in a patient with Charcot–Marie–Tooth neuropathy. *N Engl J Med* **362**(13), 1181–1191.

Metzker, M. L. (2010). Sequencing technologies: the next generation. *Nat Rev Genet* **11**(1), 31–46.

Pushkarev, D., N. F. Neff and S. R. Quake (2009). Single-molecule sequencing of an individual human genome. *Nat Biotechnol* **27**(9), 847–850.

Rosenberg, N. A., J. K. Pritchard, J. L. Weber, *et al.* (2002). Genetic structure of human populations. *Science* **298**(5602), 2381–2385.

Schuster, S. C., W. Miller, A. Ratan, *et al.* (2010). Complete Khoisan and Bantu genomes from southern Africa. *Nature* **463**(7283), 943–947.

Venter, J. C. (2010). Multiple personal genomes await. *Nature* **464**(7289), 676–677.

Venter, J. C., M. D. Adams, E. W. Myers, *et al.* (2001). The sequence of the human genome. *Science* **291**(5507), 1304–1351.

Wang, J., W. Wang, R. Li, *et al.* (2008). The diploid genome sequence of an Asian individual. *Nature* **456**(7218), 60–65.

Wheeler, D. A., M. Srinivasan, M. Egholm, *et al.* (2008). The complete genome of an individual by massively parallel DNA sequencing. *Nature* **452**(7189), 872–876.

19 Personal genomes
What's in my genome?

Be nice to the whites, they need you to rediscover their humanity.

(Archbishop Desmond Tutu)

If we have access to the genome sequence of an individual there are a very large number of questions to be posed regarding that genome. For instance, what genetic markers are present that are of a predictive value in medical terms? What trait and disease alleles are present? What genetic properties are present that make the individual sensitive or resistant to a certain drug treatment? We may also want to know about the relationship of the individual to other individuals and whether there are markers characteristic of a certain human population. These are all questions that may be addressed using bioinformatics tools. In the previous chapter we examined SNPs and used them to get an idea about the genetic differences between individuals in general. Here, we will again use SNP data to analyse genomes, but we will see how we may identify SNPs that are shared between a group of individuals. We will also illustrate how SNP data may be mapped to information regarding exons, thus identifying SNPs that are likely to be in coding regions. In this way we are able to learn about the consequences of different SNPs at the level of protein products. The genomes that we are to examine are those of a few South African individuals. These genomes also highlight interesting questions regarding the early history of man.

Human roots

Most previous studies of human evolution have been carried out using analysis of mitochondrial DNA (compare the phylogenetic analysis of mitochondrial DNA in Chapter 10). Such studies show that the root of the human tree occurs in Africa. Anatomically modern humans are thought to have developed about 200 000 years ago. This evolution took place in Africa. According to the 'Out of Africa' hypothesis, members of one branch of *Homo sapiens* migrated out of

Africa around 100 000 years ago. They colonized the other parts of the world and eventually replaced earlier human populations.

If we think about the early evolution of humans in Africa, analysis of mitochondrial sequences shows that the first lineage to branch off from the human phylogenetic tree is a mitochondrial haplogroup named L0. This group is frequent among San in South Africa and in the Sandawe of East Africa (Gonder *et al.*, 2007). The complete genomes of South African individuals to be examined in this chapter are of this old lineage (Schuster *et al.*, 2010). The individuals, named !Gubi, G/aq'o, D#kgao and !Ai[1] (Fig. 19.1), are Namibian hunter-gatherers known as Khoisan or Bushmen.[2] They will be referred to below as KB1, NB1, TK1 and MD8, respectively. In addition to the Bushmen, the genome of Archbishop Desmond Tutu (below referred to as ABT), a Bantu individual, was sequenced in the same project (Fig. 19.1).

One important reason Bushmen were selected for sequencing in this project (Schuster *et al.*, 2010) is that this group of humans represents the oldest known lineage of modern humans and that the genome sequences aid in our understanding of human diversity. In addition, Africans have previously been highly underrepresented when it comes to genome-wide data for medical applications. It is also interesting to note that analysis of the genomic sequences reveals a large extent of diversity within the Bushmen group. In fact, in terms of SNP differences, the Bushmen individuals are, on average, more different from each other than, for instance, a European and an Asian. There are a number of characteristics of the Bushmen genomes that may be associated with their particular lifestyle as hunter-gatherers. We will see one example of that in the bioinformatics section of this chapter.

BIOINFORMATICS

An SNP dataset of South African individuals

We will be working with data that originated from a dataset available from Galaxy

Tools introduced in this chapter	
Perl	constructing your own modules
	$# to return last index of array

---

[1] The characters !, / and # denote click sounds in the Khoisan languages. The ! is a palatal click, the / is a dental click and the # is an alveolar click (Schuster *et al.*, 2010). The click phonemes are unique to these African people and there is evidence that they have an old history and arose in the order of tens of thousands of years ago in sub-Saharan Africa (Tishkoff *et al.*, 2007).

[2] Some people may regard the name 'Bushmen' as derogatory and obsolete, but it will be used here partly because it is used in the publication cited (Schuster *et al.*, 2010). In no way will the word be used in a derogatory sense in this chapter.

Fig. 19.1 *South African individuals subject to genomic sequencing*. The individuals shown are G/aq'o (upper-left, here also referred to here as NB1); !Gubi (upper-right, KB1); Archbishop Desmond Tutu (lower-left, ABT); !Ai (lower-middle, MD8); and D#kgao (lower-right, TK1). NB1, KB1, MD8 and TK1 are all Namibian hunter-gatherers (belonging to a group of people referred to as Khoisan), whereas ABT is a Bantu individual. For two of the individuals, KB1 and ABT, the complete genome sequence was determined (Schuster *et al.*, 2010). Photographs of the four Khoisan were generously provided by Stephan Schuster. Photograph of ABT by permission from Dreamstime.

(http://main.g2.bx.psu.edu/library) (Goecks *et al.*, 2010). Data for chromosome 7 was extracted and the resulting file, `bushmen_chr7_all_snps.txt`, has 24 columns as listed in Table 19.1. The columns in bold are used in the example for this chapter. Column 4 is the human reference sequence, columns 6 and 8 are the KB1 and NB1 Bushmen and column 14 is ABT.

This is an example line:

```
chr7 183804 183805 C T T N T N N N N C N N Y N N Y N N N Y
```

A nomenclature is used where the second and third column numbers are the start and end positions of the feature described, in this case the SNP. In this

Table 19.1 *Contents of the data file used in analysis of Bushmen SNP data.* A file 'All SNPs in personal genomes' of the library 'Bushman' was downloaded from Galaxy (*http://main.g2.bx.psu.edu/library*). Documentation shown in the table is attached to this file and available from the same site. Items in bold were used for the analysis in this chapter.

Column	Name	Description
1	chr	chromosome
2	start	0 based start position on the chromosome
3	end	0 based end position on the chromosome
4	**hg18**	**base in the human reference sequence (NCBI Build 36.1)**
5	chimp	the corresponding chimp nucleotide
6	**kb1**	**the base in the individual KB1 using whole genome data** (454 Illumina) and exome sequencing data (454)
7	kb1g	the base in the individual KB1 using genotype data
8	**nb1**	**the base in the individual NB1 using whole genome data (454) and exome sequencing data**
9	nb1g	the base in the individual NB1 using genotype data
10	md8	the base in the individual MD8 using exome sequencing data
11	md8g	the base in the individual MD8 using genotype data
12	tk1	the base in the individual TK1 using exome sequencing data
13	tk1g	the base in the individual TK1 using genotype data
14	**bats**	**the base in the individual ABT using SOLiD data**
15	bat	the base in the individual ABT using exome sequencing data
16	abtg	the base in the individual ABT using genotype data
17	na18507	the base in the individual NA18507 (Yoruba)
18	na19240	the base in the individual NA19240 (Yoruba)
19	watson	the base in the individual Watson
20	venter	the base in the individual Venter
21	na12891	the base in the individual NA12891 (European)
22	na12892	the base in the individual NA12892 (European)
23	chinese	the base in the Chinese individual
24	korean	the base in the Korean individual

type of nomenclature the end position is excluded, so if you see an end position like 183 805, that position is not included in the feature. The start and end positions are both of a 'zero-based' index, meaning that the very first position in the chromosome has index zero. The numbers in the line above, 183 804 and 183 805, refer to a single position which is 183 804 using a zero-based index (or if you prefer, it is the position 183 805 using a 'one-based' index).

Nucleotide symbols are used as in the previous chapter. Thus, any of the letters A, T, C or G indicates that the individual is homozygous at that position, whereas an ambiguity symbol like Y means the individual is heterozygous, with T in one chromosome and C in the other. An 'N' indicates that the nucleotide in this position is unknown (and not that all four nucleotides are present).

What SNPs are unique to the Bushmen?

We will now see how we may identify SNPs that are shared by a specific collection of individuals. In the following example we want to identify SNPs that are unique to the Bushmen. Such SNPs could be a result of positive selection among this group of hunter-gatherers. We will therefore try and identify the SNPs from our dataset in which the two Bushmen are identical and at the same time different from both the reference genome sequence and that of ABT. We will use the rather simple Perl code in Code 19.1.

We start by reading the file containing the SNP data. This file contains information about 609 276 positions on chromosome 7. We split the tab-separated data up to form the array @columns in the same manner as we have done in previous chapters. To exclude all positions where at least one of the four genomes considered has an unknown composition (as represented by 'N'), we use the condition:

```
if (( $columns[3] ne 'N' )
&&  ( $columns[5] ne 'N' )
&&  ( $columns[13] ne 'N' )
&&  ( $columns[7] ne 'N' )) { }
```

Code 19.1 bushmen.pl

```perl
#!/usr/bin/perl -w

use strict;

my $infile = 'bushmen_chr7_all_snps.txt';
open(IN, $infile) or die "Could not open $infile\n";
while (<IN>) {
    my $line = $_;
    chomp;
    my @columns = split("\t");

    # columns used are 4 = reference;
    # 6 = KB1 ; 8 = NB1; 14 = ABT
    # Find instances where both reference and
    # ABT are different from KB1 and NB1

    if (
            ( $columns[3] ne 'N' )
```

```
        && ( $columns[5] ne 'N' )
        && ( $columns[7] ne 'N' )
        && ( $columns[13] ne 'N' )
        && ( $columns[3] ne $columns[5] )
        && ( $columns[3] ne $columns[7] )
        && ( $columns[13] ne $columns[5] )
        && ( $columns[13] ne $columns[7] )
      )
    {
        print "$line";
    }
}
close IN;
```

We save the result of this script:

```
% perl bushmen.pl > bushmen.out
```

The original file `bushmen_chr7_all_snps.txt` contained 609 276 lines, but in `bushmen.out` we are left with only 16 367. This is the number of SNPs in chromosome 7 that are unique to the two Bushmen individuals.

What SNPs are in coding regions?

We now move on to the problem of identifying the SNPs that are unique to the Bushmen and that are also in exonic regions of chromosome 7. We also want to know what genes these regions belong to. To solve this problem we first need information about the location of the different genes on chromosome 7. For this example we get this information using the table browser of the UCSC Genome Bioinformatics site. For details on this procedure, see Box 19.1. I chose to name the resulting file `chr7.txt`; it will have nine columns with the following type of information, as extracted from line 2 of the file:

```
1. name:        uc003sii.2
2. chrom:       chr7
3. strand:      −
4. cdsStart:    54028
5. cdsEnd:      54028
6. exonCount:   8
7. exonStarts:  54028,67562,68535,68861,69146,72018,72927,73460,
8. exonEnds:    54637,67757,68768,68919,69832,72083,73124,73584,
9. proteinID:   −
```

Hence, we see that all exon start positions are listed in one column.

Box 19.1 Downloading annotation information with the UCSC Table Browser

Information about coding sequences in the human chromosome 7 was downloaded using the UCSC Table Browser. This information included a protein sequence database accession, when present. To obtain this information:

(1) Connect to the table browser of the UCSC Genome Bioinformatics site (http://genome.ucsc.edu/cgi-bin/hgTables?command=start).
(2) Select the following:

clade:	Mammal
genome:	Human
assembly:	Mar. 2006 (NCBI36/hg18) (this was the assembly used with the Bushmen data of this chapter)
group:	Genes and Gene Prediction Tracks
track:	UCSC Genes
table:	knownGene
region:	position chr7:1–158821424
output format:	selected fields from primary and related tables
output file:	(enter a suitable name)

Click the button 'get output'.
(3) On the resulting page select from the list of fields under 'Select Fields from hg18.knownGene': name, chrom, strand, cdsStart, cdsEnd, exonCount, exonStarts, exonEnds, proteinID. Finally, click 'get output'. Save the output file.

We can now use Perl code as shown in Code 19.2 to map the SNPs to the exons as specified by the annotation information. We start by reading the annotation information from file `chr7.txt`. In this example we choose to consider only lines where there is a protein identifier. All exon start sites are in column 7, as shown above. This is a comma-separated list and we put the values in an array with the expression:

```
my @exonStarts = split( ',', $exonStarts );
```

A similar procedure is used for the exon end positions. We therefore end up with two arrays with values for the exon start and end positions, respectively. Later, we want to look up a certain position and find out whether it is in an exonic region or not. To simplify this look-up we construct a hash %genome, in which the keys to this hash are all chromosomal positions that are contained within exons. Thus, each position in the chromosome which is part of an exon will be represented as a key in the hash. The values will have information about the gene name, direction of the gene ('strand') and protein sequence database identifier. The code in this part of the script is:

```perl
for ( my $i = 0 ; $i <= $#exonStarts ; $i++ ) {
        my $start = $exonStarts[$i];
        my $stop = $exonEnds[$i];
        for ( my $j = $start ; $j <= $stop ; $j++ ) {
                # save name, strand and proteinID
                $genome{$j} = "$columns[0] $columns[2] $columns[8]";
        }
}
```

In this code we go through each of the different exons and consider their start and end positions. These positions are zero-based as described for the SNP table above. In the `for` loop with the `$i` variable the value `$#exonStarts` is the last index of the `@exonStarts` array. You could also have used `$i < scalar(@exonStarts)`, where `scalar` returns the number of elements in the `@exonStarts` array (see also Appendix III).

In the `for` loop with the `$j` variable we go through each of the different positions in an exon. As the end position is excluded we use the condition `$j < $stop`. The result of this code is that in the hash `%genome`, all positions (keys) in one specific exon will have the same value.

Code 19.2 map2exons.pl

```perl
#!/usr/bin/perl -w

use strict;

my %genome;
my $infile = 'chr7.txt';       # based on NCBI36 assembly
open(EXONS, $infile) or die "Could not open $infile\n";
while (<EXONS>) {
    unless (/^\#/) {
        chomp;
        my @columns = split("\t");
        if ( $columns[8] ) {    # if there is a protein identifier
            my $exonStarts = $columns[6];
            $exonStarts =~ s/\,$//;      # remove the trailing ','
            my @exonStarts = split( ',', $exonStarts );
            my $exonEnds = $columns[7];
            $exonEnds =~ s/\,$//;        # remove the trailing ','
            my @exonEnds = split( ',', $exonEnds );
            for ( my $i = 0 ; $i <= $#exonStarts ; $i++ ) {
                my $start = $exonStarts[$i];
```

```
            my $stop = $exonEnds[$i];
            for ( my $j = $start ; $j < $stop ; $j++ ) {
                # save name, strand and proteinID
                $genome{$j} = "$columns[0] ";
                $genome{$j} .= "$columns[2] $columns[8]";
            }
        }
      }
   }
}
close EXONS;

$infile = 'bushmen.out';
open(BUSHMEN, $infile) or die "Could not open $infile\n";
while (<BUSHMEN>) {
    my @columns = split("\t");
    my $pos     = $columns[1];
    if ( $genome{$pos} ) { print "$pos $genome{$pos}\n"; }
}
close BUSHMEN;
```

In the last part of the script in Code 19.2 we read the file bushmen.out in order to extract all the SNP positions. These are found in the second column. For each of these positions we want to find out whether they are in an exon region or not. We simply look up that information using the hash %genome. This look-up is very fast – the more time-consuming part of the script is the construction of the %genome hash.[3]

We finally run the script and save the output in a file:

```
% perl map2exons.pl > map2exons.out
```

The output has 371 SNP positions that are in exonic regions. These positions are contained within 253 different genes. We may now explore all of these genes further to better understand the significance of the different SNPs. In the following we will focus on one of the genes.

A bitter taste

If we examine the output file map2exons.out, we may note one gene with the UCSC identifier uc003vwx.1 and protein ID NP_789787. This gene has two

---

[3] While a hash, as shown in this example, is highly efficient, a potential disadvantage with this kind of approach is that the hash uses a lot of computer memory. There is a chance you will run out of memory if you store a lot of data in variables like this.

different Bushmen-specific SNPs, at locations 141 319 072 and 141 319 813, respectively. This gene is more officially known as TAS2R38. What is the function of the TAS2R38 gene product? It is a *bitter taste receptor*, one of many receptors involved in bitter taste.

The TAS2R38 receptor is known to facilitate the tasting of the synthetic organic compound phenylthiocarbamide (PTC) (Fig. 19.2) (Kim *et al.*, 2003). Not all human individuals are able to taste this compound. The difference in sensitivity to PTC was originally discovered in a laboratory incident. Arthur L. Fox was pouring PTC powder into a bottle when some of it 'flew around in the air', to quote his paper describing the discovery in 1932 (Fox, 1932). A colleague of Fox complained about a strong bitter taste, but Fox could taste nothing at all. Fox then tested a larger number of individuals and found that most of them were in two categories; either they were able to taste PTC or they were not (Wooding, 2006). There was apparently a genetic component, as individuals were much more likely to taste PTC if other members of their families did. In fact, as a PTC sensitivity test is very simple, it has been widely used in human genetics. It was even used as a paternity test before DNA tests were available.

PTC is a synthetic organic compound, but what is the biological role of the ability to taste it? It was eventually learned that PTC is chemically related to toxic alkaloids in poisonous plants. Bitter-taste receptor cells exposed to the oral cavity express TAS2 receptors, the products of a whole family of genes involved in bitter-taste reception. A large number of different plant toxins are recognized by these receptors. In terms of PTC sensitivity, the most important locus is TAS2R38, which accounts for 50–80% of variation in PTC sensitivity.

Approximately 70% of all human individuals are able to taste PTC. What do humans benefit from being able to taste PTC and related alkaloids in plants? It seems likely that it will help to avoid toxic plants, and one could expect this function to be selected for within foraging societies.

There are three different positions in the TAS2R38 protein that are subject to genetic variation. Most human individuals have either of two configurations with respect to three non-synonymous SNPs in TAS2R38 (Table 19.2). We can make use of the UCSC genome browser (http://genome.ucsc.edu) to view the location of these three SNPs (Fig. 19.3).

Individuals with 'proline–alanine–valine' (PAV) are common among 'tasters', while 'alanine–valine–isoleucine' (AVI) is common among 'non-tasters'. What does the result in the file `map2exons.out` tell us about the genetic make-up of the Bushmen? We already saw that the two SNPs at locations 141 319 072 and 141 319 813 were in that file. We can use the Unix `grep` command to find out:

```
% grep 141319813 map2exons.out
```

Fig. 19.2 *The synthetic organic compound phenylthio-carbamide (PTC). This compound may either taste very bitter, or not bitter at all, depending on your genes.*

Table 19.2 *SNPs of the gene TAS2R38*. The gene TAS2R38 encodes a bitter-taste receptor. Individuals that are sensitive to the compound PTC ('tasters') are commonly characterized by the SNPs shown here highlighted. These SNPs give rise to the amino acid configuration proline–alanine–valine (PAV). Conversely, individuals with the configuration alanine-valine-isoleucine (AVI) are much more insensitive to PTC ('non-tasters').

Chromosome position, NCBI-36 (zero-based numbering)	dbSNP entry	Reference – coding sequence nt	SNP nt	Reference codon/aa	SNP codon/aa
chr7:141 319 813	rs713598 G>C	G	C	GCA – Ala49	CCA – Pro49
chr7:141 319 173	rs1726866 C>T	C	T	GCC – Ala262	GTC – Val262
chr7:141 319 072	rs10246939 A>G	A	G	ATC – Ile296	GTC – Val296

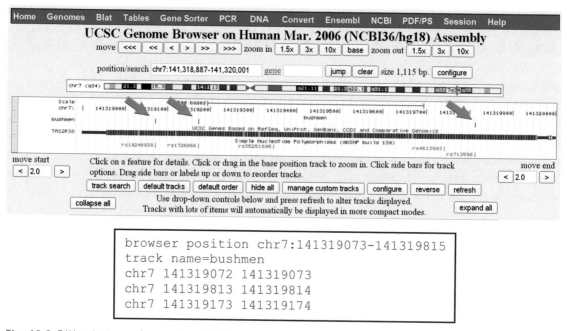

```
browser position chr7:141319073-141319815
track name=bushmen
chr7 141319072 141319073
chr7 141319813 141319814
chr7 141319173 141319174
```

Fig. 19.3 *Bitter-taste markers in the UCSC genome browser.* Gene of the TAS2R38 bitter-taste receptor and Bushmen SNPs as visualized in the UCSC genome browser. This view was obtained with the UCSC genome browser at http://genome.ucsc.edu/cgi-bin/hgGateway. The code as shown below the browser view was used with the 'add custom track' function. Red arrows have been added to indicate positions of SNPs (and are not part of the actual browser view).

The nucleotide in position 141 319 173 appears to be the same as in the reference sequence, as this SNP is not reported in the file `bushmen.out`. (It is actually present in the file `bushmen_chr7_all_snps.txt`, but was filtered out as a result of the script `bushmen.pl`.) In conclusion, both Bushmen that we analysed here have the PAV setup with respect to bitter taste. Another 12 Bushmen were analysed by Schuster *et al.* (2010) and all of them were identified as PAV. Thus, they are all sensitive to PTC. The fact that this genotype seems to be predominant in the Bushmen indicates that there has been a selective advantage for these alleles. It may have been important for these people with a foraging lifestyle to effectively avoid toxic plants.

Constructing your own modules

Now to a more technical issue. When you have written a lot of code using a language like Perl, you may discover that you are writing the same kind of code over and over again. In this situation you may want to construct your own modules. Modules will not be fully explained here, but in case you would like to have some idea, here is a simple example as a starting point. We have designed a module for reading a tab-formatted file and collecting the data from that file into a two-dimensional array. The module in this case is the file `mymodule.pm` (Code 19.3). An example of the use of this module is in Code 19.4. In that script the file `bushmen.out` is analysed and all positions will be printed where KB1 is the same as Yoruba NA18507.

Code 19.3 mymodule.pm

```perl
#!/usr/bin/perl

use strict;

package mymodule;

require Exporter;
our @ISA     = qw(Exporter);
our @EXPORT  = qw(read_tabfile);
our $VERSION = 1.00;

sub read_tabfile {
    my $infile = $_[0];
    if ( $infile eq '' ) { die "Could not find file $infile\n"; }
    my @array = ();
    my $i = 0;
    open(IN, $infile) or die "Could not open file $infile\n";
    while (<IN>) {
        unless (/^#/) {
```

```
        chomp;
        my @columns = split("\t");
        for ( my $j = 0 ; $j <= $#columns ; $j++ ) {
            $array[$i][$j] = $columns[$j];
        }
        $i++;
      }
    }
    close IN;
    return ( \@array );
  }
  1;
```

Code 19.4 use_mymodule.pl

```perl
#!/usr/bin/perl -w

use strict;

use mymodule;

my $file = 'bushmen.out';

my ($array_ref) = read_tabfile($file);
my @array = @$array_ref;
for ( my $i = 0 ; $i <= $#array ; $i++ ) {
    # is KB1 the same as NA18507 (Yoruba) ?
    if ( $array[$i][5] eq $array[$i][16] ) {
        print "pos $array[$i][1]\n";
    }
}
```

EXERCISES

19.1 Modify Code 19.1 to identify SNPs that are present in chromosome 7 of Archbishop Desmond Tutu, but that are not in NB1, KB1 or the reference genome. Make use of the dataset in `bushmen_chr7_all_snps.txt`. As in Code 19.1, exclude all positions where at least one of the four genomes is unknown ('N').

19.2 Examine all the individuals represented in the dataset `bushmen_chr7_all_snps.txt` (see Table 19.1) with respect to the three SNPs of the TAS2R38 gene. What are the predicted properties of the different individuals in terms of bitter taste?

19.3 Construct your own Perl module that reads a file with a single nucleotide sequence in FASTA format and returns two variables; the information in the identifier (definition) line and the nucleotide sequence.

REFERENCES

Fox, A. L. (1932). The relationship between chemical constitution and taste. *Proc Natl Acad Sci U S A* **18**(1), 115–120.

Goecks, J., A. Nekrutenko and J. Taylor (2010). Galaxy: a comprehensive approach for supporting accessible, reproducible, and transparent computational research in the life sciences. *Genome Biol* **11**(8), R86.

Gonder, M. K., H. M. Mortensen, F. A. Reed, A. de Sousa and S. A. Tishkoff (2007). Whole-mtDNA genome sequence analysis of ancient African lineages. *Mol Biol Evol* **24**(3), 757–768.

Kim, U. K., E. Jorgenson, H. Coon, *et al.* (2003). Positional cloning of the human quantitative trait locus underlying taste sensitivity to phenylthiocarbamide. *Science* **299**(5610), 1221–1225.

Schuster, S. C., W. Miller, A. Ratan, *et al.* (2010). Complete Khoisan and Bantu genomes from southern Africa. *Nature* **463**(7283), 943–947.

Tishkoff, S. A., M. K. Gonder, B. M. Henn, *et al.* (2007). History of click-speaking populations of Africa inferred from mtDNA and Y chromosome genetic variation. *Mol Biol Evol* **24**(10), 2180–2195.

Wooding, S. (2006). Phenylthiocarbamide: a 75-year adventure in genetics and natural selection. *Genetics* **172**(4), 2015–2023.

20 Personal genomes
Details of family genetics

Basic principles of genetic inheritance

In this chapter we will discuss inheritance within a family and address questions about the parental origin of specific sites in a genome. First we need to understand a few basic elements of genetics. All human cells contain two copies of each chromosome, one of parental and one of maternal origin. Exceptions are the reproductive cells or *gametes*, i.e. eggs in females and sperms in males, which have only one copy of each chromosome. The gametes are formed in the process of *meiosis* (Figs. 20.1 and 20.2). Meiosis starts out with a cell having two copies of each chromosome (one from the father and one from the mother; two such chromosomes are said to be *homologous*[1]). The DNA is first copied in order to generate two copies each of the paternal and maternal chromosomes. Through *crossing-over*, which occurs by homologous recombination, segments are occasionally interchanged between parental and maternal chromosomes (Fig. 20.2). Such recombination occurs about 50–100 times during meiosis, and there is evidence that in humans the recombination in maternal meiosis is about 1.7 times more frequent than in paternal meiosis (Petkov *et al.*, 2007; Roach *et al.*, 2010).

After copying of DNA and recombination events the cell divides into two daughter cells. Each of these daughter cells may receive either the maternal or paternal chromosome from each homologous pair. The segregation is random. For instance, in the special case where we consider two pairs of chromosomes there are two possible outcomes, as illustrated in Fig. 20.1.

A second round of cell division takes place in which the homologous chromosomes are separated. The resulting cells will have only one copy of each chromosome. These are the egg and sperm cells that eventually take part in sexual reproduction. During fertilization, when a sperm enters an egg cell, these two cells fuse, resulting in a genome with one copy of each chromosome. In other words, the fertilized egg will have one chromosome of paternal origin and one chromosome of maternal origin.

---

[1] The word homologous in this context does not have the same meaning as when we have previously discussed homologous genes and proteins.

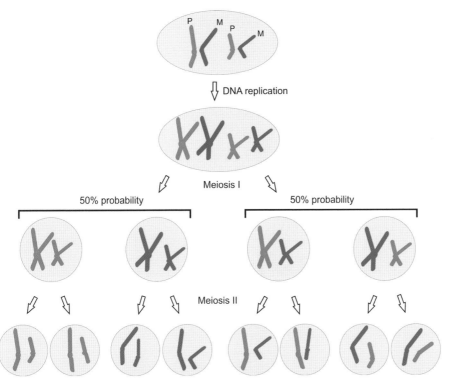

Fig. 20.1 *Meiosis and genetic variation through chromosome segregation*. Meiosis starts out with a diploid cell with chromosomes of paternal (P, red) and maternal (M, blue) origin. An example is shown for two different pairs of chromosomes, distinguished here by different sizes. The DNA is first copied to generate two copies each of the paternal and maternal chromosomes. The cell then divides into two daughter cells (Meiosis I). A second round of cell division takes place (Meiosis II) in which the homologous chromosomes are separated so the resulting cells will have only one copy of each chromosome. These are the egg and sperm cells that eventually take part in sexual reproduction. During the first cell division (Meiosis I), each of the daughter cells may receive either the maternal or paternal chromosome from each homologous pair. The segregation is random. If we consider two pairs of chromosomes there are two possible outcomes, each with 50% probability. The second cell division will generate four different types of cells. Generally, the number of possible haploid gametes from an original diploid cell with N chromosomes is 2^N. For instance, for a cell with 23 chromosomes, the number of outcomes is 2^{23}, which is about 8×10^6.

Meiosis is designed to produce the reproductive egg and sperm cells. There is an important advantage of sexual reproduction as it increases genetic diversity. There are two features of meiosis that contribute in this respect. First, there is random segregation of chromosomes in the first meiotic division. Second, there are recombination (crossing-over) events that take place before cell division. It is the recombination events that we are going to look at in this chapter.

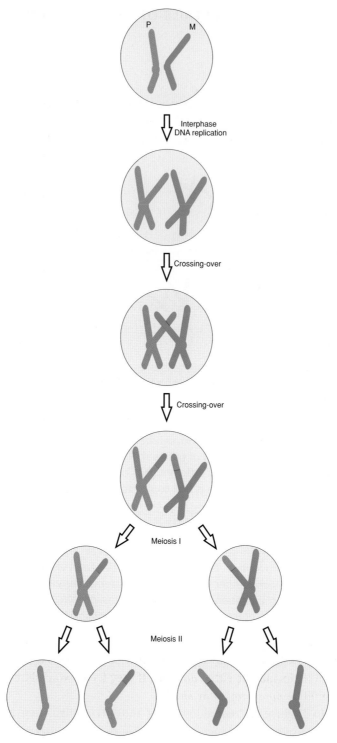

Fig. 20.2 *Meiosis and genetic variation through crossing-over.* Meiosis occurs as described in Fig. 20.1. In addition to chromosome segregation, crossing-over is another mechanism to generate genetic diversity. Through crossing-over that occurs by homologous recombination, segments are occasionally interchanged between parental and maternal chromosomes. A schematic example is shown here with only one pair of homologous chromosomes in the original diploid cell.

Analysis of a family quartet

We are to analyse the genetic relationships in a family. As in the previous chapter we will take advantage of SNPs. What are the implications of the genetic inheritance in this case? As our example we will consider the work of Roach *et al.* (2010) to examine genetic inheritance in a family quartet, i.e. two parents and their children, in this case a boy and a girl. The genomes of all family members were sequenced. Both children have two rare recessive disorders. One is Miller syndrome, an inherited disease characterized by craniofacial malformations. The other is primary ciliary dyskinesia, characterized by chronic respiratory tract infections. The results of the genome analysis of the family quartet made it possible to considerably narrow down the candidate genes for both of these disorders. (In the case of Miller syndrome, one of the identified candidate genes, DHODH, was shown to be the responsible gene in another study (Ng *et al.*, 2010).) Therefore, this study illustrates how genome sequences of a family aid in identifying genes involved in inherited disorders. Another interesting outcome of this study was that it allowed a very precise mapping of recombination (crossing-over) sites. This is the issue we are going to deal with for this chapter and we will demonstrate how SNP data may be used, with the help of a Perl script, to identify the recombination sites. Having access to genetic data from all members of the family quartet is essential in this case. If we have access to information from two children and only one parent, we end up with much poorer resolution, and analysis of two parents and only one child provides no information on recombination at all (Roach *et al.*, 2010).

In human genome sequencing, diploid cells are typically analysed, i.e. cells that have two copies of each chromosome. This was also the case for the project considered here (Roach *et al.*, 2010). Let us think of one distinct SNP position that we consider for genetic variation. We assume that there are only two variants and we name them *a* and *b*.[2] You could think of them as two different base pairs, like *a* being C–G and *b* being T–A. As each individual has two copies of each chromosome their genotype may be either *aa*, *ab* or *bb* (*ba* is the same as *ab* as we cannot distinguish the two chromosomes). There are certain basic principles of inheritance based on these alleles. If, for instance, a male is *aa* and a female is *ab*, a child of theirs may be either *aa* or *ab*. If they have two children this pair of children could be *aa/ab*, *aa/aa* or *ab/ab*.

For the family quartet sequencing that we consider here, data was collected for each individual in the family. Thus, for a number of genomic positions we have access to SNPs, and for each individual there will be one or two bases

---

[2] There are indeed more than two variants (alleles) in many positions, but for simplicity we consider here only two. We do not lose much resolution in our recombination analysis by leaving out positions with more than two alleles.

observed in that SNP position. This means the whole family quartet may be represented in each SNP position with information like 'CT/TT/TT/TC', where the order is father–mother–daughter–son. A 'CT' in one individual means that C is in one chromosome and T is in the other, and a 'TT' means that both chromosomes have a T.

To make the different SNP positions comparable and to produce a genotype of the a and b type above, we transform the base representation into a representation with the characters a and b. For instance, the genotype information 'CT/TT/TT/TC' may be represented by $ab/aa/aa/ab$. There are two different rules of symmetry in doing this transformation: (1) ab in an individual is equivalent to ba; and (2) a and b may be swapped in the whole quartet genotype. For instance, $ab/bb/bb/ba$ is equivalent to $ba/aa/aa/ab$, which in turn is equivalent to $ab/aa/aa/ab$. In the transformation into a string of the characters a and b we will assign the most common base in the two parents as a and the other base as b. We will also transform all individual ba to ab. These operations of symmetry lead to a restricted number of possible family genotypes (14 that are consistent with Mendelian laws of inheritance if we consider two alleles. For more details, see Roach *et al.* (2010)).

For the identification of recombination sites it is useful to consider four different 'states' of inheritance, as illustrated in Fig. 20.3:

(1) The children inherited the same alleles from both parents.
(2) Children share only a maternal allele.
(3) Children share only a paternal allele.
(4) Children share no allele.

To each SNP site in the genome there will be a genotype pattern with information about the parents and the two children – for example, $ab/ab/aa/aa$. Further analysis shows that there are only ten possible family genotype patterns that are consistent with the inheritance states as listed above. The other possible patterns are either not informative or they violate the rules of Mendelian inheritance. For instance, the patterns $aa/aa/aa/aa$ and $aa/bb/ab/ab$ are not informative.

In Fig. 20.3 are shown the ten different genotype patterns; the arrows indicate the inheritance states they are consistent with. In each of the boxes of Fig. 20.3 the parental genotypes are shown on top with the father to the left and mother to the right. Below are the children genotypes. The most frequent allele in the parents is denoted by a; in case of equal frequency, a denotes the most frequent allele in the children. Thus, 'aa + ab' indicates that the father is homozygous for the most frequent allele within the family, while the mother is heterozygous. A '/' symbol means that the order is not important. Most genotypes are associated with two different states. For instance, $aa/ab/ab/ab$ means that the a of the children could be either from the mother or father, whereas the b is most

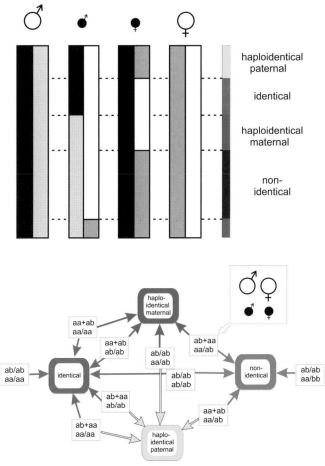

Fig. 20.3 *Inheritance states and family genotypes*. Upper panel: the four members of a family quartet are shown; left to right are father, son, daughter, mother. There are four possible states of allele inheritance, depending on whether the children inherited the same alleles from both parents (red), share only a maternal allele (green) or a paternal allele (yellow), or share none (blue). Inheritance states are observed in large contiguous blocks and transitions between blocks correspond to recombination events (dotted lines). Lower panel: ten different family genotype patterns are consistent with one or two inheritance states. We only consider biallelic positions, i.e. positions where there are exactly two different alleles. For each pattern (white boxes), the parental genotypes are shown on top (father to the left, mother to the right) and the children genotypes below. The most common allele in the parents is denoted by 'a'; in case of equal frequency, 'a' denotes the most frequent allele in the children. Thus, 'aa + ab' means the father is homozygous for the allele which is most common within the family, while the mother is heterozygous. A '/' symbol is used to indicate that the order is not important. Figures of both panels as well as accompanying legend text are adapted from Roach *et al*. (2010), by permission from the American Association for the Advancement of Science.

certainly from the mother. Thus, the family genotype is in this case either such that the children inherited the same allele from both parents, or that they share only a maternal allele.

The information in Fig. 20.3 may be exploited to examine all SNPs along a chromosome. At each SNP one or two of the four possible inheritance states may receive a score and it turns out that this approach may be used to effectively identify the sites of recombination (crossing-over).

BIOINFORMATICS

Where are the crossing-over sites?

Tools introduced in this chapter		
Perl	array of hashes	
	function `int`	

We will make use of SNP data from the family quartet to predict different inheritance states along a chromosome. The data we are using is SNP data derived from chromosome 4, and it is in the file quartet.txt.[3]

The fields in this file are separated by commas:

```
chr4,1646,T,T,C,T,T,T,T,C
```

The fields are, from left to right: chromosome, nucleotide position, base in reference genome, and then eight fields with the explicit bases of the individual genotypes that are in the order father, mother, daughter, son. Thus, for this particular line the family genotypes are TC, TT, TT and TC, respectively.

An important part of the procedure is the transformation of this genotype information to a string of the characters *a* and *b*. This first step is to identify the most common base in the parents. The following method is used:

(1) The parent string has four characters, two from the father and two from the mother. Take one of the bases of the parent string.
(2) Count how many times this base occurs in the parent string.
(3) If the result is less than two, it means it should be assigned as *b*. If it is equal to or greater than two it means it should be assigned as *a*.

With this procedure the genotype string with bases may be transformed into a string of the characters *a* and *b*. As a last step in this operation, we also replace every instance of *ba* to *ab*, as these are equivalent.

---

[3] As the actual personal genome sequencing data of this project is not freely available we will here be using simulated data. However, the simulated data is designed in such a way that it will produce exactly the same results as the real data.

The data in the file `quartet.txt` will be processed with the Perl script Code 20.1. Explanations of the code are below; the numbers refer to marked sections in the code.

(1) We divide the chromosome up into pieces (bins), 250 000 positions in length. The variable `@bins` is an *array of hashes* that stores, for each bin, the scores of the four different inheritance states. The values of `@bins` are stated as `$bins[$i]{id}`, where `$i` is the number of the bin (an integer) and 'id' is one of the inheritance states.

(2) The `int` function returns the integer part of a number. For instance, `int(3.74)` will evaluate to `3`.

(3) We choose to consider only cases where there are exactly two different alleles, i.e. two bases in the family genotype string. Hence, we want to exclude triallelic situations, i.e. cases with three or more bases, and we also exclude the non-informative case with only one base, like 'GGGGGGGG'. As a first step, therefore, we count the number of different bases in the genotype string `$quartet` with the `foreach my $base (@fourbases) {}` loop. After this loop the different bases in the string `$quartet` is the array `@bases`.

Code 20.1 quartet.pl

```perl
#!/usrc/bin/perl -w

use strict;

my $binsize = 250000;
my $bin;

# 1 #
my @bins;      # array of hashes

my @fourbases = ( 'A', 'T', 'C', 'G' );

my %genotypes = (
    'ababaaaa' => 'ID',       # identical
    'aaabaaaa' => 'IDHM',     # identical or haplomaternal
    'aaababab' => 'IDHM',     # identical or haplomaternal
    'abaaaaaa' => 'IDHP',     # identical or haplopaternal
    'abaaabab' => 'IDHP',     # identical or haplopaternal
    'ababaaab' => 'HMHP',     # haplomaternal or haplopaternal
    'abababab' => 'IDNI',     # identical or non-identical
    'abaaaaab' => 'HMNI',     # haplomaternal or non-identical
    'ababaabb' => 'NI',       # non-identical
    'aaabaaab' => 'HPNI'      # haplopaternal or non-identical
);
```

```perl
my $infile = 'quartet.txt';
open(IN, $infile) or die "Could not open $infile\n";
while (<IN>) {
   chomp;

   # input format is
   # chr4,1646,T,T,C,T,T,T,T,T,C

   my @columns = split(',');
   my $pos = $columns[1];

   # 2 #
   $bin = int( $pos / $binsize );

   my $quartet = '';
   for ( my $i = 3 ; $i < 11 ; $i++ ) {
       $quartet .= $columns[$i];
   }

   my $count = 0;
   my @bases = ();

   # 3 #
   foreach my $base (@fourbases) {
      if ( $quartet =~ /$base/ ) {
         $count++;
         push @bases, $base;
      }
    }

   if ( $count == 2 ) {    # if the genotype string has
                           # exactly two different bases

      # which is the most common base?
      my $parents = substr( $quartet, 0, 4 );

      # select one of the two bases and count how
      # many times it occurs in the parents
      my $pcount = ( $parents =~ s/$bases[0]//g );
      if ( $pcount >= 2 ) {
         $quartet =~ s/$bases[0]/a/g;
         $quartet =~ s/$bases[1]/b/g;
      }
      if ( $pcount < 2 ) {
         $quartet =~ s/$bases[1]/a/g;
         $quartet =~ s/$bases[0]/b/g;
      }
```

```perl
    # replace all ba with ab
    $quartet = swap_ab($quartet);

    # check %genotypes
    if ($genotypes{$quartet}) {
        if ( $genotypes{$quartet} =~ /ID/ ) { $bins[$bin]{id}++; }
        if ( $genotypes{$quartet} =~ /HP/ ) { $bins[$bin]{hp}++; }
        if ( $genotypes{$quartet} =~ /HM/ ) { $bins[$bin]{hm}++; }
        if ( $genotypes{$quartet} =~ /NI/ ) { $bins[$bin]{ni}++; }
    }
  }
}

close IN;

# Some of the elements in @bins are undefined.
# This is to set these to zero instead.
for ( my $i = 0 ; $i <= $#bins ; $i++ ) {
   unless ( $bins[$i]{id} ) { $bins[$i]{id} = 0; }
   unless ( $bins[$i]{ni} ) { $bins[$i]{ni} = 0; }
   unless ( $bins[$i]{hp} ) { $bins[$i]{hp} = 0; }
   unless ( $bins[$i]{hm} ) { $bins[$i]{hm} = 0; }
}

# Finally print all the data in @bins
print "bin\tidentical\thaplopaternal\thaplomaternal\tnonidentical\n";
for ( my $i = 0 ; $i <= $#bins ; $i++ ) {
   print "$i\t$bins[$i]{id}\t$bins[$i]{hp}\t";
   print "$bins[$i]{hm}\t$bins[$i]{ni}\n";
}

sub swap_ab {
   my $str    = $_[0];
   my $newstr = '';
   for ( my $i = 0 ; $i < 8 ; $i = $i + 2 ) {
      my $temp = substr( $str, $i, 2 );
      if ( $temp eq 'ba' ) { $temp = 'ab'; }
      $newstr .= $temp;
   }
   return $newstr;
}
```

The output of Code 20.1 has been plotted in Fig. 20.4. For each of the bins, the *most supported* type of inheritance state – i.e. haploidentical paternal (yellow), haploidentical maternal (green), identical (red) or non-identical (blue) – is shown. From this graph we are able to identify large blocks of consistent inheritance, and these blocks are bounded by recombination events. It should

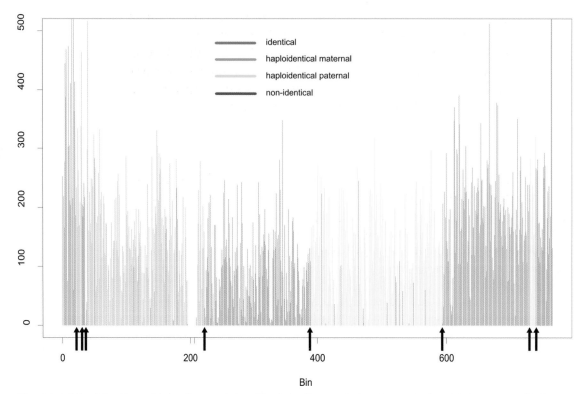

Fig. 20.4 *Identification of inheritance states*. The inheritance states shown in Fig. 20.3 were studied. Thus, the states are that children (1) inherited the same alleles from both parents (red), (2) share only a maternal allele (green), (3) share only a paternal allele (yellow) or (4) share none (blue). Chromosome 4 was studied and divided up into segments of 250 000 nucleotide positions each. For each segment, the counts of inferred inheritance states were obtained; shown in the plot are the counts of the most frequent state. The important conclusion from the plot is that large blocks of consistent inheritance may be identified and these are bounded by recombination events. Thus, in this plot of chromosome 4, at least eight different crossing-over events may be identified, as indicated by the arrows. A region close to bin 200 that has very low counts corresponds to the centromeric region of chromosome 4. The plot was obtained using an R script (see the web resources for this book).

be noted that these sites represent the sum of the two children and none of the individual sites may be attributed to a specific child. (Only through analysis of a third child, if available within the family, can we gather information about child-specific recombination.)

The analysis performed by our Perl script allows a very precise mapping of recombination sites. Thus, for chromosome 4 we may identify at least eight recombination sites (Fig. 20.4). As indicated previously, this result and the high resolution in Fig. 20.4 is dependent on analysis of all four members of the family. Had we analysed only the mother and two children, the resolution would have been much poorer.

We said previously that crossing-over in maternal meiosis is thought to be about 1.7 times more frequent than in paternal meiosis. Our results are consistent with this relationship. We may deduce from Fig. 20.4 that there are five maternal and three paternal crossing-over sites on chromosome 4. Analysis of all chromosomes identified 98 and 57 crossing-over points in maternal and paternal meiosis, respectively, as shown by Roach *et al.* (2010).

EXERCISES

20.1 In Code 20.1 the variable `@bins` is an *array of hashes*. Modify the code to instead use a *hash of hashes*. See Appendix III for these data structures.

20.2 Construct a graph based on the output from Code 20.1 which is similar to that in Fig. 20.3, but in which all four inheritance states are plotted, and where the states are sorted by decreasing level of support. (You will have to use R or other plotting software to do this. For a brief introduction to R, see Appendix IV.)

20.3 Design a Perl script that will try and elucidate the recombination breakpoints from the output of Code 20.1.

REFERENCES

Ng, S. B., K. J. Buckingham, C. Lee, et al. (2010). Exome sequencing identifies the cause of a Mendelian disorder. *Nat Genet* **42**(1), 30–35.

Petkov, P. M., K. W. Broman, J. P. Szatkiewicz and K. Paigen (2007). Crossover interference underlies sex differences in recombination rates. *Trends Genet* **23**(11), 539–542.

Roach, J. C., G. Glusman, A. F. Smit, et al. (2010). Analysis of genetic inheritance in a family quartet by whole-genome sequencing. *Science* **328**(5978), 636–639.

APPENDIX I
Brief Unix reference

Operating systems

Whenever you are using a computer you interact with it with the help of an *operating system* (OS), a vital interface between the hardware and the user. The operating system does a number of different things. For example, multiple programs are often run at the same time and in this situation the operating system allocates resources to the different programs or may be able to appropriately interrupt programs. Another common feature of an operating system is a graphical user interface (GUI), originally developed for personal computers. Examples of popular operating systems are Microsoft Windows, Mac OS X and Linux.

Linux is an example of a Unix (or 'Unix-like') operating system. Unix was originally developed in 1969 at Bell Laboratories in the United States. Many different flavours of the Unix OS have been developed, such as Solaris, HP-UX and AIX, and there are a number of freely available Unix or Unix-like systems such as GNU/Linux in different distributions such as Red Hat Enterprise Linux, Fedora, SUSE Linux Enterprise, openSUSE and Ubuntu.

If you are at a personal computer your access to Unix depends on what operating system you are using (Fig. AI.1). For instance, Microsoft Windows is not based on Unix and does not provide a Unix interface. If you would like to have a Unix environment within Windows, a possible choice is to install Cygwin (http://www.cygwin.com). Cygwin has a lot of Unix functionality and useful Unix programs; it also includes Perl (see also Appendix III). The Mac OS X operating system is based on Unix and all you need to do to communicate with Unix commands is to open a terminal window (available under Applications > Utilities). The same applies if you are using a computer with Linux or any other Unix flavour; just open a terminal window to get started.

Accessing a Unix computer

Even though you do not have Unix at your personal computer, you can connect to a Unix-based server and operate from there. In fact, this is a common mode of working in bioinformatics. You typically make use of an SSH (secure shell) client program to connect. SSH is a network protocol that allows data to be transferred between two networked computers. The SSH client program communicates with an SSH daemon running on the server side. A typical application of the SSH

```
        $geneName = $1;
        $geneName =~ s/ENST0000(.*)\|ENSG.*/ENST$1/;
        $geneName = $abbrev . "." . $geneName;
    } else {
        $seq .= $_;
    }
}
    close(IN);
} else {
    die "Can't get gene source info from filename $fname\n";
}
}
bio%
bio%
bio%
bio%
bio%
bio%
bio%
bio%
bio%
bio%
bio%
```

```
Main Options  VT Options  VT Fonts
AFRRSWAERKCSVINSQTFATCHSKVYHLPYYEACVRDACGCDSGGDCECLCDAVAAYAQACLDKGVCVD
WRTPAFCPIYCGFYNTHTQDGHGEYQYTQEANCTWHYQPCLCPSQPQSVPGSNIEGCYNCSQDEYFDHEE
GVCVPCMPPTTPQPPTTPQLPTTGSRPTQVMPMTGTSTTIGLLSSTGPSPSSNHTPASPTQTPLLPATLT
SSKPTASSGEPPRPTTAVTPQATSGLPPTATLRSTATKPTVTQATTRATASTASPATTSTAQSTTRTTMT
LPTPATSGTSPTLPKSTNQELPGTTATQTTGPRPTPASTTGPTTPQPGQPTRPTATETTQTRTTTEYTTP
QTPHTTHSPPTAGSPVPSTGPVTATSFHATTTYPTPSHPETTLPTHVPPFSTSLVTPSTHTVITPTHAQM
ATSASNHSAPTGTIPPPTTLKATGSTHTAPPITPTTSGTSQAHSSFSTNKTPTSLHSHTSSTHHPEVTPT
STTTITPNPTSTRTRTPVAHTNSATSSRPPPPFTTHSPPTGSSPFSSTGPMTATSFKTTTTYPTPSHPQT
TLPTHVPPFSTSLVTPSTHTVITPTHAQMATSASIHSMPTGTIPPPTTLKATGSTHTAPTMTLTTSGTSQ
ALSSLNTAKTSTSLHSHTSSTHHAEATSTSTTNITPNPTSTGTPPMTVTTSGTSQSRSSFSTAKTSTSLH
SHTSSTHHPEVTSTSTTSITPNHTSTGTRTPVAHTTSATSSRLPTPFTTHSPPTGTTPISSTGPVTATSF
QTTTTYPTPSHPHTTLPTHVPSFSTSLVTPSTHTVIIPTHTQMATSASIHSMPTGTIPPPTTIKATGSTH
TAPPMTPTTSGTSQSPSSFSTAKTSTSLPYHTSSTHHPEVTPTSTTNITPKHTSTGTRTPVAHTTSASSS
RLPTPFTTHSPPTGSSPFSSTGPMTATSFQTTTTYPTPSHPQTTLPTHVPPFSTSLVTPSTHTVIITTHT
QMATSASIHSTPTGTVPPPTTLKATGSTHTAPPMTVTTSGTSQTHSSFSTATASSSFISSSWLPQNSSS
RPPSSPITTQLPHLSSATTPVSTTNQLSSSFSPSPSAPSTVSSYVPSSHSSPQTSSPSVGTSSSFVSAPV
HSTTLSSGSHSSLSTHPTTASVSASPLFPSSPAASTTIRATLPHTISSPFTLSALLPISTVTVSPTPSSH
LASSTIAFPSTPRTTASTHTAPAFSSQSTTSRSTSLTTRVPTSGFVSLTSGVTGIPTSPVTNLTTRHPGP
TLSPTTRFLTSSLTAHGSTPASAPVSSLGTPTPTSPGVCSVREQQEEITFKGCMANVTVTRCEGACISAA
SFNIITQQVDARCSCCRPLHSYEQQLELPCPDPSTPGRRLVLTLQVFSHCVCSSVACGD
bash-3.2$
bash-3.2$
bash-3.2$
bash-3.2$
```

```
Terminal — bash — 80×24
$
$
$
$
$
$
$
$
$
$
$
$
$
$
$
$ perl find_cag.pl | head
Repeat found in >gi|157151758|ref|NM with length 30
Repeat found in >gi|116875847|ref|NM with length 18
Repeat found in >gi|223646108|ref|NM with length 30
Repeat found in >gi|114431247|ref|NM with length 27
Repeat found in >gi|125346191|ref|NM with length 21
Repeat found in >gi|154350223|ref|NM with length 21
Repeat found in >gi|157168352|ref|NM with length 18
Repeat found in >gi|154350245|ref|NM with length 21
Repeat found in >gi|209862781|ref|NM with length 24
Repeat found in >gi|197383729|ref|NR with length 18
$
```

Fig. AI.1 *Terminal windows*. No matter what computer platform you are comfortable with, you have access to terminal windows where you are able to operate in command-line mode for bioinformatics applications. Examples shown here are: Gnome terminal window of Unix Redhat (top), Cygwin xterm window running under Windows (middle) and Mac OS X terminal (bottom).

client program is to login to a remote computer and execute commands at the remote computer. Examples of freely available implementations of SSH are openSSH (http://www.openssh.com), copSSH (http://www.itefix.no/i2/copssh) and PuTTY (http://www.chiark.greenend.org.uk/∼sgtatham/putty).

From now on, we assume you have access to a Unix terminal window. We will be focusing on operations that are possible at the command line, as this is where you are most effective in the long run.

File system

Examples of directories and their contents

/	root of file system
/bin	executable binary files
/dev	special files used to represent real physical devices
/etc	commands and files used for system administration
/home	contains a home directory for each user of the system
/home/joe	home directory of user joe
/lib	libraries used by various programs and programming languages
/tmp	a 'scratch' area where any user can store files temporarily
/usr	system files and directories that you share with other users

Moving around in the file system

To find out what directory you are currently in, use:

% pwd   'present working directory'

Move to a specific directory with cd (change directory):

% cd /tmp   change directory to /tmp
% cd ..   go up one level in directory tree
% cd   (without argument) go to your home directory

Find out what files are in the current directory:

% ls   show files in current directory
% ls -al   show files in current directory

The ls -al command will result in a more detailed output, such as:

```
-rw-r--r--   1 joe users    383269  2007-11-25 16:54  PF02854.txt
drwxr-xr-x   9 joe users       656  2003-04-04 20:12  scripts/
-rw-r--r--   1 joe users      4898  2006-09-12 09:12  README.txt
-rwxr-xr-x   1 joe users    120635  2004-08-03 01:47  dnapars*
```

In this listing there is in each line a set of characters describing the file status. In a string like -rw-r--r--, the first '-' means that it is a regular file (a 'd' says that it is a directory). The next symbols are three groups of three in which the first is what the owner can do, the second what group members can do, and the third what other users can do. In each group of three, the symbols are 'r' = readable, 'w' = writable, 'x' = executable. To change the file attributes, see documentation on the Unix command chmod.

As with many other Unix commands, the wildcard * may be used with ls. The following command will list files with names beginning with 'HIV':

```
% ls HIV*
```

Manipulating files and directories

Copying files

```
cp [source] [destination]
```

Examples:

(1) Copy the file /tmp/seq.fa to the current working directory. The current directory is represented by a dot '.':

```
% cp /tmp/seq.fa .
```

(2) Copy the file /tmp/seq.fa to another file in /tmp named seq2.fa:

```
% cp /tmp/seq.fa /tmp/seq2.fa
```

(3) Copy all files (*) in the directory /home/joe/seqfiles to the directory /tmp:

```
% cp /home/joe/seqfiles/* /tmp
```

Moving and renaming files

```
mv [source] [destination]
```

Examples:

(1) Rename the file seq.fa in /tmp to seq2.fa:

```
% mv /tmp/seq.fa /tmp/seq2.fa
```

(2) Move all files (*) in the directory /home/joe/seqfiles to the directory /tmp:

```
% mv /home/joe/seqfiles/* /tmp
```

Removing files

```
% rm seq.fa
```

Creating and removing directories

```
% mkdir dirname
% rmdir dirname
```

Viewing and editing files

`% cat seq.fa` will show the contents of `seq.fa` on the screen. The `cat` command may also be used to merge (concatenate) files. The following command will merge three different files – `file1`, `file2` and `file3` – into a new file `newfile`:

```
% cat file1 file2 file3 > newfile
```

The symbol > means we are redirecting the output of the `cat` program to a file instead of the standard output (which is the screen). We can also append to an existing file by using the symbols >>:

```
% cat file4 >> newfile
```

Viewing a text file on the screen one page at a time

To view a text file on the screen, use either `more` or `less`. With `less`:

```
% less seq.fa
```

Some useful keys for `less` are:

space	Move down one page
enter	Go down one line
u	Go up (back)
/HIV	Search for 'HIV'
q	Quit program

Viewing or extracting the first or last lines of a file

The first and last lines of a file may be displayed using the commands `head` and `tail`, respectively.

```
% head seq.fa          (by default head will show the first ten lines of the file)
% head -1000 seq.fa (the first 1000 lines of the file will be extracted)
```

Text-mode editor vi

```
% vi seq.fa
```

The editor vi is very useful whenever you do not have access to a graphical editor. An adequate description of it, however, requires a book on its own. For more information on vi, the reader is referred to http://en.wikibooks.org/wiki/ Learning_the_vi_Editor and http://unixhelp.ed.ac.uk/vi.

Graphical editors

Examples of graphical editors are emacs (http://www.gnu.org/software/emacs), gedit (http://www.gedit.org) and nedit (http://www.nedit.org).

Extracting file components with cut

Consider the content of a file dat.txt, in which the columns are separated with tabs:

```
1       12      1300       1306
2       11      1500       1458
3       17      1620       1700
```

We may extract columns 1 and 3 with cut:

```
% cut -f1,3 dat.txt
```

which produces:

```
1       1300
2       1500
3       1620
```

The fields or columns to be extracted are specified with the -f option. The default separator is a tab, but we may use any separator. The separator is specified with the -d option. Consider the file dat2.txt, which contains:

```
A;2;4500
B;5;4505
F;4;4510
```

We use the cut command:

```
% cut -f1,2 -d ';' dat2.txt
```

The output will be:

```
A;2
B;5
F;4
```

Sorting

Consider the file `dat3.txt`, which contains:

```
A     12     1300     1306
C     11     1500     1458
B     17     1620     1700
```

The lines may be sorted using `sort`:

```
% sort dat3.txt
```

The output will be:

```
A     12     1300     1306
B     17     1620     1700
C     11     1500     1458
```

The `sort` utility sorts lines alphabetically by default. Sorting is done numerically if we use the option `-n`. In addition, we may specify sorting with respect to a specific column, using the parameter `-k`:

```
% sort -n -k2 dat3.txt
```

The output is then:

```
C     11     1500     1458
A     12     1300     1306
B     17     1620     1700
```

Note that now the values in column 2 are in numerical order. We may also reverse the order of sorting with the `-r` parameter:

```
% sort -n -k2 -r dat3.txt
```

Unique lines

The `uniq` command is used to identify the unique lines in a file:

```
% uniq sortedfile
```

For this to work well the lines in the file first need to be sorted with `sort`. A useful option to `uniq` is `-c`. The effect is to list the number of times each line occurs:

```
% uniq -c sortedfile
```

Comparing files

To report differences between the two (sorted) files:

```
% diff sortedfile1 sortedfile2
```

To show lines that are shared between two (sorted) files:

```
% comm -12 sortedfile1 sortedfile2
```

Counting words

To count lines, words and number of bytes:

```
% wc filename
```

To count lines only:

```
% wc -l filename
```

Redirection and pipes

For the > and >> redirection symbols, see above under 'Viewing and editing files'.

The output of a file may be directed as input to another program with a pipe '|' symbol, like:

```
% sort somefile | uniq | wc -l
```

In the above command line the output of sort will be sent to uniq and the output of uniq will in turn be sent to wc. The final result is the number of unique lines in the file somefile.

Finding text with grep

A highly useful Unix utility is the grep command, used to search files for text strings or regular expression matching:

```
% grep ">" seq.fa | wc
```

In this example grep will identify all lines in seq.fa that contain '>'. The output of grep will then be directed to wc. The final output is therefore the number of lines with '>'. This is a simple way of counting the number of sequences in a FASTA format file. Some useful parameters of the grep command are illustrated here:

```
% grep -v -i -l "AACGTA" seqfile
```
-v   report lines where AACGTA does *not* match
-i   ignore case, i.e. we consider also 'aacgta'
-l   show only the file name, not the matching text

Finding files

The `find` command is very useful to locate files. To simplify it somewhat, the `find` command has the following syntax:

```
find [path or list of files ] [expression]
```

To locate files with the extension '`.fa`':

```
% find -name "*.fa"
```

By default, `find` is recursive so it will search all subdirectories as well. Use `'-maxdepth 1'` to restrict the search only to the current directory.

You may also want to locate files with a certain content. The following command will show all lines in all files with the extension `.fa` that contain the string 'HIV':

```
% find -name "*.fa" -exec grep HIV {} \;
```

The `-exec` parameter means that any program following `exec` will be executed on the files found by `find`. There are some peculiar symbols towards the end of the command line. The pair of curly brackets `{}` is a placeholder to indicate where the names of files found by `find` should be placed. The backslashed semicolon indicates the end of the command specified by `-exec`.

As another example, you may instead want to list the files that contain the string 'HIV':

```
% find -name "*.fa" -exec grep -l HIV {} \;
```

Useful features of Unix shells when typing commands

When typing a command, the *tab key* may be used for command-line completion. Thus, you can type the first few characters of a program or file name and press the tab key to fill in the rest of the item. This function saves a lot of typing.

Arrows on the keyboards are used to recall and edit previous commands. Thus, there is no need to type a long and complex command that you already typed before.

When editing a command, *Ctrl-E* is used to move to the end of the line and *Ctrl-A* is used to move to the beginning of the line.

More information about Unix commands

For more information on the different Unix commands, use the `man` command at the Unix command prompt. Many systems also have `info`. For example:

```
% man cut
% info cut
```

Interrupting a program

On many occasions you will find that you want to interrupt a program – for instance, you may find that it takes too long to run and you do not want to wait for the results, or you discover you typed the wrong parameters to a program. In many cases you can interrupt a program with the Ctrl-C key combination.

Programs running in the background

When you invoke a program with a GUI, or when you initiate a program that takes a while to complete, you typically want to be able to return to the Unix command prompt while the program is still running. The way to do that is to use an `&` (ampersand) after the program command. The program is said to 'run in the background':

```
% xclock &
```

For further studies on the interruption of a program and on programs running in the background, the reader is encouraged to read documentation on the Unix `bg` (background), `fg` (foreground) and `kill` commands.

Utilities when retrieving data over the network

The program `wget` is a useful Unix utility to retrieve a specific URL without making use of a web browser. Here is how to retrieve records of *Bacillus anthracis* CDC 684 from the NCBI FTP site Genbank (the following is to be typed on one line only):

```
wget ftp://ftp.ncbi.nih.gov/genomes/Bacteria/
Bacillus_anthracis_CDC_684_uid59303/NC_012581.gbk
```

In some cases data has been compressed and archived and may have the extension 'tar.gz'. Consider `wget` with this file containing a distribution of the Linux BLAST program:

```
wget
ftp://ftp.ncbi.nih.gov/blast/executables/release/2.2.25/blast-2.2.25-
ia32-linux.tar.gz
```

Having downloaded the file you need to uncompress it:

```
gunzip blast-2.2.25-ia32-linux.tar.gz
```

The resulting file is `blast-2.2.25-ia32-linux.tar`.
Then you need to unpack the contents of the tar archive:

```
tar -xvf blast-2.2.25-ia32-linux.tar
```

You can also skip the `gunzip` step to use this option of `tar`:

```
% tar -zxvf blast-2.2.25-ia32-linux.tar.gz
```

Some files are compressed with `bzip2`, which generates files with the extension `.bz2`. They can be unpacked with:

```
% bunzip2 some_archive.tar.bz2
```

or

```
% tar -xjvf some_archive.tar.bz2
```

Learning more about Unix

There is a whole range of books and free web resources about Unix. For example, see *Unix in a Nutshell* by Arnold Robbins (http://oreilly.com/catalog/9780596100292) and *UNIXhelp for Users* (http://unixhelp.ed.ac.uk).

A selection of biological sequence analysis software

A selection of bioinformatics software focused on biological sequence analysis is presented in this section. There are a lot more programs out there that deserve a mention, but all programs listed here are well documented and most of them are mentioned earlier in this book. For more software, including next-generation sequencing software, see web resources to this book.

EMBOSS

EMBOSS is the 'European Molecular Biology Open Software Suite'. EMBOSS is a free open source software analysis package with nearly 200 programs for biological sequence analysis (Rice *et al.*, 2000). For more details, see http://emboss.open-bio.org. A few examples of EMBOSS programs are listed below.

degapseq	Removes gap characters from sequences
distmat	Creates a distance matrix from multiple alignment
dottup	Displays a dotplot of two sequences
infoseq	Displays selected information about sequences
needle	Needleman–Wunsch global alignment
plotorf	Plot potential open reading frames
prettyplot	Displays aligned sequences; with colouring and boxes
revseq	Reverse and complement a sequence
seqret	Reads and writes sequences, converts between formats
shuffleseq	Shuffles a set of sequences while maintaining composition
sixpack	Translates nucleotide sequences
water	Smith–Waterman local alignment

The EMBOSS program `sixpack` will produce translation products of an input nucleotide sequence. The output will be two files with the extensions 'fasta' and 'sixpack', respectively. The sixpack file lists the input nucleotide sequence, its complementary sequence and all possible translation products. Help information for all EMBOSS programs is available by adding `-help` to the name of the program on the command line:

```
% sixpack -help
```

NCBI BLAST

BLAST is a frequently used program to search databases for sequence similarity to a query sequence (Altschul *et al.*, 1990, 1997). In the following we are referring to the NCBI blast+ software, which contains the programs `blastp`, `blastn`, `tblastn`, `blastx` and `tblastx`, as listed in Table 11.1. In addition, there are tools such as `makeblastdb` and `blastdbcmd`. Help information is available for these programs by adding `-help` to the program name at the command line.

```
% makeblastdb -help
% blastp -help
```

makeblastdb

The program `makeblastdb` is used for formatting sequences for use with BLAST. An example command line is:

```
% makeblastdb -in swissprot -dbtype prot -parse_seqids
```

where `-dbtype prot` indicates that we are formatting a protein sequence database. (For a nucleotide database, use `-dbtype nucl`.) The parameter `-in` specifies the FASTA-formatted database to be used, in this case in the file `swissprot`. The option `-parse_seqids` is an instruction to parse sequence identifiers in the FASTA definition line. If this option is selected we can retrieve sequences from the database using `blastdbcmd`, as described below. As a result of the `makeblastdb` command, a number of binary files with extensions .pnd, .psi, .phr, .pni, .psq, .pin and .psd will be created.

blastdbcmd

The program `blastdbcmd` is used to retrieve sequences from a database. Important parameters are:

`-db` [BLAST database name]	BLAST database name, default is 'nr'
`-dbtype` [`guess`, `nucl` or `prot`]	Default is 'guess'
`-entry` [sequence identifier(s)]	Comma-delimited string(s) of sequence identifiers
`-range` [start-stop]	Range of sequence to extract
`-strand` [`minus` or `plus`] Default is 'plus'	Strand of nt sequence to extract
`-out` [output file name]	Output file name

Here is an example to retrieve the first 20 amino acids of the ABL1_HUMAN protein sequence:

```
% blastdbcmd -entry ABL1_HUMAN -db swissprot -range 1-20
```

blastp, blastn, etc.

Important parameters to the programs such as `blastp`, `blastn`, `tblastn` and `tblastx` are:

`-query` [File_In]	Input file name
`-db` [database name]	BLAST database name
`-out` [File_out]	Output file name
`-word_size` <integer>=2>	Word size for wordfinder algorithm
`-outfmt` [string]	Alignment view options:
	0 = pairwise,
	1 = query-anchored, showing identities,
	2 = query-anchored, no identities,
	3 = flat query-anchored, show identities,
	4 = flat query-anchored, no identities,
	5 = XML BLAST output,
	6 = tabular,
	7 = tabular with comment lines,
	8 = Text ASN.1,
	9 = Binary ASN.1,
	10 = Comma-separated values
`-num_descriptions` [integer]	Number of database sequences to show one-line descriptions for; default is '500'
`-num_alignments` [integer]	Number of database sequences to show alignments for; default is '250'
`-seg` [string]	Filter query sequence with SEG (Format: 'yes', 'window locut hicut', or 'no' to disable); default is 'no'. (Parameter does not apply to `blastn`.)

An example command line is:

```
% blastp -query bcrabl_human.fa -db swissprot -out bcrabl.blastp
```

Older versions of NCBI BLAST

Older versions of NCBI BLAST use other program names and parameters to those described above for blast+. For example, to initiate a blastp search, use:

```
% blastall -i query -d database -p blastp -o outputfile
```

To format a collection of FASTA-formatted sequences, use:

```
% formatdb -i swissprot
```

For help on these commands use:

```
% blastall --help
% formatdb --help
```

PSI-BLAST

Searches with PSI-BLAST are carried out with the program `psiblast`. For help information use:

```
% psiblast -help
```

Many parameters such as `-query`, `-db`, `-out` and `-outfmt` are used as described for `blastp`. Specific to `psiblast` is, for instance, `-num_iterations`, specifying the number of iterations to perform. An example command is:

```
% psiblast -query input -db swissprot -num_iterations 8
```

Blat

Blat (Kent, 2002) is a local alignment tool like BLAST. However, it has been optimized with respect to speed, and a key element as compared to BLAST is that it keeps an index of all non-overlapping k-mers in the genome searched. Help information is provided by:

```
% blat -help
```

An example command line is:

```
% blat humandb query output
```

where `humandb` is the database to be searched, `query` is a sequence used as the query and `output` is the name of the output file. The database and query sequences are each either `.fa`, `.nib` or `.2bit` files.

ClustalW

ClustalW (http://www.clustal.org) (Thompson *et al.*, 2002) is a program for multiple sequence alignments.

Help information is obtained with

```
% clustalw2 -help
```

Some examples of useful command-line parameters are:

- `options`
- `bootstrap`
- `tree`
- `convert`
- `profile1=filename`
- `profile2=filename`
- `gapopen=f`
- `gapext=f`
- `output=gcg OR gde OR pir OR phylip OR nexus OR fasta`
- `outputtree=nj OR phylip OR dist OR nexus`
- `outfile=filename`
- `outorder=input OR aligned`

The input to ClustalW is typically a collection of sequences in a FASTA format. An example command line is:

```
% clustalw2 sequences.fa
```

Two different output files are produced. One has the extension 'aln' and is the actual alignment. The other file has the extension 'dnd' and is a file with information on the tree that was used in the construction of the alignment. The tree may be viewed with programs such as `njplot` (Perriere and Gouy, 1996).

T-Coffee

T-Coffee (Notredame *et al.*, 2000) (http://www.tcoffee.org) is another example of a multiple alignment tool. For help information, use:

```
% t_coffee -help
```

Muscle

Muscle (Edgar, 2004) (http://www.drive5.com) is another program to carry out multiple alignments. Help information is obtained by typing the name of the program without parameters:

```
% muscle
```

Jalview

Jalview (Waterhouse *et al.*, 2009) is a Java-based multiple sequence alignment editor and viewer with a GUI interface (http://www.jalview.org).

PHYLIP

PHYLIP (the PHYLogeny Inference Package) is a package of more than 30 programs for inferring phylogenies (evolutionary trees) (Felsenstein, 1989, 2005) (http://evolution.genetics.washington.edu/phylip). Specific details will be provided here to show how the PHYLIP Dnapars program was used to create the phylogenetic tree in Fig. 10.4.

Bootstrap replicates are first formed with the program `seqboot`. The file `mito.fa` contains mitochondrial genome sequences. ClustalW may be used to convert this file to the format used by PHYLIP. The file with PHYLIP format, `mito.phy`, is copied to a file named `infile`:

```
% cp mito.phy infile
```

This file is then used as input to `seqboot`.

```
% seqboot
```

As you start `seqboot`, as in all other PHYLIP programs, you are presented with a menu which allows you to modify different parameters of the program. In this case, type 'Y' to accept the settings (including the default 100 replicates). Also type an odd number as a seed to a random-number generator. The output file from `seqboot`, named `outfile`, now has 100 random alignments in PHYLIP format. As the next step this file will be used as input to Dnapars. Therefore, copy it to a file named `infile`.

```
% cp outfile infile
```

Now run Dnapars:

```
% dnapars
```

In the menu resulting from this program you need to change the option 'Analyse multiple data sets?'. Therefore, type 'M'. You are then prompted for 'Multiple data sets or multiple weights? (type D or W)'. Type 'D' and '100' as the number of datasets. Again, type a random-number seed and a number for 'Number of times to jumble'. Then type 'Y' to start the analysis. Now each of the 100 replicates will be analysed. The output files from Dnapars will be `outtree` and `outfile`; `outtree` has a number of trees in Newick format. The information in this file is finally used to produce a consensus tree, using the Consense program of the PHYLIP package.

First, copy the `outtree` file to `intree`:

```
% cp outtree intree
```

Then run the Consense program:

```
% consense
```

Accept the settings by typing 'Y'. The output from Consense will end up in the file `outtree`. This is the tree shown in Fig. 10.4. The text format is shown below:

```
(Canis_lupu:100.0,((((((Dasyurus_h:100.0,Phascogale:100.0):100.0,
Sminthopsi:100.0):100.0,Myrmecobiu:100.0):72.5,Thylacinus:100.0):100.0,
(((((Phalanger_:100.0,Trichosuru:100.0):100.0,Vombatus_u:100.0):63.7,
(Potorous_t:100.0,Macropus_r:100.0):100.0):68.0,Dactylopsi:100.0):98.0,
(Monodelphi:100.0,(Isoodon_ma:100.0,Echymipera:100.0):100.0):89.5):
69.7):93.0,Notoryctes:100.0):100.0);
```

MrBayes

MrBayes is a program for Bayesian estimation of phylogeny (Ronquist and Huelsenbeck, 2003). It was used to generate one of the trees in Fig. 10.4. The sequences to be analysed need to be in the Nexus format. ClustalW may be used to convert the sequences into this format. The file generated by ClustalW may still need some manual editing. The first lines of the file should have this format to work well with MrBayes:

```
#NEXUS
BEGIN DATA;
dimensions ntax=16 nchar=15577;
format interleave=yes datatype=DNA gap= -;
```

We assume that the file with mitochondrial sequences in Nexus format is named `mito.nxs`. To start the MrBayes program under Unix, type 'mb'. The following commands may then be used:

(1) `execute mito.nxs`
(2) `lset nst=6 rates=invgamma`
(3) `mcmc ngen=10000 samplefreq=10`

if standard deviation > 0.01 then repeat the command 3.
if standard deviation < 0.01 then:

(4) `sump burnin = 250 (25% of 10000/10)`
(5) `sumt burnin = 250`

Convergence was reached after 56 000–100 000 generations, depending on the run. One can also design a script to make things more automatic with MrBayes. For example:

```
begin mrbayes;
        set autoclose=yes nowarn=yes;
        execute mito.nxs;
        lset nst=6 rates=invgamma;
        mcmc nruns=2 ngen=100000 samplefreq=10;
        sump burnin=2500;
        sumt burnin=2500;
        quit;
end;
```

The final tree shown in Fig. 10.4 is shown here in Nexus format:

```
(Dasyurus_hallucatus:0.259794,Phascogale_tapoatafa:0.240719,
(Sminthopsis_crassicaudata:0.234747,(Myrmecobius_fasciatus:0.314110,
(Thylacinus_cynocephalus:0.420154,((((Isoodon_macrourus:0.164670,
Echymipera_rufescens_australis:0.148569)1.00:0.157430,Monodelphis_
domestica:0.464971)1.00:0.054760,(((Trichosurus_vulpecula:0.135767,
Phalanger_interpositus:0.178641)1.00:0.090922,(Vombatus_ursinus:
0.258777,Dactylopsila_trivirgata:0.324118)0.54:0.023144)0.55:0.022629,
(Macropus_robustus:0.173136,Potorous_tridactylus:0.171551)1.00:0.105689)
1.00:0.059132)1.00:0.029939,(Notoryctes_typhlops:0.474563,Canis_lupus:
1.441093)0.82:0.041982)1.00:0.126314)1.00:0.025187)1.00:0.072658)
1.00:0.074571);
```

For documentation on MrBayes and further explanation of the commands above, see the manual available at http://mrbayes.csit.fsu.edu.

HMMER

Many of the practical details of HMMER (Eddy, 1998; http://hmmer.org) are described in Chapter 12, including Table 12.1. To get help information, consult the *HMMER User's Guide* (http://hmmer.org), or use the -h option of the program:

```
% hmmsearch -h
```

Here is an example command line, in which the first argument is a Pfam model of the VWD protein domain and the second argument is the database to be searched:

```
% hmmsearch vwd.hmm swissprot
```

A utility that is part of the HMMER package is `sreformat`, which reads a sequence file in any supported format, reformats it into a new format specified, then prints the reformatted text. Output format choices are:

```
Unaligned       Aligned
---------       -------
fasta           stockholm
embl            msf
genbank         a2m
gcg             phylip
gcgdata         clustal
pir             selex
raw             eps
```

Infernal

The Infernal ('INFERence of RNA ALignment') software (Nawrocki *et al.*, 2009; http://infernal.janelia.org) is for searching DNA sequence databases for RNA structure and sequence similarities. It is briefly introduced in Chapter 17. The programs as described in the user guide are:

`cmalign`	Align sequences to an existing model.
`cmbuild`	Build a model from a multiple sequence alignment.
`cmcalibrate`	Determine expectation value scores (E-values) for more sensitive searches and appropriate HMM filter score cutoffs for faster searches.
`cmemit`	Emit sequences probabilistically from a model.
`cmscore`	Test the efficacy of different alignment algorithms (mainly useful for development and testing).
`cmsearch`	Search a sequence database for matches to a model.
`cmstat`	Report statistics on a model.

A typical procedure when using Infernal is:

(1) Build a model with `cmbuild`. The input is a multiple sequence alignment in Stockholm format.
(2) Calibrate that model with `cmcalibrate`.
(3) Search a collection of sequences with `cmsearch`.

Help information is available through the user guide, as well as by typing `-h` at the command line:

```
% cmsearch -h
```

More reading on bioinformatics software

For more information on bioinformatics software, see the following sources:

- Network Science website (http://www.netsci.org)
- Bioinformatics Organization (http://www.bioinformatics.org)
- The journal *Briefings in Bioinformatics* (http://bib.oxfordjournals.org), which publishes reviews for the users of databases and analytical tools of bioinformatics
- Many tools and algorithms relevant to next-generation sequencing applications have been published in *Bioinformatics*, and these are gathered together in a 'Bioinformatics for next generation sequencing' virtual issue, which is continually updated (http://www.oxfordjournals.org/our_journals/bioinformatics/nextgenerationsequencing.html)
- A listing of software is available in Markel and León's (2003) *Sequence Analysis in a Nutshell.*

REFERENCES

Altschul, S. F., W. Gish, W. Miller, E. W. Myers and D. J. Lipman (1990). Basic local alignment search tool. *J Mol Biol* **215**(3), 403–410.

Altschul, S. F., T. L. Madden, A. A. Schaffer, *et al.* (1997). Gapped BLAST and PSI-BLAST: a new generation of protein database search programs. *Nucleic Acids Res* **25**(17), 3389–3402.

Eddy, S. R. (1998). Profile hidden Markov models. *Bioinformatics* **14**(9), 755–763.

Edgar, R. C. (2004). MUSCLE: a multiple sequence alignment method with reduced time and space complexity. *BMC Bioinformatics* **5**, 113.

Felsenstein, J. (1989). PHYLIP: Phylogeny Inference Package (Version 3.2). *Cladistics* **5**, 164–166.

Felsenstein, J. (2005). PHYLIP (Phylogeny Inference Package) version 3.6. Distributed by the author. Department of Genetics, University of Washington, Seattle.

Kent, W. J. (2002). BLAT: the BLAST-like alignment tool. *Genome Res* **12**(4), 656–664.

Markel, S. and D. León (2003). *Sequence Analysis in a Nutshell: A Guide to Tools and Databases*. Farnham and Sebastopol, CA, O'Reilly.

Nawrocki, E. P., D. L. Kolbe and S. R. Eddy (2009). Infernal 1.0: inference of RNA alignments. *Bioinformatics* **25**(10), 1335–1337.

Notredame, C., D. G. Higgins and J. Heringa (2000). T-Coffee: a novel method for fast and accurate multiple sequence alignment. *J Mol Biol* **302**(1), 205–217.

Perriere, G. and M. Gouy (1996). WWW-query: an on-line retrieval system for biological sequence banks. *Biochimie* **78**(5), 364–369.

Rice, P., I. Longden and A. Bleasby (2000). EMBOSS: the European Molecular Biology Open Software Suite. *Trends Genet* **16**(6), 276–277.

Ronquist, F. and J. P. Huelsenbeck (2003). MrBayes 3: Bayesian phylogenetic inference under mixed models. *Bioinformatics* **19**(12), 1572–1574.

Thompson, J. D., T. J. Gibson and D. G. Higgins (2002). Multiple sequence alignment using ClustalW and ClustalX. *Curr Protoc Bioinformatics* Chapter 2, Unit 2.3.

Waterhouse, A. M., J. B. Procter, D. M. Martin, M. Clamp and G. J. Barton (2009). Jalview Version 2: a multiple sequence alignment editor and analysis workbench. *Bioinformatics* **25**(9), 1189–1191.

APPENDIX III
A short Perl reference

In short, when the genome project was foundering in a sea of incompatible data formats, rapidly changing techniques, and monolithic data-analysis programs, Perl saved the day. Though it's not perfect, Perl seems to meet the needs of the genome centers remarkably well, and is usually the first tool we turn to when we have a problem to solve.

(Lincoln Stein, from 'How Perl saved the genome project'[1])

The Perl programming language was invented by Larry Wall; his version 1.000 was presented in 1987. Perl is said not to be an acronym, but still you occasionally see it said to represent 'Practical Extraction and Report Language' or 'Pathologically Eclectic Rubbish Lister'. The word 'Perl' (with a capital P) refers to the programming language as such, whereas 'perl' refers to the interpreter (implementation). Larry Wall originally invented the language to help out in system administration and in the analysis of huge text files, but through the years Perl has been continuously developed and has for many years been widely used in areas such as web programming and bioinformatics.

The information found in this appendix is very far from a complete description of Perl. The focus here is on the features of Perl that are mentioned in this book. There are certain important elements that are left out, such as variable references and object-oriented approaches. For more extensive information the reader is referred to other sources as listed towards the end of this appendix – for example, *Perl Programming*, for which Larry Wall is one of the coauthors.

To execute Perl programs you need a Perl interpreter. Perl is available for different computer platforms. As with Unix, you can select to run Perl on your own personal computer or, alternatively, you can connect with SSH to a remote Unix-based server computer (see Appendix I). If you choose to carry out your work at a Unix computer, Perl is most likely already installed and you may execute a Perl program by typing in a terminal window:

```
% perl someprogram.pl
```

[1] http://www.bioperl.org/wiki/How_Perl_saved_human_genome.

Also, Macs running OS X or later operating systems are based on Unix. You may thus access Perl by opening a terminal window (available under Applications > Utilities).

Perl is not available by default in the Windows operating systems (alas), but you can install Cygwin (http://www.cygwin/com), which is a useful Unix environment for Windows.[2] Perl is one of many Unix components that are available using Cygwin. The are also other ways of running Perl under Windows, such as ActivePerl (http://www.activestate.com/activeperl) and Strawberry Perl, the latter an open source distribution for Windows (http://strawberryperl.com).

Once Perl is installed on a computer, you eventually need to install additional modules that you need for specific tasks. Towards the end of this appendix there is more information about how to install modules.

Data types and variables

Scalars

The scalar is a fundamental data type, typically strings or numbers from which more complicated structures are built. Scalars are named with an initial $:

```
$dna = 'gctatatat';
$pi = 3.14;
```

A string is specified by single quotes. Double quotes are used when variable interpolation is desired:

```
$dna = 'gctatatat';
print "The DNA is $dna"; # will print The DNA is gctatatat
```

Arrays

An array is an ordered list of scalars. Arrays are named with an initial @:

```
@dna = ('A', 'G', 'C', 'T');
@numbers = (3, 6, 12, 13, 16);
```

Elements in arrays are referred to by number (0, 1, 2, etc.):

```
@dna = ('A', 'G', 'C', 'T');
print "Third element in array dna is $dna[2]"; # will print C
```

---

[2] In fact, the author used Cygwin while writing the Perl code for this book. Just to make things more difficult.

Hashes

A hash is an unordered set of key–value pairs:

```
%id2description = (
        AC988823 => 'protein kinase'
        BC887682 => 'dehydrogenase'
        NX772123 => 'hexokinase'
)
```

In the listing above, a string like AC988823 is a *key*, and the corresponding *value* for that key is 'protein kinase'. Hashes are named with an initial %. Here is how to obtain the value for the key AC988823:

```
print "$id2description{AC988823}"; # will print protein kinase
```

To print all keys and values of a hash, you can use:

```
foreach $key (keys %id2description) {
        print "$key $id2description{$key}\n";
}
```

Variable $_

The variable $_ is the default input and pattern-search space. For example:

while (<IN>) {...}	is the same as	while ($_ =<IN>) {...}
/^Subject:/;	is the same as	$_ =~ /^Subject:/;
tr/a/A/;	is the same as	$_ =~ tr/a/A/;
print;	is the same as	print $_;

Operators

Assignment operators

A basic assignment operator is =, as in $a = 3. More examples are:

$a += 2	short for $a = $a + 2
$dna .= 'gcg'	short for $dna = $dna . 'gcg'

Arithmetic operators

Below are the available arithmetic operators:

+	addition
−	subtraction
*	multiplication
/	division
%	modulus
**	exponentiation

String operators

Available string operators are:

.   concatenation

x   repeat operator (not to be confused with the numeric multiplication operator
      * shown above)

For example:

```
$codon1 = 'gct';
$codon2 = 'gcc';
print "$codon1.$codon2";  # prints gctgcc
print "$codon1 x 4";      # prints gctgctgctgct
```

Logical operators

The three logical operators are OR, AND and NOT. In Perl they are represented
by the symbols ||, && and !, respectively. Alternatively, you can use the closely
related operators or, and and not (with lower case, not upper case).

Numeric and string comparison operators

Comparison	Numeric	String
Equal	==	eq
Not equal	!=	ne
Less than	<	lt
Greater than	>	gt
Less than or equal	<=	le
Greater than or equal	>=	ge
Comparison*	<=>	cmp

* 0 if equal, 1 if $a is greater, −1 if $b is greater

Conditional statements
if/else/elsif

```
if ($a < 10) {print "less than 10";}
elsif ($a == 10) {print "10";}
else {print "greater than 10";}
```

unless

```
unless ($a < 10) {print "10 or greater";}
```

Loop constructs

while

```
$seq = 'gcgggat';
while (length($seq) > 3) {
        $seq =~ s/^.//;
        print "$seq ";
} # will print cgggat gggat ggat gat
```

until

```
$a = 10;
until ($a == 5) {$a--; print "$a ";} # will print 9 8 7 6 5.
```

for

```
for ($a = 0; $a < 4; $a++) {
        print "$a ";
} # will print 0 1 2 3
```

In a for loop there are three expressions within parentheses: the initial value of the loop variable, the condition to be tested and an expression to change the loop variable.

foreach

The foreach statement is used to go through a set of scalars.

```
@dna = (A, G, C, T);
foreach $letter(@dna) {
        print "$letter ";
}       # will print A G C T
```

Breaking out of a loop

```
@dna = (A, G, C, T);
foreach $letter(@dna) {
        print "$letter ";
```

```
        if ($letter eq 'G') {last;}
}       # will print A G
```

You can also add a label to a loop and specify that `last` refers to that specific loop.

```
DNA:
for ($i = 0; $i < 3; $i++) {
        foreach $letter(@dna) {
                print "$i $letter ";
                if ($letter eq 'G') {last DNA;}
        }
}
```

Avoiding a specific iteration

```
@dna = (A, G, C, T);
foreach $letter(@dna) {
        if ($letter eq 'G') {next;}
        else {print "$letter ";}
}       # will print A C T
```

True and false in Perl

Everything in Perl counts as 'true' except:

(1) the strings `' '` (the empty string) and `'0'` (a string containing only the character `'0'`), or any string expression that evaluates to either `""` (the empty string) or `"0"`;

(2) any numeric expression that evaluates to a numeric 0;

(3) any undefined value.

Pattern matching

A 'regular expression' or 'pattern' is used to describe a set of strings, without having to list all elements of that set. For instance, all sequences cleaved by the enzyme BfmI may be described by the pattern CT[AG][CT]AG.

```
$seq = 'GGACTATAGCT';
if ($seq =~ m/CT[AG][CT]AG/){  # is there a match of CT[AG][CT]AG
                               # to the string in $seq?
        print "$&";            # print what is matched
}
```

=~   is the 'binding operator'

/ /   is the regular expression delimiter

The m may be left out. m/CT[AG][CT]AG/ is the same as /CT[AG][CT]AG/. In addition to the m/// match operator, there are also the *substitution* and *transliteration* operators as described further below.

Pattern modifiers

```
$seq =~ /A/g;    # globally find all matches
$seq =~ /a/i;    # ignore case (in this case find A and a)
```

Special characters used in patterns

Most characters in patterns match themselves. The special characters that do not are:

. ^ $ | ( ) [ ] \ { } * + ?

. *(match one character)*

```
$seq = 'GCAT';
if ($seq =~ /G.A/ ) {print "G, any, A";}
```

^ *(true at beginning of string)*

```
$seq = 'GCAT';
if ($seq =~ /^G/ ) {
    print "G is at the beginning of the string";
    }
```

$ *(true at end of string)*

```
$seq = 'GCAT';
if ($seq =~ /$T/ ) {print "T is at the end of the string";}
```

...|... *(match one or the other)*

```
$seq = 'GCAT';
if ($seq =~ /G|C/) {print 'match';}
# match either G or C
```

(. . .) (grouping)

```
$seq = 'GCAT';
if ($seq =~ /(GC)|(CA)/) {print 'match';}
# match either GC or CA
```

[. . .] (group of characters)

```
/[AGCT]/    matches any of the characters A G C and T
/[^AGCT]/   matches any character which is not A G C or T
```

\ (backslash)

Backslash is used if you want to match any of the special characters – for example, \(matches a left parenthesis.

For the characters { } * + ?, see under 'Quantifiers' below.

Metasymbols

```
\d    digit (same as [0-9])
\D    non-digit ([^0-9])
\s    whitespace ([\t\n\qr\f])
\S    non-whitespace ([^\t\n\qr\f])
\w    word character ([a-zA-Z0-9_])
\W    non-word character ([^a-zA-Z0-9_])
\n    new line
\t    tab
```

Quantifiers

```
*              Match zero or more times
+              Match one or more times
?              Match one or zero times
{COUNT}        Match exactly COUNT times
{MIN,}         Match at least MIN times
{MIN,MAX}      Match at least MIN times but at most MAX times
```

For example:

```
/G{1,2}/    matches a series of Gs occurring in the range 1–2, i.e. 'G' or 'GG'
/G+/        matches one or more Gs
/G*/        matches zero, one or more Gs
```

By default, matching in Perl is 'greedy'. Consider this example:

```
$seq = 'GCAGCTTA';
$seq =~ /G.*A/;
```

Here, the pattern will try to grab as much as possible. It grabs the whole string GCAGCTTA. If you instead want to grab as small a portion as possible, you have to use a ? after the quantifer:

*?   Match zero or more times

+?   Match one or more times

For example:

```
$seq = 'GCAGCTTA';
$seq =~ /G.*?A/; # matches GCA
```

Capturing matched patterns

If you place parentheses around a pattern, the matched string will be stored in the variables $1, $2, etc.

```
$seq = 'GGACTTTCTG';
$seq =~ /(G.A).*(...)$/ ;
print "$1\n"; # prints GGA
print "$2\n"; # prints CTG
```

Matching, substitution and transliteration operators

We have encountered the m// match operator above to describe features of patterns. It is of the form:

```
m/PATTERN/
```

or

```
m/PATTERN/modifier
```

In addition, there are the substitution s/// and transliteration tr/// operators. The s/// operator has the form

```
s/PATTERN/REPLACEMENT/modifier
```

For example:

```
$dna = 'GCATTT';
$rna = $dna;
$rna =~ s/T/U/g;   # g = global modifier
                   # replace all Ts with Us
                   # $rna is now GCAUUU
```

The `s///` operator may be used for counting characters or patterns in a string, like

```
$count = ($dna =~ s/T//g);
```

The transliteration operator `tr///` has the form

```
tr/SEARCHLIST/REPLACEMENTLIST/modifier
```

Note that the SEARCHLIST is a list of characters and *not* a pattern (regular expression).

For example:

```
$seq =~ tr/T/U/; # replace all Ts with Us
$reverse =~ tr/ATCG/TAGC/; # replacing bases according to
                           # the base complementarity rules
$string =~ tr/a-z/A-Z/;    # from lower case to upper case
$p = ($protein =~ tr/P//); # counting the occurrence of P
```

Perl functions used in this book
String functions

`reverse`	reverse order of string characters
`substr`	extract substring from string

```
$dna = 'GCAAAA';
print substr($dna, 2, 3); # 3 characters starting from
# position 2 ; prints AAA
```

`length`	return length of string
`uc`	make upper case
`lc`	make lower case
`split`	split string as specified by the separator into a list of substrings (compare join)

```
$line = 'G,C,A,T';
@bases = split(',', $line); # G C A and T are now
# elements of the array @bases
$seq = 'GCATTT';
@seqarray = split('', $seq); # sequence string is
# converted into an array.
$seq = 'GCATTT';
@seqarray = split(/G|C/, $seq);
# the elements of @seqarray are
# $segarray[0]: empty string
```

```
              # $seqarray[1]: empty string
              # $seqarray[2]: ATTT
join          join a set of strings such as those in an array (compare split).
              @seqarray = ('A', 'G', 'C', 'T');
              $seq = join('', @seqarray);
              # $seq is now the string 'AGCT'
```

Numerical functions

```
int    return integer portion of a number
```

General functions

```
print         print to standard output
open/close    file open and close operations
sort          sorts a list
```

Examples of `sort` are:

```
@array = ('A', 'C', 'B');
@sortedarray = sort(@array); # @sortedarray is now
                                  ('A', 'B', 'C')
```

By default `sort` is based on string comparison:

```
@array = ('20', '3', '1');
@sortedarray = sort(@array); # @sortedarray is now
                                  ('1', '20', '3')
```

To do a numerical sort:

```
@array = (20, 3, 1);
@sortedarray = sort {$a <=> $b} (@array);
# The elements of @sortedarray are now in
# numerical order ('1', '3', '20')
```

Remember, hashes are *unordered* list of scalars. Consider this example:

```
%hash = (
'3' => 'CAG',
'5' => 'CAA',
'35' => 'CAC',
'1' => 'CAT');
foreach $key (keys %hash) {
    print "$key $hash{$key}\n";
}
```

This will print:

```
35 CAC
1 CAT
3 CAG
5 CAA
```

Thus, keys and their values are printed in a non-predictable order. If we want the output to be *numerically* sorted according to the key values, we can replace the `foreach` loop with this code, where the '`<=>`' is a comparison operator which is very useful for sorting:

```
foreach $key (sort {$a <=> $b} keys %hash) {
        print "$key $hash{$key}\n";
}
```

This will print:

```
1 CAT
3 CAG
5 CAA
35 CAC
```

Making your own functions

A user-defined function is described with the following type of code. The word `sub` (subroutine) is followed by the name of the function, and within curly brackets is a block of code. An example without arguments:

```
sub dna {
        print "It looks like a DNA sequence\n";
}
```

This function is called with an expression like:

```
dna ();
```

or

```
dna;
```

Here is an example of a function with one argument:

```
$dna = 'GATTTTT';
$rna = dna_to_rna($dna);
print "$rna\n"; # will print GAUUUUU
sub dna_to_rna {
        my $rna = $_[0];
```

```
        $rna =~ tr/T/U/;
        return ($rna);
}
```

The $_[0] in the function definition is the first element in an array named @_. This array is a list of the arguments passed to the function. The individual elements of this array are referred to by $_[0], $_[1], etc. In this case there is only one element in that array, the $dna variable. Therefore, the second line means that the variable $rna receives its contents from the first element in the @_ array, which is $_[0]. The return function by the end of the function specifies what is to be returned from this function, in this case the $rna variable.

An example with three arguments:

```
$newseq = merge ('ATTA', 'TTTA', 'CGGA');
print "$newseq\n";
sub merge {
        my ($seq1, $seq2, $seq3)= @_;
        return ($seq1.$seq2.$seq3);
}
```

If we want to check whether our function has the expected number of arguments, we could use the following:

```
if (@_ != 3) {
        print "Warning, function should get exactly three arguments!\n";
}
```

Filehandles

<FILEHANDLE> in a scalar context reads a single line from the file opened by FILEHANDLE. Examples are:

```
open(IN, "file"); # read from a file
open(OUT, ">file"); # create a file and write to it
open(FILE, ">>file"); # append to an existing file
```

Examples of use are:

```
open(IN, "file");
while (<IN>) {print;}
close IN;
open(OUT, ">newfile");
print OUT "This is a new file\n";
close OUT;
```

In array context, filehandle FILEHANDLE reads the whole file:

```
open(IN, 'seq.fa');
@dna = <IN>;
close IN;
print "@dna";
```

More complex data structures

Arrays of arrays

A two-dimensional array may be constructed is this way:

```
@matrix = (
['G', 'C', 'A', 'T'],
['AG', 'GA', 'CT', 'TC'],
['CGA', 'GAA', 'ATT', 'TTT']
);
```

To access an element:

```
print $matrix[2][0]; # prints "CGA"
```

To print the whole array:

```
for ($i = 0; $i <= $#matrix ; $i++) {
    for ($j = 0; $j <= $#{$matrix[$i]}; $j++) {
        print "row $i col $j = $matrix[$i][$j]\n"
    }
}
```

Here, `$#` is the last array index. An alternative is to use `scalar`, as in `scalar(@matrix)`, which is the same as `$#matrix + 1`.

Hashes of arrays

A hash of arrays may be constructed in this way:

```
%code = (
    glycine => ['GGA', 'GGT', 'GGC', 'GGG'],
    valine => ['GTA', 'GTT', 'GTC', 'GTG'],
    met => ['ATG']
);
```

To access an element:

```
print "$code{glycine}[1]"; # prints GGT
```

To print the whole hash:

```
for $aminoacid (keys %code) {
    print "$aminoacid : @{$code{$aminoacid}}\n";
}
```

Arrays of hashes

An array of hashes may be constructed in this way:

```
@symbols = (
    {
        Y => 'CT',
        R => 'AG',
        N => 'AGCT',
    },
    {
        Y => 'Tyr',
        R => 'Arg',
        N => 'Asn',
    }
);
```

To access an element:

```
print "$symbols[1]{R}\n"; # prints Arg
```

To print the whole array:

```
for ($i = 0; $i<= $#symbols; $i++) {
    print "index = $i : ";
    for $string (keys %{ $symbols[$i] } ) {
        print "$string = $symbols[$i]{$string}\n";
    }
}
```

Hashes of hashes

A hash of hashes may be constructed in this way:

```
%bases = (
    T => { type => 'pyrimidine', pairswith => 'A' },
    C => { type => 'pyrimidine', pairswith => 'G' },
    A => { type => 'purine',     pairswith => 'T' },
    G => { type => 'purine',     pairswith => 'C' },
);
```

To access an element:

```
print "$bases{T}{pairswith}\n"; # prints A
```

To print the whole hash:

```
for $base ( keys %bases ) {
    print "$base : ";
    for $property ( keys %{ $bases{$base} } ) {
        print "$property = $bases{$base}{$property} ";
    }
    print "\n";
}
```

Readable Perl code and comments

It is good practice to comment your code so that it will be easier to understand when you later return to it. In particular, well-commented code will be a lot easier to understand for other programmers. For instance, as a beginner you may not immediately realize what the following code accomplishes (code of Stefan Washietl, University of Vienna):

```
map {print " "x$_->[0],$xx=(keys(%x))[int(rand 4)],"-"x$_->
[1],$x{$xx},"\n"}
(@_=([5,0],[4,2],[3,3],[2,4],[1,4],[0,3],[0,2],[1,0]),reverse @_)
while (%x=(a,t,t,a,g,c,c,g))
```

I will not reveal here the output of this code, but the interested reader may find out, either by working out the details of the code, or by simply running the program. Speaking of code that is difficult to read, there was for some time the 'Obfuscated Perl Contest', where entries were judged on 'aesthetics, output and incomprehensibility'. Here we will have less of obfuscation and return to the issue of commenting code. There are two methods of introducing comments in Perl. One method uses the # character. Anything in a line that follows such a character is a comment and will be disregarded by the Perl interpreter. An example line:

```
# This line is a comment
```

If you need to use the # character in some other way, you need to put a backslash in front of it:

```
print "\#"; # will print '#'
```

If you have a lot of lines you want to comment out – for instance, when you are tracing a program bug – putting a # in front of many lines is tedious.

You can then make use of the following syntax (which is actually part of 'POD', a mark-up language used for writing documentation for Perl), where the code between the lines =sometext and =cut is disregarded by the Perl interpreter.

```
=sometext
$seq = 'gcgggagt';
print "$seq\n";
=cut
```

Executing Perl programs

We assume that you have Perl code in a file named program.pl. In order to execute this program you use:

```
% perl program.pl
```

or:

```
% program.pl
```

The second alternative assumes that there is a line #!/usr/bin/perl at the head of the program file and that the file has been made executable. If it is not, you can make it executable with

```
chmod +x program.pl
```

Using arguments with a Perl script

It is often convenient to supply information to a Perl script on the command line. Assume we have a Perl script named file.pl, which has the following content:

```
open(IN, $ARGV[0]) or die "There is no file with that name\n";
print "This is the content of file $ARGV[0]:\n";
while (<IN>) {
    print;
}
close IN;
```

When we run this script we can supply the name of the file to be analysed on the command line, like this:

```
% perl file.pl notes.txt
```

Perl stores the arguments at the command line in the array @ARGV. It contains all the words listed on the command line after the name of the Perl script. The variable $ARGV[0] has the first element of this array, in this case notes.txt.

Another way to pass on information to a Perl script is to use 'piping'. As an example, we have a Perl script named `analyse.pl` with the following content:

```
while (<>) { # default input
        if (/^AC(.*)/) {print "accession number is $1\n"; } }
```

We can then analyse the file `seq.fa` with this command:

```
% cat seq.fa | perl analyse.pl
```

Perl one-liners

Once you get more familiar with Perl and you find you are frequently writing relatively short scripts you may consider a 'Perl one-liner', meaning that you run Perl code using a single Unix command line. A simple example:

```
% perl -e 'print "Hello, RNA world\n";'
```

Here, the parameter `-e` is to indicate that everything that follows after it should be run as a Perl script. Please note the use of single quotes surrounding the Perl code. We can also use recursive procedures with the `-n` parameter. In the example below, a file will be searched for lines beginning with 'ID' and in case such a line is found the text following 'ID' will be printed:

```
% cat seqfile | perl -ne 'if (/^ID(.*)/) {print "$1\n";} '
```

Other programs and Perl

It is often useful to be able to execute an external program from within a Perl script. Two different methods are shown here, one using the `system` function, and one using the 'backtick' method.

Using the system function

```
$id = 'brca1_human';
system ("blastdbcmd -entry $id -db swissprot > brca1_human.fa");
```

Everything within the double quotes as an argument to the `system` function is a Unix command. Note that variable interpolation is possible, as with `$id` here. The result of this code is that the sequence with accession 'brca1_human' is retrieved from Swiss-Prot and stored in the file named `brca1_human.fa`.

Ideally we should check that the `system` function is successful. We could do that with

```
system ("blastdbcmd -entry $id -db swissprot
> brca1_human.fa") == 0 or die "blastdbcmd failed : $!";
```

Here, the `$!` variable has information about the system error.

Using backticks

```
$id = `brca1_human`;
$brca1_human = `blastdbcmd  -entry $id  -db swissprot`;
print $brca1_human;
```

The Swiss-Prot entry 'brca1_human' is now in the variable `$brca1_human`. As with the `system` function, variable interpolation is possible.

Perl modules and BioPerl

Perl modules are extra pieces of code that are not within the core of the Perl programming language. Some of these modules are part of the standard installation of Perl and some need to be installed (see below). You may also design your own modules. Modules come in two different categories. One category is 'traditional' modules that define functions (subroutines) and variables to be used in a program. Another category is object-oriented modules. Object-oriented approaches in Perl will not be described in this book, but are widely used in Perl programming.[3] Of particular interest for bioinformatics is *BioPerl* (http://www.bioperl.org), a collection of modules that are typically object-oriented and that aid in developing Perl programs for bioinformatics applications.

As an example of the use of an object-oriented module, consider the example in Code AIII.1, which is designed to parse the output from the BLAST program. In Chapter 11 we analysed a BLAST report in tabular format with the help of a Perl script. Here we assume that the BLAST report is in its standard format, and that the report is in the file `bcrabl.blastp`, the same file as we considered in Chapter 7. The module we are using is a BioPerl module named `Bio::SearchIO`, designed to parse output files from a number of different sequence analysis programs. We use it in this case to print all hits with an E-value less than 1E–20 and where the hit name matches the string 'BCR_'.

---

[3] But you can *use* object-oriented modules without understanding too much of objected-oriented programming (although some Perl programmers would hate me for such a statement).

Code AIII.1 searchio.pl

```perl
#!/usr/bin/perl -w

use strict;
use Bio::SearchIO;

my $in = new Bio::SearchIO(
    -format => 'blast',
    -file   => 'bcrabl.blastp'
);

while ( my $result = $in->next_result ) {
   while ( my $hit = $result->next_hit ) {
      while  ( my $hsp = $hit->next_hsp ) {
          if ( ( $hsp->evalue < 1e-20 ) && ( $hit->name =~ /BCR_/ ) ) {
              print "Query=", $result->query_name,
              " Hit=", $hit->name,
              " Percent_id=", $hsp->percent_identity,
              "\n";
          }
      }
   }
}
```

The use statement specifies the module to use. We start by creating an object
with the my $in = new ... statement. The SearchIO module has three different
objects, Result, Hit and HSP. Each of these objects has a number of methods. For
instance, the object Result has the method 'query_name', the Hit object has the
method 'name' and the HSP object has the method 'evalue', as shown in Code
AIII.1. Further examples of elements that may be extracted from the BLAST
report are shown in Table AIII.1.

Installation of Perl modules

Most modules are available from CPAN – the Comprehensive Perl Archive Network. There are two different methods that are typically used to install Perl
modules, one 'manual' approach and one 'automatic'. We start by describing
the manual method.

You often download modules in an archived, compressed format. For instance,
BioPerl version 1.6.1 may be downloaded as a file named BioPerl-1.6.1.tar.gz.
Once having downloaded this file you need to decompress it with:

```
% gunzip BioPerl-1.6.1.tar.gz
```

> **Table AIII.1** *Selected elements of a BLAST report that may be extracted using the Perl module SearchIO. There are three objects of SearchIO, Result, Hit and Hsp. Each of these objects has a number of methods. A 'hit' refers to the database sequence identified as being similar to the query sequence. See http://www.bioperl.org/wiki/ HOWTO:SearchIO for more parameters and documentation.*
>
> | | |
> |---|---|
> | $result->query_name | Name of query sequence |
> | $result->query_length | Length of query sequence |
> | $hit->name | Name of the hit |
> | $hit->length | Length of hit sequence |
> | $hsp->frac_identical | Fraction identical |
> | $hsp->length('total') | Total length of HSP (including gaps) |
> | $hsp->hit_string | Sequence of the hit from the alignment |
> | $hsp->evalue | E-value of HSP |
> | $hsp->query_string | Sequence of the query sequence of the alignment |
> | $hsp->rank | Rank of HSP |
> | $hsp->strand('hit') | Strand of hit sequence, i.e. + or - |
> | $hsp->strand('query') | Strand of query sequence, i.e. + or - |
> | $hsp->start('hit') | Start position of the hit sequence of the alignment |
> | $hsp->end('hit') | End position of the hit sequence of the alignment |
> | $hsp->start('query') | Start position of the query sequence of the alignment |
> | $hsp->end('query') | End position of the query sequence of the alignment |

The result of this operation is a file `BioPerl-1.6.1.tar`. Unpack the tar file with:

```
% tar -xvf BioPerl-1.6.1.tar
```

Or you may skip the gunzip step and instead do:

```
% tar -zxvf BioPerl-1.6.1.tar.gz
```

The result of the `tar` operation is that new directories are created and files in the tar archive end up in these directories. To move further in the installation you therefore have to navigate to the relevant directory, one containing a file named `Makefile.PL`. The standard procedure is then:

```
% perl Makefile.PL
% make
% make test
% make install
```

If you are lucky there will be no error messages, meaning the installation was successful.

Another method which is more automated is one that uses CPAN. There is a module named CPAN.pm that is designed to install modules. One use of it is this command line:

```
% perl -MCPAN -e 'shell'
```

The `-M` parameter with Perl allows you to use a module from the Unix command line, in this case the CPAN module. The command line above opens a cpan shell; you should have a prompt that looks something like this:

```
cpan[1]>
```

If you do not know the name of the installation file (and that is often quite impossible to know), try the 'i' command. For instance, if you want to install BioPerl, you can try the search:

```
i /bioperl/
```

Here, 'i' is a command to search for authors, bundles, distribution files and modules, and in the example above we are searching with a regular expression. The result of the query is a list of modules matching the regular expression. For instance, as a result of this search you may have identified the file CJFIELDS/BioPerl-1.6.1.tar.gz. Then the installation is initiated with the command `install`.

```
install CJFIELDS/BioPerl-1.6.1.tar.gz
```

Making your own modules

See Chapter 19 for an example of how to construct your own module.

More about Perl

For more information on Perl, see the following resources.

Online information

The official Perl website is at http://www.perl.org

CPAN, the Comprehensive Perl Archive Network is at http://www.cpan.org

A collection of Perl tools for Bioinformatics is available through The BioPerl Project (http://bioperl.org).

The Perl Reference Guide (http://www.vromans.org/johan/perlref.html) is available online in several formats.

Further motivation to study Perl and Unix is provided by the article 'How Perl saved the Human Genome Project', by Lincoln Stein.

BOOKS

Christiansen, T. and N. Torkington (2003). *Perl Cookbook*. Sebastopol, CA, O'Reilly.

Schwartz, R. L., T. Phoenix and B. D. Foy (2008). *Learning Perl*. Sebastopol, CA, O'Reilly.

Tisdall, J. D. (2001). *Beginning Perl for Bioinformatics*. Sebastopol, CA, O'Reilly.

Wall, L., T. Christiansen and J. Orwant (2000). *Programming Perl*, 3rd edition. Cambridge, MA, O'Reilly.

BOOKS ON MORE ADVANCED TOPICS SUCH AS PERL OBJECTS

Friedl, J. E. F. (2006). *Mastering Regular Expressions*. Sebastopol, CA, O'Reilly.

Schwartz, R. L. and T. Phoenix (2003). *Learning Perl Objects, References, and Modules*. Sebastopol, CA, O'Reilly.

Tisdall, J. D. (2003). *Mastering Perl for Bioinformatics*. Sebastopol, CA, O'Reilly.

A brief introduction to R

R is a free software environment for statistical computing and graphics. Some of the figures in this book were created with R, and for this reason this appendix has some introductory information about this software. The R scripts related to this book are available for download from the web resources of this book.

The home page for R is http://www.r-project.org and versions for Windows, Mac OS X and Linux may be downloaded from CRAN (the Comprehensive R Archive Network) mirror site.

First time with R

Having installed and started R, you will find a command-line window with the prompt '>'. You communicate with R by typing commands or expressions at this prompt. An example of an expression to enter is:

```
> 2*3
[1] 6
```

Here, 2 multiplied by 3 is evaluated to 6. Thus, R functions as a calculator. The result is, in this case, labelled [1]. This means it is the first element of the result (and there is only one element in this case).

Installing packages

As in Perl, some pieces of code are not part of the standard distribution and need to be installed separately. If we are to install additional modules, this is done with the command `install.packages()`. For instance, it you want to install seqinr, a package for exploratory data analysis and data visualization for biological sequence information, use:

```
> install.packages ("seqinr")
```

However, the bioconductor (http://www.bioconductor.org) package useful for many bioinformatics applications is installed by

```
> source("http://bioconductor.org/biocLite.R")
> biocLite()
```

When a package is to be used it needs to be attached with the `library()` function.

```
> library ("seqinr")
```

Getting help

To start a browser window with help information:

```
> help.start()
```

To have help information open up in a separate window:

```
> ?help
```

or

```
> help()
```

For help on a specific program (plot):

```
> help(plot)
```

To list all functions where 'color' is part of the name:

```
> apropos ("color")
```

To search the help system for the word 'color':

```
> help.search("color")
```

Quitting R

To quit R, use:

```
> quit();
```

or

```
> q();
```

You will be asked whether the workspace is to be saved on not. The workspace has information about what variables have been defined in the current R session.

Defining simple variables and vectors

The assignment operator in R is '<-'. To assign a numerical value, use:

```
> y <- 5
```

To assign a string:

```
> str <- "AUGGUGCACCCUG"
```

Defining a vector (comparable to a Perl array) is done via the `c()` function, which joins a set of values into a vector or list:

```
> x <- c(1,3,7,11)
```

An element in the array x above is referred to like this:

```
> x[4]
[1] 11
```

Patterned data

An expression like 1:6 is used to generate all integers in a range, in this case 1–6:

```
> 1:6
[1] 1 2 3 4 5 6
```

The print() function

To print a value or object use either of the two methods in these examples:

```
> x <- 12
> x
[1] 12
> print(x)
[1] 12
```

Conditional statements

```
> i <- 2
> j <- 3
> if (i < j) {k <- 1}
```

Loop structures

```
for (i in (1:4)) {
  j <- i * 3
  print (i)
  print (j)
}
```

Reading data from a file and data frames

Assume we have a data file with the following contents and with tab-separated columns:

```
x       BRCA1    MGAM
1       10.34    9.56
2       11.05    10.80
3       12.35    11.87
4       10.05    8.76
5       9.34     8.02
```

The data is in a file named `genes.txt` (available from the web resources for this book). We may read the contents of this file with the function `read.table()`:

```
> genes <- read.table("genes.txt", sep="\t", header=TRUE)
```

The option `header=TRUE` assumes that each column in the file has header information, in this case 'x', 'BRCA1' and 'MGAM'. We can test that the data has been read from the file by typing `genes`:

```
> genes
  x BRCA1 MGAM
1 1 10.34  9.56
2 2 11.05 10.80
3 3 12.35 11.87
4 4 10.05  8.76
5 5  9.34  8.02
```

The object `genes` is an example of a data frame. A data frame is like a matrix, but one difference is that in a data frame the columns may be of different types (like 'logical', 'numeric' or 'character'). Using the header information we may refer to the columns read from the table as `genes$x`, `genes$BRCA1` and `genes$MGAM`.

```
> genes$x
[1] 1 2 3 4 5
```

Element 2 in the BRCA1 column:

```
> genes$BRCA1[2]
[1] 11.05
```

or

```
> genes[2,2]
[1] 11.05
```

The `edit()` function opens up a new window in which editing of the data frame is possible:

```
> genes <- edit(genes)
```

Plots

We can plot the data in the data frame `genes`. To plot column BRCA1 against column x:

```
> plot(genes$x, genes$BRCA1)
```

or

```
> plot(genes$BRCA1 ~ genes$x)
```

To plot with both lines and points:

```
> plot(genes$x, genes$BRCA1, type="b")
```

To define the x and y ranges to show:

```
> plot(genes$x, genes$BRCA1, xlim= c(0, 6), ylim=c(0, 15))
```

Additional options are possible with `plot`. A title for the plot may be specified using `main` (`main="GENES"`). Labels to the axes may be changed like in this example: `xlab="x"`, `ylab="Expression"`.

Colour may be added to the plot with the `col` parameter, for instance, `col="red"`. The command `colors()` shows all colours available with the current palette. See also the `palette()` function for more information on colours.

An empty plot may first be created using the option `type = "n"`. Points and lines may be added as appropriate:

```
> plot(0, type = "n", xlim = c(0, 6), ylim = c(3, 15),
main = "GENES", xlab = "x", ylab = "Expression")
> lines(genes$x, genes$BRCA1, col = "red")
> points(genes$x, genes$MGAM, col = "blue")
```

To go through all values in a column, use:

```
for (i in (1:5)) {
     if (genes$BRCA1[i] > 11) {
          lines(c(i - 0.5, i + 0.5), c(8, 8))
     }
}
```

Adding a legend to the plot above is done by:

```
> legend(2, 7, c("BRCA1", "MGAM"), col = c("red", "blue"), lwd = 2)
```

There are other useful commands to add text and shapes to plots. For instance, the following is an example of the `text()` function; the numbers 1 and 14 are the x and y coordinates:

```
> text(1, 14, "This is text")
```

Text may be added to one of four margins of the figure by using the function `mtext()`. The values for the parameter side are 1 = bottom, 2 = left, 3 = top, 4 = right:

```
> mtext("Margin text", side = 4)
```

To draw a rectangle, use the following command, in which the first arguments are the xleft, ybottom, xright and ytop values:

```
> rect(3, 13.5, 3.5, 14, col = "green")
```

The `arrow()` function may be used to draw an arrow in the plot. This function has the syntax `arrows(x0, y0, x1, y1)`, in which an arrow is drawn from the point x0, y0 to x1, y1. For example:

```
> arrows (1, 5, 5, 5)
```

Fig. AIV.1 shows the plot generated using the commands above, starting from the empty plot created with the option `type = "n"`.

Printing to a PDF

```
> pdf("genes.pdf")
> genes <- read.table("genes.txt", sep="\t", header=TRUE)
> plot(genes$x, genes$BRCA1)
> dev.off()
```

Batch procedures

Assume that the file `R.txt` has the following content:

```
pdf("genes.pdf")
genes <- read.table("genes.txt", sep="\t", header=TRUE)
plot(genes$x, genes$BRCA1)
dev.off()
```

R may then be run in batch mode, as in this example for R being run in a Linux environment:

```
% R --vanilla < R.txt
```

GENES

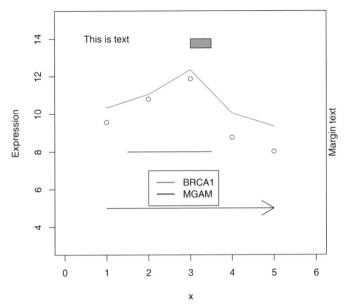

Fig. AIV.1 *Plotting with R*. This plot was obtained by using the commands as described in the text, and shows examples of the functions plot(), lines(), points(), legend(), text(), mtext(), rect() and arrows().

Arguments may also be passed to R through the command line:

```
% R --args ARG1 ARG2
> cmd_args = commandArgs();
> print (cmd_args[3])
[1] "ARG1"
> print (cmd_args[4])
[1] "ARG2"
```

FURTHER READING

Gentleman, R. (2009). *R Programming for Bioinformatics*. Boca Raton, FL, CRC Press.
Maindonald, J. H. and J. W. Braun (2007). *Data Analysis and Graphics Using R: An Example-Based Approach*. Cambridge, Cambridge University Press.
R Development Core Team, An Introduction to R. http://cran.r-project.org/doc/manuals/ R-intro.pdf.

INDEX